Siegfried Selberherr

Analysis and Simulation of Semiconductor Devices

Springer-Verlag Wien New York

Dipl.-Ing. Dr. Siegfried Selberherr
Institut für Allgemeine Elektrotechnik und Elektronik, Technische Universität Wien, Austria

Printed in Austria

With 126 Figures

ISBN 3-211-81800-6 Springer-Verlag Wien-New York
ISBN 0-387-81800-6 Springer-Verlag New York-Wien

To Margit

Preface

The invention of semiconductor devices is a fairly recent one, considering classical time scales in human life. The bipolar transistor was announced in 1947, and the MOS transistor, in a practically usable manner, was demonstrated in 1960. From these beginnings the semiconductor device field has grown rapidly. The first integrated circuits, which contained just a few devices, became commercially available in the early 1960s. Immediately thereafter an evolution has taken place so that today, less than 25 years later, the manufacture of integrated circuits with over 400.000 devices per single chip is possible.

Coincident with the growth in semiconductor device development, the literature concerning semiconductor device and technology issues has literally exploded. In the last decade about 50.000 papers have been published on these subjects.

The advent of so called Very-Large-Scale-Integration (VLSI) has certainly revealed the need for a better understanding of basic device behavior. The miniaturization of the single transistor, which is the major prerequisite for VLSI, nearly led to a breakdown of the classical models of semiconductor devices.

The characteristic feature of early (classical) device modeling is primarily the separation of the interior of the device into different regions, treated by closed form solutions based on restrictive and sometimes drastic assumptions. The solutions in the independently treated regions are simply connected and matched at boundaries to produce a global solution. Any other approach is obviously prohibitive if results with an analytic appearance are intended. For the purpose of analysis, however, this classical approach has turned out to be only limitedly applicable, particularly when a technically acceptable prediction of device performance is desired.

As a consequence numerical analysis and simulation based on comparatively fundamental differential equations has become necessary and popular. This trend has been supported considerably by the enormous progress in technology and performance of digital computers. Contemporary modeling of semiconductor devices has attained such a high level of sophistication that two-dimensional simulation of the static behavior is almost standard in the development stage of device prototypes. Even three-dimensional transient simulations have been reported very recently, but, due to a still too extensive consumption of computer resources, these are at the moment more of academic importance than of practical relevance.

Numerical analysis of semiconductor devices can be expected to become a basic methodology of research and development engineers. However, one must not expect that people using computer programs as numerical analysis tools are specialists considering the complexity of the assumptions, algorithms and implementation details of the programs they use. In particular, this book has been written with two primary objectives kept in view: First, the interested device engineer should be introduced to the physical and mathematical problems to be solved by an analysis program. This category of readers should gain a more fundamental understanding of the applicability of device simulation programs. Secondly, this book will benefit authors of device simulation programs by providing a compact reference with many citations and a critical overview of the various physical and mathematical approaches which are used worldwide today.

The chapters in this book are arranged in a logical sequence without many crossreferences. Each chapter is more or less self-contained. Readers who are only interested in particular subjects will be able to extract easily the information they require.

In preparing the material for this book many people have assisted me to a considerable extent. I am extremely grateful to Prof. H. Pötzl for reviewing my manuscripts and the resulting endless discussions and suggestions. I am indebted to my colleagues at the university for many discussions and the friendly atmosphere: Drs. W. Agler, J. Demel, A. Franz, G. Franz, W. Griebel, E. Guerrero, W. Jüngling, M. Kowatsch, H. Lafferl, E. Langer, W. Mader, P. Markowich, Prof. F. Paschke, P. Pichler, C. Ringhofer, A. Schütz, Prof. F. Seifert, Prof. H. Stetter, F. Straker, Doz. Ch. Überhuber, Prof. R. Weiß. I would like to express my sincere appreciation to Dr. S. E. Laux, IBM T. J. Watson Research Center, for proofreading my manuscript. I would like to thank the Austrian "Fonds zur Förderung der wissenschaftlichen Forschung" and the Research Laboratory of Siemens AG, Munich, FRG, for supporting many projects which have evolved into much of the material presented in this book. Last but not least I would like to gratefully acknowledge the generous amount of computer resources provided by Dipl.-Ing. D. Schornböck, and the excellent computer access made possible by the whole staff of the local computer center.

I hope my book will be used by many engineers and scientists who wish to gain insight into the subject of numerical device modeling. It is my sincere wish that this book will contribute to bridging the gaps between solid-state physicists, numerical analysists, computer scientists and device engineers.

Vienna, April 1984

Siegfried Selberherr

Contents

Notation

$\vec{A}$	vector potential
$\vec{B}$	magnetic induction vector
C	net ionized impurity concentration
CI	total ionized impurity concentration
CN	total neutral impurity concentration
C_c^{OPT}	optical capture coefficient
Cc_i	electrically inactive concentration of *i*-th impurity
C_{cv}^{AU}	Auger capture coefficient
C_{cv}^{SRH}	Shockley-Read-Hall capture coefficient
C_e^{OPT}	optical emission coefficient
C_{ev}^{AU}	Auger emission coefficient
C_{ev}^{SRH}	Shockley-Read-Hall emission coefficient
Ct_i	total concentration of *i*-th species
$\vec{D}$	electric displacement vector
D_i^+	diffusivity of *i*-th impurity due to singly positive charged vacancies
D_i^-	diffusivity of *i*-th impurity due to singly negative charged vacancies
$D_i^=$	diffusivity of *i*-th impurity due to doubly negative charged vacancies
D_i^0	diffusivity of *i*-th impurity due to neutral vacancies
D_i	effective diffusivity of *i*-th impurity
D_v^T	thermal diffusion coefficient
D_v	effective diffusivity
$\vec{E}$	electric field vector
$\vec{E}_v$	effective field
E	energy
E	electric field
E_{ac}	acoustic deformation potential of conduction band
E_{av}	acoustic deformation potential of valence band
E_c	conduction band energy
E_{co}	conduction band edge
E_{fv}	quasi-Fermi energy
E_g	band gap
E_i	ionization energy
E_i	intrinsic Fermi energy
E_v^{crit}	critical field

E_ν	driving force
E_r	average energy loss per high energetic collision
E_v	valence band energy
E_{vo}	valence band edge
$E_\perp$	electric field component perpendicular to current flow direction
$E_{\|\|}$	electric field component parallel to current flow direction
$\vec{F}$	force vector
$\vec{F}_{\nu e}$	external force
$\vec{F}_{\nu i}$	internal force
$F_{1/2}$	Fermi integral of order 1/2
$\vec{H}$	magnetic field vector
H	thermal generation
$\vec{J}$	total electric current density
$\vec{J}_i$	flux of *i*-th impurity
$\vec{J}_\nu$	carrier current density
N_A^-	concentration of singly ionized acceptors
N_D^+	concentration of singly ionized donors
N_c	effective density of states in conduction band
N_d	implantation dose
N_t	concentration of traps
N_v	effective density of states in valence band
R	net carrier generation/recombination
R^{AU}	net Auger generation/recombination
R^{II}	net impact ionization generation rate
R^{OPT}	net optical generation/recombination
R^{SRH}	net Shockley-Read-Hall generation/recombination
R^{SURF}	net surface generation/recombination
R_p	projected range
S_ν	scattering probability
T	lattice temperature
T_ν	carrier temperature
Ut	thermal voltage ($=k \cdot T/q$)
V^+	normalized concentration of singly positive charged vacancies
V^-	normalized concentration of singly negative charged vacancies
$V^=$	normalized concentration of doubly negative charged vacancies
Z_i	charge state of *i*-th impurity
a	crystal lattice constant
a_B	Bohr radius ($=5.2917706 \cdot 10^{-11}$ m)
α_ν	ionization rate
β_2	kurtosis
β_c	equilibrium cluster coefficient
c	speed of light in vacuum ($=2.99792458 \cdot 10^8$ ms^{-1})
c	specific heat
δE_c	shift energy for conduction band edge
δE_v	shift energy for valence band edge
df	field enhancement factor
ε	absolute permittivity

ε_0	permittivity constant in vacuum ($=8.854187818 \cdot 10^{-12}$ As V^{-1} m^{-1})
ε_r	relative permittivity
f_v	distribution function
φ_v	quasi-Fermi potential
f_t	fraction of occupied traps
γ_1	skewness
h	Planck constant ($=6.626176 \cdot 10^{-34}$ V As^2)
$\vec{k}$	momentum vector
k	thermal conductivity
k	Boltzmann constant ($=1.380662 \cdot 10^{-23}$ V As K^{-1})
kc	clustering rate
kd	declustering rate
λ	mean free path between high energetic collisions
λ	screening length
m	cluster size
μ_i	i-th central moment
m_v^*	effective mass
μ_v^Σ	carrier mobility
μ_0	permeability constant in vacuum ($=4 \cdot \pi$)
m_0	electron rest mass ($=9.109534 \cdot 10^{-31}$ V As^3 m^{-2})
n	electron concentration
n_i	intrinsic carrier concentration
n_{ie}	effective intrinsic carrier concentration
n_0	equilibrium concentration of electrons
p	hole concentration
p_0	equilibrium concentration of holes
q	elementary charge ($=1.6021892 \cdot 10^{-19}$ As)
ρ	specific mass density
ρ	space charge
ρ_A	density of states in acceptor band
ρ_D	density of states in donor band
ρ_c	density of states in conduction band
ρ_v	density of states in valence band
σ_{DA}	standard deviation for donor and acceptor band
σ_{cv}	standard deviation for conduction and valence band tails
σ_p	standard deviation
t	time
t_{mask}	mask thickness
τ_v	relaxation time
τ_v	lifetime
$\vec{u}_v$	group velocity
ψ	electrostatic potential
ψ_b	built-in potential
$\vec{v}_v$	drift velocity
v_v^{sat}	saturation velocity
$\vec{x}$	space vector
x_{ox}	oxide thickness

Subscript "v" stands for "n" or "p" denoting the respective quantity for electrons or holes.

Superscript "Σ" in the carrier mobility stands for any combination of the following list.

- C carrier-carrier impurity scattering
- E velocity saturation
- I ionized impurity scattering
- L lattice scattering
- N neutral impurity scattering
- S surface scattering

A superscript "*" or no superscript indicates the effective mobility which is comprised of all above given effects.

Landau Symbols

A) $f(x)=O(g(x))$ as $x \to x_0$ means that

$$\left|\frac{f(x)}{g(x)}\right| < \text{const.}$$

for x sufficiently close to x_0.

B) $f(x)=o(g(x))$ as $x \to x_0$ means that

$$\lim_{x \to x_0} \frac{f(x)}{g(x)} = 0$$

C) Sometimes we say (sloppily) that "a quantity f is $O(g)$" which means that $|f|$ is of approximate order of magnitude $|g|$.

Introduction 1

1.1 The Goal of Modeling

At the outset it seems necessary to clarify the frequently used terms analysis, simulation and modeling. By tracing the literature one often has the impression that authors use these terms in a fairly arbitrary manner. A while ago I picked up a heavy dictionary and, among many others, I have found the following interpretations to be quite appropriate:

Analysis

- separation of a whole into its component parts, possibly with comment and judgement
- examination of a complex, its elements, and their relations in order to learn about

Simulation

- imitative representation of the functioning of one system or process by means of the functioning of another
- examination of a problem not subject to experimentation

Modeling

- to produce a representation or simulation of a problem or process
- to make a description or analogy used to help visualize something that cannot be directly observed

Therefore, as difficult as it might be to decide in an individual case, analysis is at least intended to mean "exact analysis" and simulation must mean "approximate simulation" by inference. Modeling is obviously a necessity for analysis and simulation.

With a model one can analyse some phenomena, provided that the effects one wants to extract are built in the model, possibly in a very complex manner. A model for the purpose of pure simulation (like a curve fitting model) is usually much more simple than a model for analysis. Many effects can be treated in a very heuristic manner for the purpose of simulation, just reflecting the underlying physics in a qualitative way.

An excellent example to highlight these aspects can be found in the application of a Monte Carlo method. "Modeling" with a Monte Carlo method is equivalent to "producing an imitative representation of the functioning of a system". But the

purpose of a Monte Carlo model is strictly analysis and not just simulation, because the underlying basis is "a separation of a whole into its component parts".
However, one has to keep foremost in mind the limitations of any model in order not to interprete too naively results which are just obtained by improper application of a model.
I feel obliged to explicitly state my personal opinion about the quality of the results which can be obtained by contemporary device modeling. The development of devices involves several iterations of trial and error in fabrication until a specified goal in terms of design conditions is reached. The application of device models can now, and sometimes fairly substantially, decrease the number of trial and error steps during the development. A serious speculation about the average savings in development effort could be on the order of forty percent. Obviously, this number depends strongly on the individual conditions of a specific project. The total elimination of trial and error in device development is not possible nowadays, because the uncertainties of several parameters of the available models, although they are already very sophisticated, are still too large. I absolutely expect not being wrong in claiming that device modeling will become more and more important in the near future. This assumption is also supported by the fact that computer resources are going to be cheaper compared to drastically increasing costs for experimental investigations. Hence, many more engineers will have to face the problem of numerical device modeling in order to stay competitive.
It remains to say that the main power of higher dimensional device model lies in their capability to provide insight into the functioning of devices by means of distributions of the various physical quantities in the interior of a device. However, many device engineers are not at all used to interpreting those results; they prefer global quantities like current-voltage characteristics. A properly tuned higher dimensional device model is certainly able to predict global device parameters with a desired accuracy, but much simpler and cheaper (in terms of computer resources) models will often be able to deliver global results with equally good reliability. For miniaturized devices, however, higher dimensional models are often the only existing and imaginable tool for the accurate prediction of device performance.

1.2 The History of Numerical Device Modeling

Fully numerical modeling of a semiconductor device based on partial differential equations [1.87] which describe all different regions of a device in one unified manner was first suggested by Gummel [1.29] in 1964 for the one dimensional bipolar transistor. This approach was further developed and applied to *pn*-junction theory by De Mari [1.18], [1.19] and to IMPATT diodes by Scharfetter and Gummel [1.75]. A two dimensional solution of Poisson's equation with application to a MOS structure was first published by Loeb et al. [1.49] and Schroeder and Muller [1.76] in 1968. Kennedy and O'Brien [1.39] investigated in 1969 the junction field effect transistor by means of a two dimensional numerical solution of Poisson's equation and one continuity equation. At the same time Slotboom [1.82] presented a two dimensional analysis of the bipolar transistor solving Poisson's equation and both continuity equations. Since then two dimensional modeling has been applied to nearly all important devices. It is not possible to cite here all relevant papers in the field; however, to present at least a comprehensive menu of key papers is worthwile.

The junction field effect transistor has been investigated in two dimensions by solving the Poisson equation and one continuity equation by, e.g., Himsworth [1.34], Kennedy and O'Brien [1.40] and Yamaguchi et al. [1.91]. The transient behavior in two dimensions of those devices has been simulated by, e.g., Reiser [1.69].

MESFET's have been analyzed also by Reiser [1.70] and by, e.g., Barnes et al. [1.8], [1.9]. More sophisticated equations for the physical and mathematical model of MESFET's have been solved by Cook and Frey [1.17] (energy transport equations) and by Moglestue [1.58], [1.59] and Pone et al. [1.65] (particle equations).

Many activities have been concentrated on the simulation of MOS devices due to their intrinsically two dimensional nature, e.g.: in 1971 [1.88], in 1972 [1.89], in 1973 [1.41], [1.55], in 1976 [1.35], in 1977 [1.24], in 1978 [1.63], [1.83], in 1979 [1.43], [1.77], in 1980 [1.16], [1.64], [1.78], [1.84], in 1981 [1.44], [1.71], [1.92], in 1982 [1.67], [1.73], [1.74], [1.90] and in 1983 [1.60]. Two dimensional transient simulations of MOSFET's have been carried out by, e.g., Mock [1.56], Oh et al. [1.62] and Yamaguchi [1.93]. Three dimensional static modeling has been published in, e.g.: [1.13], [1.15], [1.81].

Out of the many papers which have been published on modeling bipolar transistors and thyristors it seems worthwhile to cite, e.g.: in 1970 [1.21], in 1971 [1.42], in 1973 [1.33], [1.94], in 1974 [1.51], [1.80], in 1975 [1.53] (two dimensional transient simulation), in 1976 [1.27], in 1978 [1.2], in 1979 [1.1], [1.3], in 1981 [1.45], [1.47], [1.85] and in 1983 [1.26], [1.50] (nonlinear small signal simulation).

Several non-standard and unusual devices have been simulated during their development, e.g.: the permeable base transistor [1.10], the insulating gate rectifier [1.6], and the "dielectric surface loaded GaAs bulk element" [1.38]. However, only a very few computer programs which allow the simulation of a fairly arbitrary device structure have been published, e.g.: [1.14], [1.25], [1.28], [1.30], [1.31].

As it can be obviously expected, many dissertations on numerical modeling of semiconductor devices have been undertaken, e.g. (in order of appearance): [1.68], [1.32], [1.52], [1.7], [1.36], [1.66], [1.48], [1.79], [1.72].

Kurata [1.46] and Mock [1.57] have published a monograph in 1982 and 1983 respectively. Various conferences with proceedings published as books, e.g.: [1.11], [1.12], [1.54] have taken place, and summer courses, e.g.: [1.5], [1.20], [1.86] have been held.

Among many more the following outstanding review papers have been published [1.22], [1.23], [1.37] and [1.61]. In 1975 Agajanian [1.4] has published a bibliography on device modeling (not only numerical modeling) with about 500 references selected from the most important papers of the preceeding four years.

1.3 References

[1.1] Adachi, T., Yoshii, A., Sudo, T.: Two-Dimensional Semiconductor Analysis Using Finite-Element Method. IEEE Trans. Electron Devices *ED-26,* 1026 – 1031 (1979).

[1.2] Adler, M. S.: Accurate Calculations of the Forward Drop and Power Dissipation in Thyristors. IEEE Trans. Electron Devices *ED-25,* No. 1, 16 – 22 (1978).

[1.3] Adler, M. S.: A Method for Achieving and Choosing Variable Density Grids in Finite Difference Formulations and the Importance of Degeneracy and Band Gap Narrowing in Device Modeling. Proc. NASECODE I Conf., pp. 3 – 30. Dublin: Boole Press 1979.

[1.4] Agajanian, A. H.: A Bibliography on Semiconductor Device Modeling. Solid-State Electron. *18,* 917 – 929 (1975).

[1.5] Antognetti, P., Antoniadis, D. A., Dutton, R. W., Oldham, W. G.: Process and Device Simulation for MOS-VLSI Circuits. The Hague: Martinus Nijhoff 1983.

[1.6] Baliga, B. J., Adler, M. S., Gray, P. V., Love, R. P., Zommer, N.: The Insulated Gate Rectifier (IGR): A New Power Switching Device. Proc. International Electron Devices Meeting, pp. 264 – 267 (1982).

[1.7] Barnes, J. J.: A Two-Dimensional Simulation of MESFET's. Dissertation, University of Michigan, 1976.

[1.8] Barnes, J. J., Lomax, R. J., Haddad, G. I.: Finite-Element Simulation of GaAs MESFET's with Lateral Doping Profiles and Submicron Gates. IEEE Trans. Electron Devices *ED-23,* No. 9, 1042 – 1048 (1976).

[1.9] Barnes, J. J., Lomax, R. J.: Finite-Element Methods in Semiconductor Device Simulation. IEEE Trans. Electron Devices *ED-24,* 1082 – 1089 (1977).

[1.10] Bozler, C. O., Alley, G. D.: Fabrication and Numerical Simulation of the Permeable Base Transistor. IEEE Trans. Electron Devices *ED-27,* 1128 – 1141 (1980).

[1.11] Browne, B. T., Miller, J. J. H.: Numerical Analysis of Semiconductor Devices. Dublin: Boole Press 1979.

[1.12] Browne, B. T., Miller, J. J. H.: Numerical Analysis of Semiconductor Devices and Integrated Circuits. Dublin: Boole Press 1981.

[1.13] Buturla, E. M., Cottrell, P. E., Grossman, B. M., Salsburg, K. A., Lawlor, M. B., McMullen, C. T.: Three-Dimensional Finite Element Simulation of Semiconductor Devices. Proc. Int. Solid-State Circuits Conf., pp. 76 – 77 (1980).

[1.14] Buturla, E. M., Cottrell, P. E., Grossman, B. M., Salsburg, K. A.: Finite-Element Analysis of Semiconductor Devices: The FIELDAY Program. IBM J. Res. Dev. *25,* 218 – 231 (1981).

[1.15] Chamberlain, S. G., Husain, A.: Three-Dimensional Simulation of VLSI MOSFET's. Proc. Int. Electron Devices Meeting, pp. 592 – 595 (1981).

[1.16] Colak, S., Singer, B., Stupp, E.: Lateral DMOS Power Transistor Design. IEEE Electron Dev. Lett. *EDL-1,* 51 – 53 (1980).

[1.17] Cook, R. K., Frey, J.: Two-Dimensional Numerical Simulation of Energy Transport Effects in Si and GaAs MESFET's. IEEE Trans. Electron Devices *ED-29,* No. 6, 970 – 977 (1982).

[1.18] DeMari, A.: An Accurate Numerical Steady-State One-Dimensional Solution of the P-N Junction. Solid-State Electron. *11,* 33 – 58 (1968).

[1.19] DeMari, A.: An Accurate Numerical One-Dimensional Solution of the P-N Junction under Arbitrary Transient Conditions. Solid-State Electron. *11,* 1021 – 2053 (1968).

[1.20] DeMeyer, K. M.: VLSI Process and Device Modeling. Katholieke Universiteit Leuven, 1983.

[1.21] Dubock, P., D.C. Numerical Model for Arbitrarily Biased Bipolar Transistors in Two Dimensions. Electron. Lett. *6,* 53 – 55 (1970).

[1.22] Engl, W. L., Dirks, H. K., Meinerzhagen, B.: Device Modeling. Proc. IEEE *71,* No. 1, 10 – 33 (1983).

[1.23] Engl, W. L., Manck, O., Wieder, A. W.: Device Modeling. In: Process and Device Modeling for Integrated Circuit Design, pp. 3 – 17. Leyden: Noordhoff 1977.

[1.24] Fortino, A. G., Nadan, J. S.: An Efficient Method for the Small-Signal AC Analysis of MOS Capacitors. IEEE Trans. Electron Devices *ED-24,* No. 9, 1137 – 1147 (1977).

[1.25] Franz, A. F., Franz, G. A., Selberherr, S., Ringhofer, C., Markowich, P.: Finite Boxes – A Generalization of the Finite Difference Method Suitable for Semiconductor Device Simulation. IEEE Trans. Electron Devices *ED-30,* No. 9, 1070 – 1082 (1983).

[1.26] Franz, G. A., Franz, A. F., Selberherr, S., Markowich, P.: A Quasi Three Dimensional

Semiconductor Device Simulation Using Cylindrical Coordinates. Proc. NASECODE III Conf., pp. 122 – 127. Dublin: Boole Press 1983.

[1.27] Gaur, S. P., Navon, D. H.: Two-Dimensional Carrier Flow in a Transistor Structure under Nonisothermal Conditions. IEEE Trans. Electron Devices *ED-23,* 50 – 57 (1976).

[1.28] Greenfield, J. A., Dutton, R. W.: Nonplanar VLSI Device Analysis Using the Solution of Poisson's Equation. IEEE Trans. Electron Devices *ED-27,* 1520 – 1532 (1980).

[1.29] Gummel, H. K.: A Self-Consistent Iterative Scheme for One-Dimensional Steady State Transistor Calculations. IEEE Trans. Electron Devices *ED-11,* 455 – 465 (1964).

[1.30] Hachtel, G. D., Mack, M. H., O'Brien, R. R., Speelpennig, B.: Semiconductor Analysis Using Finite Elements – Part 1: Computational Aspects. IBM J. Res. Dev. *25,* 232 – 245 (1981).

[1.31] Hachtel, G. D., Mack, M. H., O'Brien, R. R.: Semiconductor Analysis Using Finite Elements – Part 2: IGFET and BJT Case Studies. IBM J. Res. Dev. *25,* 246 – 260 (1981).

[1.32] Heimeier, H. H.: Zweidimensionale numerische Lösung eines nichtlinearen Randwertproblems am Beispiel des Transistors im stationären Zustand. Dissertation, Technische Hochschule Aachen, 1973.

[1.33] Heimeier, H. H.: A Two-Dimensional Numerical Analysis of a Silicon N-P-N Transistor. IEEE Trans. Electron Devices *ED-20,* 708 – 714 (1973).

[1.34] Himsworth, B.: A Computer Aided Two-Dimensional Analysis of Gallium Arsenide and Silicon Junction Field Effect Transistors. Int. J. Electronics *31,* No. 4, 365 – 371 (1971).

[1.35] Hori, R., Masuda, H., Minato, O., Nishimatu, S., Sato, K., Kubo, M.: Short Channel MOS-IC Based on Accurate Two Dimensional Device Design. Jap. J. Appl. Phys. *15,* 193 – 199 (1976).

[1.36] Jesshope, C. R.: Bipolar Transistor Modelling with Numerical Solutions to the 2-Dimensional DC and Transient Problems. Dissertation, University of Southampton, 1976.

[1.37] Kani, K.: A Survey of Semiconductor Device Analysis in Japan. Proc. NASECODE I Conf., pp. 104 – 119. Dublin: Boole Press 1979.

[1.38] Kataoka, S., Tateno, H., Kawashima, M.: Two-Dimensional Computer Analysis of Dielectric-Surface-Loaded GaAs Bulk Element. Electron. Lett. *6,* No. 6, 169 – 171 (1970).

[1.39] Kennedy, D. P., O'Brien, R. R.: Two-Dimensional Mathematical Analysis of a Planar Type Junction Field-Effect Transistor. IBM J. Res. Dev. *13,* 662 – 674 (1969).

[1.40] Kennedy, D. P., O'Brien, R. R.: Two-Dimensional Analysis of JFET Structures Containing a Low-Conductivity Substrate. Electron. Lett. *7,* No. 24, 714 – 717 (1971).

[1.41] Kennedy, D. P., Murley, P. C.: Steady State Mathematical Theory for the Insulated Gate Field Effect Transistor. IBM J. Res. Dev. *17,* 2 – 12 (1973).

[1.42] Kilpatrick, J. A., Ryan, W. D.: Two-Dimensional Analysis of Lateral-Base Transistors. Electron. Lett. *7,* No. 9, 226 – 227 (1971).

[1.43] Kotani, N., Kawazu, S.: Computer Analysis of Punch-Through in MOSFET's. Solid-State Electron. *22,* 63 – 70 (1979).

[1.44] Kotani, N., Kawazu, S.: A Numerical Analysis of Avalanche Breakdown in Short-Channel MOSFET's. Solid-State Electron. *24,* 681 – 687 (1981).

[1.45] Kurata, M.: Hybrid Two-Dimensional Device Modelling. Proc. NASECODE II Conf., pp. 88 – 112. Dublin: Boole Press 1981.

[1.46] Kurata, M.: Numerical Analysis for Semiconductor Devices. Lexington, Mass.: Lexington Press 1982.

[1.47] Latif, M., Bryant, P. R.: Network Analysis Approach to Multi-Dimensional Modeling of Transistors Including Thermal Effects. Proc. Int. Symp. Circuits and Systems 1981.

[1.48] Laux, S. E.: Two-Dimensional Simulation of Gallium-Arsenide MESFET's Using the Finite-Element Method. Dissertation, University of Michigan, 1981.

[1.49] Loeb, H. W., Andrew, R., Love, W.: Application of 2-Dimensional Solutions of the Shockley-Poisson Equation to Inversion-Layer M.O.S.T. Devices. Electron Lett. *4,* 352 – 354 (1968).

[1.50] Machek, J., Fulop, W.: Harmonic Distortion in a One-Dimensional p-n-p Transistor. Solid-State Electron. *26,* No. 6, 525 – 536 (1983).

[1.51] Manck, O., Heimeier, H. H., Engl, W. L.: High Injection in a Two-Dimensional Transistor. IEEE Trans. Electron Devices *ED-21,* 403 – 409 (1974).

[1.52] Manck, O.: Numerische Analyse des Schaltverhaltens eines zweidimensionalen bipolaren Transistors. Dissertation, Technische Hochschule Aachen, 1975.

[1.53] Manck, O., Engl, W. L.: Two-Dimensional Computer Simulation for Switching a Bipolar Transistor Out of Saturation. IEEE Trans. Electron Devices *ED-22,* No. 6, 339-347 (1975).

[1.54] Miller, J. J. H.: Numerical Analysis of Semiconductor Devices and Integrated Circuits. Dublin: Boole Press 1983.
[1.55] Mock, M. S.: A Two-Dimensional Mathematical Model of the Insulated-Gate Field-Effect Transistor. Solid-State Electron. *16,* 601 – 609 (1973).
[1.56] Mock, M. S.: A Time-Dependent Numerical Model of the Insulated-Gate Field-Effect Transistor. Solid-State Electron. *24,* 959 – 966 (1981).
[1.57] Mock, M. S.: Analysis of Mathematical Models of Semiconductor Devices. Dublin: Boole Press 1983.
[1.58] Moglestue, C., Beard, S. J.: A Particle Model Simulation of Field Effect Transistors. Proc. NASECODE I Conf., pp. 232 – 236. Dublin: Boole Press 1979.
[1.59] Moglestue, C.: A Monte-Carlo Particle Model Study of the Influence of the Doping Profiles on the Characteristics of Field-Effect Transistors. Proc. NASECODE II Conf., pp. 244 – 249. Dublin: Boole Press 1981.
[1.60] Navon, D. H., Wang, C. T.: Numerical Modeling of Power MOSFET's. Solid-State Electron. *26,* No. 4, 287 – 290 (1983).
[1.61] Newton, A. R.: Computer-Aided Design of VLSI Circuits. Proc. IEEE *69,* 1189 – 1199 (1981).
[1.62] Oh, S. Y., Ward, D. E., Dutton, R. W.: Transient Analysis of MOS Transistors. IEEE Trans. Electron Devices *ED-27,* 1571 – 1578 (1980).
[1.63] Oka, H., Nishiuchi, K., Nakamura, T., Ishikawa, H.: Two-Dimensional Numerical Analysis of Normally-Off Type Buried Channel MOSFET's. Proc. Int. Electron Devices Meeting, pp. 30 – 33 (1979).
[1.64] Oka, H., Nishiuchi, K., Nakamura, T., Ishikawa, H.: Computer Analysis of a Short-Channel BC MOSFET. IEEE Trans. Electron Devices *ED-27,* 1514 – 1520 (1980).
[1.65] Pone, J. F., Castagne, R. C., Courat, J. P., Arnodo, C.: Two-Dimensional Particle Modeling of Submicrometer Gate GaAs FET's Near Pinchoff. IEEE Trans. Electron Devices *ED-29,* No. 8, 1244 – 1255 (1982).
[1.66] Price, C. H.: Two-Dimensional Numerical Simulation of Semiconductor Devices. Dissertation, Stanford University, 1980.
[1.67] Rahali, F.: Analyse Numerique a 2 Dimensions de Transistors MOS par la Methode des Elements Finis. Laboratoire d'electronique generale, Lausanne, 1982.
[1.68] Regier, F.: A New Analysis of Field Effect Transistors. Dissertation, Yale University, 1968.
[1.69] Reiser, M.: Difference Methods for the Solution of the Time-Dependent Semiconductor Flow-Equations. Electron. Lett. *7,* 353 – 355 (1971).
[1.70] Reiser, M.: A Two-Dimensional Numerical FET Model for DC, AC, and Large-Signal Analysis. IEEE Trans. Electron Devices *ED-20,* 35 – 44 (1973).
[1.71] Schütz, A., Selberherr, S., Pötzl, H. W.: Numerical Analysis of Breakdown Phenomena in MOSFET's. Proc. NASECODE II Conf., pp. 270 – 274. Dublin: Boole Press 1981.
[1.72] Schütz, A.: Simulation des Lawinendurchbruchs in MOS-Transistoren. Dissertation, Technische Universität Wien, 1982.
[1.73] Schütz, A., Selberherr, S., Pötzl, H. W.: A Two-Dimensional Model of the Avalanche Effect in MOS Transistors. Solid-State Electron. *25,* 177 – 183 (1982).
[1.74] Schütz, A., Selberherr, S., Pötzl, H. W.: Analysis of Breakdown Phenomena in MOSFET's. IEEE Trans. Computer-Aided-Design of Integrated Circuits *CAD-1,* 77 – 85 (1982).
[1.75] Scharfetter, D. L., Gummel, H. K.: Large-Signal Analysis of a Silicon Read Diode Oscillator. IEEE Trans. Electron Devices *ED-16,* 64 – 77 (1969).
[1.76] Schroeder, J. E., Muller, R. S.: IGFET Analysis Through Numerical Solution of Poisson's Equation. IEEE Trans. Electron Devices *ED-15,* No. 12, 954 – 961 (1968).
[1.77] Selberherr, S., Fichtner, W., Pötzl, H. W.: MINIMOS – a Program Package to Facilitate MOS Device Design and Analysis. Proc. NASECODE I Conf., pp. 275 – 279. Dublin: Boole Press 1979.
[1.78] Selberherr, S., Schütz, A., Pötzl, H. W.: MINIMOS – a Two-Dimensional MOS Transistor Analyzer. IEEE Trans. Electron Devices *ED-27,* 1540 – 1550 (1980).
[1.79] Selberherr, S.: Zweidimensionale Modellierung von MOS-Transistoren. Dissertation, Technische Universität Wien, 1981.
[1.80] Seltz, D., Kidron, I.: A Two-Dimensional Model for the Lateral p-n-p Transistor. IEEE Trans. Electron Devices *ED-21,* No. 9, 587 – 592 (1974).

[1.81] Shigyo, N., Konaka, M., Dang, R. L. M.: Three-Dimensional Simulation of Inverse Narrow-Channel Effect. Electron. Lett. *18*, No. 6, 274–275 (1982).

[1.82] Slotboom, J. W.: Iterative Scheme for 1- and 2-Dimensional D.C.-Transistor Simulation. Electron. Lett. *5*, 677–678 (1969).

[1.83] Toyabe, T., Yamaguchi, K., Asai, S., Mock, M. S.: A Numerical Model of Avalanche Breakdown in MOSFET's. IEEE Trans. Electron Devices *ED-25*, 825–832 (1978).

[1.84] Toyabe, T., Yamaguchi, K., Asai, S., Mock, M. S.: A Two-Dimensional Avalanche Breakdown Model of Submicron MOSFET's. Proc. Int. Electron Devices Meeting, pp. 432–435 (1980).

[1.85] Toyabe, T., Mock, M. S., Okabe, T., Ujiie, K., Nagata, M.: A Two-Dimensional Analysis of I2L with Multi-Stream Function Technique. Proc. NASECODE I Conf., pp. 290–292. Dublin: Boole Press 1979.

[1.86] VanDeWiele, F., Engl, W. L., Jespers, P. G.: Process and Device Modeling for Integrated Circuit Design. Leyden: Noordhoff 1977.

[1.87] Van Roosbroeck, W. V.: Theory of Flow of Electrons and Holes in Germanium and Other Semiconductors. Bell Syst. Techn. J. *29*, 560–607 (1950).

[1.88] Vandorpe, D., Xuong, N. H.: Mathematical 2-Dimensional Model of Semiconductor Devices. Electron. Lett. *7*, 47–50 (1971).

[1.89] Vandorpe, D., Borel, J., Merckel, G., Saintot, P.: An Accurate Two-Dimensional Numerical Analysis of the MOS Transistor. Solid-State Electron. *15*, 547–557 (1972).

[1.90] Wilson, C. L., Blue, J. L.: Two-Dimensional Finite Element Charge-Sheet Model of a Short Channel MOS Transistor. Solid-State Electron. *25*, No. 6, 461–477 (1982).

[1.91] Yamaguchi, K., Toyabe, T., Kodera, H.: Two-Dimensional Analysis of Triode-Like Operation of Junction Gate FET's. IEEE Trans. Electron Devices *ED-22*, 1047–1049 (1975).

[1.92] Yamaguchi, K., Takahashi, S.: Theoretical Characterization and High-Speed Performance Evaluation of GaAs IGFET's. IEEE Trans. Electron Devices *ED-28*, No. 5, 581–587 (1981).

[1.93] Yamaguchi, K.: A Time Dependent and Two-Dimensional Numerical Model for MOSFET Device Operation. Solid-State Electron. *26*, No. 9, 907–916 (1983).

[1.94] Zaluska, E. J., Dubock, P. A., Kemhadhan, H. A.: Practical 2-Dimensional Bipolar-Transistor-Analysis Algorithm. Electron. Lett. *9*, 599–600 (1973).

2 Some Fundamental Properties

To accurately analyze an arbitrary semiconductor structure which is intended as a self contained device under various operating conditions, a mathematical model has to be given. The equations which form this mathematical model are commonly called the basic semiconductor equations. They can be derived from Maxwell's equations (2-1), (2-2), (2-3) and (2-4), several relations obtained from solid-state physics knowledge about semiconductors and various – sometimes overly simplistic – assumptions.

$$\mathrm{rot}\ \vec{H} = \vec{J} + \frac{\partial \vec{D}}{\partial t} \tag{2-1}$$

$$\mathrm{rot}\ \vec{E} = -\frac{\partial \vec{B}}{\partial t} \tag{2-2}$$

$$\mathrm{div}\ \vec{D} = \rho \tag{2-3}$$

$$\mathrm{div}\ \vec{B} = 0 \tag{2-4}$$

$\vec{E}$ and $\vec{D}$ are the electric field and displacement vector; $\vec{H}$ and $\vec{B}$ are the magnetic field and induction vector, respectively. $\vec{J}$ denotes the conduction current density, and ρ is the electric charge density.
The next sections will be devoted entirely to an outline of the procedures which have to be carried out in order to derive the basic semiconductor equations.

2.1 Poisson's Equation

Poisson's equation is essentially the third Maxwell equation (2-3). However, to make this equation directly applicable to semiconductor problems, some manipulations have to be undertaken. We first introduce a relation for the electric displacement vector $\vec{D}$ and the electric field vector $\vec{E}$ (2.1-1).

$$\vec{D} = \varepsilon \cdot \vec{E} \tag{2.1-1}$$

ε denotes the permittivity tensor. This relation is valid for all materials which have a time independent permittivity. Furthermore, polarization by mechanical forces is neglected [2.44]. Both assumptions hold relatively well considering the usual applications of semiconductor devices. However, an investigation of piezoelectric phenomena, ferroelectric phenomena and nonlinear optics is impossible when using only (2.1-1).

As the next step it is desirable to relate the electric field vector $\vec{E}$ to the electrostatic potential ψ. For that purpose we solve (2-4) by introducing a vector field $\vec{A}$ and remembering that "div rot" applied to any vector is always zero.

$$\vec{B} = \text{rot}\,\vec{A}, \quad \text{div}\,\vec{A} = 0 \tag{2.1-2}$$

Note that we assume for the applicability of (2.1-2) that the speed of light is large compared to all velocities which are relevant for a device [2.44].

We substitute (2.1-2) into (2-2) and we obtain readily (2.1-3).

$$\text{rot}\left(\vec{E} + \frac{\partial \vec{A}}{\partial t}\right) = 0 \tag{2.1-3}$$

If "rot $\vec{z} = 0$" holds for a vector field $\vec{z}$ we know from basic differential calculus that $\vec{z}$ can be expressed as a gradient field. Therefore, the electric field vector $\vec{E}$ can be expressed as:

$$\vec{E} = -\frac{\partial \vec{A}}{\partial t} - \text{grad}\,\psi \tag{2.1-4}$$

Now we substitute (2.1-4) into (2.1-1) and then the result into (2-3).

$$\vec{D} = -\varepsilon \cdot \frac{\partial \vec{A}}{\partial t} - \varepsilon \cdot \text{grad}\,\psi \tag{2.1-5}$$

$$\text{div}\left(\varepsilon \cdot \frac{\partial \vec{A}}{\partial t}\right) + \text{div}\,(\varepsilon \cdot \text{grad}\,\psi) = -\rho \tag{2.1-6}$$

The first term in (2.1-6) is zero if the permittivity ε can be considered to be homogenous. Thus, we finally end up with (2.1-7) which is the well known form of Poisson's equation.

$$\text{div}\,(\varepsilon \cdot \text{grad}\,\psi) = -\rho \tag{2.1-7}$$

The space charge density ρ can be further broken apart (2.1-8) into the product of the elementary charge q times the sum of the positively charged hole concentration p, the negatively charged electron concentration n and an additional concentration C which will be subject of later investigations.

$$\rho = q \cdot (p - n + C) \tag{2.1-8}$$

From a purely mathematical point of view (2.1-8) represents a substitution only, without introducing any assumptions. However, additional assumptions are brought about by modeling the quantities n, p etc. as will become clearly apparent in Sections 2.3 and 2.4.

The permittivity ε will be treated here in all further investigations as a scalar quantity. In principle it has to be represented as a tensor of rank two. However, the materials currently in use for device fabrication do not show a significant anistropy of the permittivity owing to their special composition, e.g. cubic lattice or amorphous structure. Inhomogeneity effects of the permittivity have been neglected in (2.1-7). There does not exist pronounced experimental evidence for inhomogeneity effects. For some materials the relative permittivity constants $\varepsilon_r = \varepsilon/\varepsilon_0$ are summarized in Table 2.1-1.

Table 2.1-1. *Relative permittivity constants*

material	ε_r [ ]
Si	11.7
SiO_2	3.9
Si_3N_4	7.2 typical
GaAs	12.5
Ge	16.1

In particular for Si_3N_4 the value of ε_r depends strongly on the individual processing conditions; it can vary quite significantly.
If we introduce (2.1-8) and the assumption of a homogeneous scalar permittivity into (2.1-7), we obtain the final form of Poisson's equation to be used for semiconductor device modeling.

$$\operatorname{div} \operatorname{grad} \psi = \frac{q}{\varepsilon} \cdot (n - p - C) \tag{2.1-9}$$

2.2 Continuity Equations

The continuity equations can be derived in a straightforward manner from the first Maxwell equation (2-1). If we apply the operator "div" on this equation we obtain:

$$\operatorname{div} \operatorname{rot} \vec{H} = \operatorname{div} \vec{J} + \frac{\partial \rho}{\partial t} = 0 \tag{2.2-1}$$

Now we split the conduction current density $\vec{J}$ into a component $\vec{J}_p$ caused by holes and a component $\vec{J}_n$ caused by electrons:

$$\vec{J} = \vec{J}_p + \vec{J}_n \tag{2.2-2}$$

Furthermore, we assume that all charges in the semiconductor, except the mobile carriers electrons and holes, are time invariant. Thus we neglect the influence of charged defects, e. g. vacancies, dislocations, deep recombination traps, which may change their charge state in time.

$$\frac{\partial C}{\partial t} = 0 \tag{2.2-3}$$

If we substitute (2.1-8) and (2.2-2) into (2.2-1) and if we make use of (2.2-3) we obtain:

$$\operatorname{div} (\vec{J}_p + \vec{J}_n) + q \cdot \frac{\partial}{\partial t} (p - n) = 0 \tag{2.2-4}$$

This result is interpreted fairly trivially. It just means that sources and sinks of the total conduction current are fully compensated by the time variation of the mobile charge. In order to obtain two continuity equations a few formal steps have to be carried out. We first define a quantity R in (2.2-5) and, secondly, we rewrite (2.2-4) by making use of the definition R.

$$\operatorname{div} \vec{J}_n - q \cdot \frac{\partial n}{\partial t} = q \cdot R \tag{2.2-5}$$

$$\operatorname{div} \vec{J}_p + q \cdot \frac{\partial p}{\partial t} = -q \cdot R \tag{2.2-6}$$

It is obvious that we can not gain information by writing one equation (2.2-4) in two different ways (2.2-5), (2.2-6). However, these formal steps enable us to interprete the equation more easily. The quantity R can be understood as a function describing the net generation or recombination of electrons and holes. Positive R means recombination and negative R means generation. So far we have no information about the structure of R except equations (2.2-5) and (2.2-6). R has to be modeled carefully (cf. Section 4.2) using knowledge from the solid-state physics of semiconductors. If we have a model for R, equations (2.2-5) and (2.2-6) can really be considered as two equations. It seems worthwhile to note explicitly here that there is no necessity or even evidence that R can be expressed as a function depending only upon local quantities and not upon integral quantities; non-local generation or recombination phenomena may certainly take place in semiconductor devices considering only the derivation of the continuity equations.

2.3 Carrier Transport Equations

The derivation of current relations for the semiconductor equations is a very cumbersome task. It is not the intention of this book to cover the extraordinarily wide field of physics behind all the considerations necessary to derive the current relations in detail. Therefore, some of the required relations will be given without proof, but with reference to a text more specialized in that field.
Without loss of generality the current density of charged particles is the product of the charge constant per particle, the particle concentration and the average velocity (drift velocity) of the particles. So the hole current density and the electron current density can be written as (2.3-1) and (2.3-2), respectively.

$$\vec{J}_p = q \cdot p \cdot \vec{v}_p \tag{2.3-1}$$

$$\vec{J}_n = -q \cdot n \cdot \vec{v}_n \tag{2.3-2}$$

The major problem is to find expressions which relate the average carrier velocities to the electric field vector $\vec{E}$ and to the carrier concentration. In order to obtain information about the drift velocity we have to describe the carrier concentration by means of a distribution function f_v in phase space which is the space of spatial coordinates $\vec{x} = (x, y, z)^T$, momentum coordinates $\vec{k} = (k_x, k_y, k_z)^T$ and time t, thus a seven dimensional space. The distribution function determines the carrier concentration per unit volume of phase space. By integrating the distribution function over the entire momentum volume V_k we obtain the carrier concentration $\nu(\vec{x}, t)$. ν stands for n or p, denoting electrons or holes.

$$\frac{1}{4 \cdot \pi^3} \cdot \int_{Vk} f_\nu(\vec{x}, \vec{k}, t) \cdot d\vec{k} = \nu(\vec{x}, t) \tag{2.3-3}$$

This normalization (2.3-3) defines f_ν as a probability. In the literature various different normalizations can be found, e.g. [2.42], [2.49].
The distribution function has the property that its derivative along a particle trajectory $\vec{x}_\nu(t)$, $\vec{k}_\nu(t)$ with respect to time vanishes in the entire phase space in compliance with the Liouville theorem about the invariance of the phase volume for a system moving along the phase paths or an account of the conservation of the number of states [2.49].

$$\frac{d}{dt} f_\nu(\vec{x}_\nu(t), \vec{k}_\nu(t), t) = 0 \tag{2.3-4}$$

Equation (2.3-4) is the Boltzmann transport equation in implicit form. By expanding the total derivative we obtain:

$$\frac{\partial f_\nu}{\partial t} + \mathrm{grad}_k\, f_\nu \cdot \frac{d\vec{k}_\nu}{dt} + \mathrm{grad}_x\, f_\nu \cdot \frac{d\vec{x}_\nu}{dt} = 0 \tag{2.3-5}$$

Here grad_k denotes the gradient operator with respect to the momentum coordinates $\vec{k}$; grad_x is the gradient operator with respect to the spatial coordinates $\vec{x}$. Equation (2.3-5) shows that the variation of the distribution function at each point of phase space $(\vec{x}, \vec{k})$ with time is caused by the motion of particles in normal space $(\vec{x})$ and in momentum space $\vec{k}$.
The derivative of $\vec{k}_\nu$ with respect to time multiplied with Planck's constant $\hbar$ equals the sum of all forces $\vec{F}$. These forces have to be devided into two classes (2.3-7).

$$\frac{d\vec{k}_\nu}{dt} = \frac{\vec{F}_\nu}{\hbar}, \quad \hbar = \frac{h}{2 \cdot \pi} \tag{2.3-6}$$

$$\vec{F}_\nu = \vec{F}_{\nu e} + \vec{F}_{\nu i} \tag{2.3-7}$$

$\vec{F}_{\nu e}$ comprises forces due to macroscopic external fields and $\vec{F}_{\nu i}$ denotes forces due to internal localized crystal attributes like impurity atoms or ions, vacancies, and thermal lattice vibrations. It is quite impossible to calculate the effect of internal forces $\vec{F}_{\nu i}$ upon the distribution function from the laws of dynamics [2.49]. Statistical laws have to be invoked instead. By introducing the quantity $S_\nu(\vec{k}, \vec{k}') \cdot d\vec{k}'$ which is the probability per unit time that a carrier in the state $\vec{k}$ will be scattered into the momentum volume $d\vec{k}'$, we can write the internal collision term as follows:

$$\begin{aligned} \mathrm{grad}_k\, f_\nu \cdot \frac{\vec{F}_{\nu i}}{\hbar} = \int_{Vk'} & (f_\nu(\vec{x}, \vec{k}, t) \cdot (1 - f_\nu(\vec{x}, \vec{k}', t)) \cdot S_\nu(\vec{k}, \vec{k}') - \\ & - f_\nu(\vec{x}, \vec{k}', t) \cdot (1 - f_\nu(\vec{x}, \vec{k}, t)) \cdot S_\nu(\vec{k}', \vec{k})) \cdot d\vec{k}' \end{aligned} \tag{2.3-8}$$

(2.3-8) is termed the collision integral. The first term in the integrand describes the number of carriers scattered from the state $\vec{k}$ into the volume element $d\vec{k}'$ per unit time. $f_v(\vec{x}, \vec{k}, t)$ gives the probability that a carrier initially occupies the state $\vec{k}$. $[1-f_v(\vec{x}, \vec{k}', t)]$ gives the probability that the volume element $d\vec{k}'$ is initially unoccupied and can, therefore, accept a carrier. $S_v(\vec{k}, \vec{k}')$ gives an a priori probability of the scattering event. Correspondingly, the second term in the integrand of (2.3-8) equals the number of electrons scattered from volume element $d\vec{k}'$ into state $\vec{k}$ per unit time. Thorough investigations about the scattering probability $S_v(\vec{k}, \vec{k}')$ can be found in, e. g., [2.21], [2.95].

The derivative of $\vec{x}_v$ with respect to time represents the group velocity of the carriers.

$$\frac{d\vec{x}_v}{dt} = \vec{u}_v \tag{2.3-9}$$

We have now to substitute the relations (2.3-6) to (2.3-9) into (2.3-5), and we obtain the Boltzmann transport equation in explicit form.

$$\begin{aligned} &\frac{\partial f_v}{\partial t} + \frac{\vec{F}_{ve}}{\hbar} \cdot \operatorname{grad}_k f_v + \vec{u}_v \cdot \operatorname{grad}_x f_v = \\ &= -\int\limits_{Vk'} \big(f_v(\vec{x}, \vec{k}, t) \cdot (1 - f_v(\vec{x}, \vec{k}', t)) \cdot S_v(\vec{k}, \vec{k}') - \\ &\quad - f_v(\vec{x}, \vec{k}', t) \cdot (1 - f_v(\vec{x}, \vec{k}, t)) \cdot S_v(\vec{k}', \vec{k})\big) \cdot d\vec{k}' \end{aligned} \tag{2.3-10}$$

A fairly accurate approach would be to directly solve (2.3-10) in order to calculate carrier concentrations and drift velocities. However, this is an extraordinarily difficult task to accomplish. (2.3-10) represents an integro-differential equation with seven independent variables. This equation does not admit a closed solution. It rather requires the use of iterative procedures which, moreover, are scarcely suitable for numerical approaches [2.10], or additionally, invoke very stringent assumptions [2.42].

An alternative approach to solving the Boltzmann equation consists in simulating the motion of one or more carriers at microscopic level with Monte Carlo methods, e. g. [2.43]. However, this category of simulations is very computationally intensive [2.64], [2.65] and therefore, with a few exceptions only, not suitable for engineering application.

One should be aware of the fact that the validity of the Boltzmann equation (2.3-10) implies already several assumptions (cf. [2.10], [2.17]).

- The scattering probability is independent of external forces.
- The duration of a collision is much shorter than the average time of motion of a particle; collisions are instantaneous.
- Carrier-carrier interaction is negligible. This effect would change the integrand of the right hand side integral in (2.3-10) highly nonlinear in f_v [2.4].
- External forces are almost constant over a length comparable to the physical dimensions of the wave packet describing the motion of a carrier.

- The band theory and the effective mass theorem apply to the semiconductor under consideration [2.76].

However, it is my intention here to outline the derivation of the classical current relations and only to pinpoint the problems associated with much more basic and error-prone models.
By assuming that all scattering processes are elastic and by neglecting all effects by degeneracy the scattering integral can be approximated and the Boltzmann equation is reduced to a pure differential equation [2.21], [2.42], [2.76].

$$\frac{\partial f_\nu}{\partial t}+\frac{\vec{F}_{\nu e}}{\hbar}\cdot \mathrm{grad}_k\, f_\nu+\vec{u}_\nu\cdot \mathrm{grad}_x\, f_\nu=-\frac{f_\nu-f_{\nu o}}{\tau_\nu} \tag{2.3-11}$$

The physical motivation for the right hand side of (2.3-11) is as follows: Suppose that at some moment of time $t=0$ all external forces are switched off and f_ν is homogeneous.

$$\frac{\vec{F}_{\nu e}}{\hbar}\cdot \mathrm{grad}_k\, f_\nu+\vec{u}_\nu\cdot \mathrm{grad}_x\, f_\nu=0 \tag{2.3-12}$$

It follows from (2.3-11) that the distribution function will change as a result of collisions only. (2.3-13) will reduce to:

$$\frac{\partial f_\nu}{\partial t}=-\frac{f_\nu-f_{\nu o}}{\tau_\nu} \tag{2.3-13}$$

The solution of this differential equation is quite simple.

$$f_\nu(\vec{x},\vec{k},t)=f_{\nu o}(\vec{x},\vec{k})+\big(f_\nu(\vec{x},\vec{k},0)-f_{\nu o}(\vec{x},\vec{k})\big)\cdot e^{-t/\tau_\nu} \tag{2.3-14}$$

$f_{\nu o}$ is the equilibrium distribution function, and the quantity τ_ν shows the rate of return to the state of equilibrium from the disturbed state, therefore, it is termed the relaxation time. Under the very restrictive assumptions stated above the problem of solving the Boltzmann equation can be eased drastically by modeling the relaxation time as only a function of energy [2.42].
In order to obtain the current relations from (2.3-11) we multiply this equation with the group velocity $\vec{u}_\nu$, and then we integrate the equation over momentum space.

$$\int\limits_{Vk}\vec{u}_\nu\cdot\frac{\partial f_\nu}{\partial t}\cdot d\vec{k}+\int\limits_{Vk}\vec{u}_\nu\cdot\left(\frac{\vec{F}_{\nu e}}{\hbar}\cdot \mathrm{grad}_k\, f_\nu\right)\cdot d\vec{k}+\int\limits_{Vk}\vec{u}_\nu\cdot(\vec{u}_\nu\cdot \mathrm{grad}_x\, f_\nu)\cdot d\vec{k}=$$

$$=-\int\limits_{Vk}\vec{u}_\nu\cdot\frac{f_\nu-f_{\nu o}}{\tau_\nu}\cdot d\vec{k} \tag{2.3-15}$$

For the solution of (2.3-15) we make use of the following four solutions to integrals, the verification and discussion of which is not necessarily trivial, but well established in the literature, e. g. [2.13], [2.21], [2.71].

$$\int_{Vk} \vec{u}_\nu \cdot \frac{\partial f_\nu}{\partial t} \cdot d\vec{k} = 4 \cdot \pi^3 \cdot \frac{\partial}{\partial t}(\nu \cdot \vec{v}_\nu) \tag{2.3-16}$$

$$\int_{Vk} \vec{u}_\nu \cdot \left(\frac{\vec{F}_{\nu e}}{\hbar} \cdot \mathrm{grad}_k\, f_\nu\right) \cdot d\vec{k} = \frac{\vec{F}_{\nu e}}{\hbar} \cdot \int_{Vk} \vec{u}_\nu \cdot \mathrm{grad}_k\, f_\nu \cdot d\vec{k} = -4 \cdot \pi^3 \cdot \vec{F}_{\nu e} \cdot \frac{\nu}{m_\nu^*} \tag{2.3-17}$$

$$\int_{Vk} \vec{u}_\nu \cdot (\vec{u}_\nu \cdot \mathrm{grad}_x\, f_\nu) \cdot d\vec{k} = \frac{4 \cdot \pi^3}{m_\nu^*} \cdot \mathrm{grad}_x (\nu \cdot k \cdot T) \tag{2.3-18}$$

$$\int_{Vk} \vec{u}_\nu \cdot \frac{f_\nu - f_{\nu o}}{\tau_\nu} \cdot d\vec{k} = \frac{1}{\tau_\nu} \cdot \int_{Vk} \vec{u}_\nu \cdot (f_\nu - f_{\nu o}) \cdot d\vec{k} = 4 \cdot \pi^3 \cdot \frac{\nu \cdot \vec{v}_\nu}{\tau_\nu} \tag{2.3-19}$$

ν denotes the carrier concentration, $\vec{v}_\nu$ is the drift velocity, m_ν^* represents the effective mass, T denotes the lattice temperature and τ_ν in the right hand sides of (2.3-19) is an average collision time. The constant k on the right hand side of (2.3-18) denotes the Boltzmann constant. The external forces $\vec{F}_{\nu e}$ can be expressed in terms of the electric field $\vec{E}$ if the magnetic induction $\vec{B}$ (Lorentz force) is neglected which is a requirement also for the validity of (2.3-16).

$$\vec{F}_{ne} = -q \cdot \vec{E}, \quad \vec{F}_{pe} = q \cdot \vec{E} \tag{2.3-20}$$

In (2.3-18) it has been assumed that the drift energy of the carriers is negligibly small compared to the thermal energy. Therefore, relation (2.3-18) is invalid for hot carriers (cf. [2.67]).
We obtain ordinary differential equations for the drift velocities of electrons and holes using the above given integrals and the force relations (2.3-20).

$$\frac{\partial}{\partial t}(n \cdot \vec{v}_n) + \frac{q}{m_n^*} \cdot n \cdot \vec{E} + \frac{1}{m_n^*} \cdot \mathrm{grad}\,(n \cdot k \cdot T) = -\frac{n \cdot \vec{v}_n}{\tau_n} \tag{2.3-21}$$

$$\frac{\partial}{\partial t}(p \cdot \vec{v}_p) - \frac{q}{m_p^*} \cdot p \cdot \vec{E} + \frac{1}{m_p^*} \cdot \mathrm{grad}\,(p \cdot k \cdot T) = -\frac{p \cdot \vec{v}_p}{\tau_p} \tag{2.3-22}$$

These equations can be regarded also as macroscopic force balance equations. A "closed solution" of these equations is, unfortunately, not possible. In order to obtain an approximate solution we introduce effective carrier mobilities μ_n and μ_p.

$$\mu_n = \frac{q \cdot \tau_n}{m_n^*} \tag{2.3-23}$$

$$\mu_p = \frac{q \cdot \tau_p}{m_p^*} \tag{2.3-24}$$

We rewrite (2.3-21) and (2.3-22) after multiplication with the corresponding average collision times τ_ν and charge constant $\pm q$, and – remembering (2.3-1) and (2.3-2) – we end up with:

$$\tau_n \cdot \frac{\partial \vec{J}_n}{\partial t} + \vec{J}_n = q \cdot \mu_n \cdot n \cdot \left(\vec{E} + \frac{1}{n} \cdot \text{grad} \left(n \cdot \frac{k \cdot T}{q} \right) \right) \tag{2.3-25}$$

$$\tau_p \cdot \frac{\partial \vec{J}_p}{\partial t} + \vec{J}_p = q \cdot \mu_p \cdot p \cdot \left(\vec{E} - \frac{1}{p} \cdot \text{grad} \left(p \cdot \frac{k \cdot T}{q} \right) \right) \tag{2.3-26}$$

The average collision times τ_v are very small, typically in the order of tenth of picoseconds. Therefore, equations (2.3-25) and (2.3-26) can be understood as being singularly perturbed. This suggests to expand the solution into powers of the perturbation parameter which is the collision time.

$$\vec{J}_n(\tau_n) = \sum_{i=0}^{\infty} \vec{J}_{ni} \cdot (\tau_n)^i \tag{2.3-27}$$

$$\vec{J}_p(\tau_p) = \sum_{i=0}^{\infty} \vec{J}_{pi} \cdot (\tau_p)^i \tag{2.3-28}$$

We have an algebraic equation for the zero order term of the current density.

$$\vec{J}_{no} = q \cdot \mu_n \cdot n \cdot \left(\vec{E} + \frac{1}{n} \cdot \text{grad} \left(n \cdot \frac{k \cdot T}{q} \right) \right) \tag{2.3-29}$$

$$\vec{J}_{po} = q \cdot \mu_p \cdot p \cdot \left(\vec{E} - \frac{1}{p} \cdot \text{grad} \left(p \cdot \frac{k \cdot T}{q} \right) \right) \tag{2.3-30}$$

These equations are formally approximations of order τ_v.

$$\vec{J}_n = \vec{J}_{no} + O(\tau_n) \tag{2.3-31}$$

$$\vec{J}_p = \vec{J}_{po} + O(\tau_p) \tag{2.3-32}$$

We further assume that the lattice temperature is constant.

$$T = \text{const.} \tag{2.3-33}$$

Then we can use the substitutions (2.3-34) and (2.3-35) which by means of physical interpretation are termed the Einstein relations

$$D_n = \mu_n \cdot \frac{k \cdot T}{q} \tag{2.3-34}$$

$$D_p = \mu_p \cdot \frac{k \cdot T}{q} \tag{2.3-35}$$

to define the diffusion constants D_v, and, finally, we are able to write down the current relations in the well known, established form as sums of a drift and a diffusion component.

$$\vec{J}_n \cong q \cdot n \cdot \mu_n \cdot \vec{E} + q \cdot D_n \cdot \text{grad}\, n \tag{2.3-36}$$

$$\vec{J}_p \cong q \cdot p \cdot \mu_p \cdot \vec{E} - q \cdot D_p \cdot \text{grad}\, p \tag{2.3-37}$$

In the following I should like to summarize the most important assumptions which had to be performed over and above to the ones necessary for the validity of the Boltzmann equation to obtain the current relations (2.3-36) and (2.3-37).

- All scattering processes have been assumed to be elastic. Therefore, for instance, polar optical phonon scattering, which is a major scattering mechanism in GaAs, has been neglected.
- The spatial variations of the collision time and the band structure are neglected. This implies a slowly varying impurity concentration over a carrier mean free path.
- Effects of degeneracy have been neglected in the approximation for the scattering integral.
- The spatial variation of the external forces is neglected which implies a slowly varying electric field vector.
- The influence of the Lorentz force is ignored by assuming zero magnetic induction.
- The time and spatial variation of carrier temperature is neglected and, furthermore, lattice and carrier temperature are assumed to be equal. Therefore, the diffusion of hot carriers is improperly described. Several authors have tried to overcome this problem by using modified Einstein relations [2.9], [2.52], [2.53], [2.67], [2.68], [2.69], [2.85].
- Parabolic energy bands are assumed which is an additional reason why degenerate semiconductor materials cannot be treated properly. Calculations of the realistic band structure of various semiconductors can be found in, e.g., [2.20]. However, for a realistic band structure it can become necessary to use a system of Boltzmann equations to describe the carrier distribution instead of just one (cf. [2.96]).
- The zero order term of the series expansions of $\vec{J}_n$ and $\vec{J}_p$ into powers of the collision time only has been taken into account. Thus, all time dependent conductivity phenomena, like velocity overshoot, are not included.
- The semiconductor has been assumed to be infinitely large. In a real device the distribution function is changed in a complex, highly irregular manner in the vicinity of boundaries, for instance contacts [2.58] and interfaces [2.36]. It can be expected that the drift-diffusion approximation fails within a few carrier mean free paths of boundaries.

In the literature we can find quite a few papers and books whose authors use a different form of the current relations. These are based upon special assumptions equivalent to the drift-diffusion approximations. The procedure to derive these slightly different relations will be outlined next. We know from semiconductor statistics that the equilibrium distribution functions for electrons and holes are Fermi functions.

$$f_{no}(\vec{x},\vec{k})=\frac{1}{1+\exp\left(\dfrac{E_c(\vec{x},\vec{k})-E_{fn}(\vec{x})}{k\cdot T(\vec{x})}\right)} \tag{2.3-38}$$

$$f_{po}(\vec{x},\vec{k})=\frac{1}{1+\exp\left(\dfrac{E_{fp}(\vec{x})-E_v(\vec{x},\vec{k})}{k\cdot T(\vec{x})}\right)} \tag{2.3-39}$$

E_c and E_v denote the conduction and the valence band energy, respectively.

$$E_c(\vec{x},\vec{k}) = E_{co} - q\cdot\psi(\vec{x}) + \frac{\hbar^2\cdot\vec{k}\cdot\vec{k}}{2\cdot m_n^*} \tag{2.3-40}$$

$$E_v(\vec{x},\vec{k}) = E_{vo} - q\cdot\psi(\vec{x}) - \frac{\hbar^2\cdot\vec{k}\cdot\vec{k}}{2\cdot m_p^*} \tag{2.3-41}$$

E_{co} and E_{vo} are the conduction band and the valence band edge, respectively; $\psi(\vec{x})$ is the electrostatic potential as defined in Section 2.1.
E_{fn} and E_{fp} in (2.3-38) and (2.3-39) determine the Fermi energy for electrons and holes. We shall try now to calculate a correction term (2.3-42) to the equilibrium distribution function.

$$f_v(\vec{x},\vec{k},t) = f_{vo}(\vec{x},\vec{k}) + f_{v1}(\vec{x},\vec{k},t) \tag{2.3-42}$$

We recall the Boltzmann equation with the relaxation time approximation:

$$\frac{\partial f_v}{\partial t} + \frac{\vec{F}_{ve}}{\hbar}\cdot \mathrm{grad}_k\, f_v + \vec{u}_v\cdot \mathrm{grad}_x\, f_v = -\frac{f_v - f_{vo}}{\tau_v} \tag{2.3-43}$$

By assuming

$$\left|\frac{\partial f_v}{\partial t}\right| \ll \left|\frac{\vec{F}_{ve}}{\hbar}\cdot \mathrm{grad}_k\, f_v + \vec{u}_v\cdot \mathrm{grad}_x\, f_v\right| \tag{2.3-44}$$

we obtain a simplified form of the Boltzmann equation.

$$\frac{\vec{F}_{ve}}{\hbar}\cdot \mathrm{grad}_k\, f_v + \vec{u}_v\cdot \mathrm{grad}_x\, f_v = -\frac{f_v - f_{vo}}{\tau_v} \tag{2.3-45}$$

(2.3-45) is valid only for small perturbations from equilibrium. Then we estimate the correction term f_{v1} to the equilibrium distribution:

$$f_{v1} \cong -\tau_v\cdot\left(\frac{\vec{F}_{ve}}{\hbar}\cdot \mathrm{grad}_k\, f_{vo} + \vec{u}_v\cdot \mathrm{grad}_x\, f_{vo}\right) \tag{2.3-46}$$

The spatial gradients of the equilibrium distribution functions for electrons and holes are:

$$\mathrm{grad}_x\, f_{no} = f_{no}\cdot(1-f_{no})\cdot \mathrm{grad}_x\left(\frac{q\cdot\psi + E_{fn}}{k\cdot T}\right) \tag{2.3-47}$$

$$\mathrm{grad}_x\, f_{po} = -f_{po}\cdot(1-f_{po})\cdot \mathrm{grad}_x\left(\frac{q\cdot\psi + E_{fp}}{k\cdot T}\right) \tag{2.3-48}$$

The gradient of the equilibrium distribution function with respect to $\vec{k}$ evaluates to:

$$\mathrm{grad}_k\, f_{vo} = -f_{vo}\cdot(1-f_{vo})\cdot\frac{\hbar^2\cdot\vec{k}}{m_v^*\cdot k\cdot T} \tag{2.3-49}$$

The group velocity can be expressed for a parabolic, isotropic band as:

$$\vec{u}_v = \frac{\hbar\cdot\vec{k}}{m_v^*} \tag{2.3-50}$$

By assuming vanishing variation of temperature $\text{grad}_x\, T=0$ and substituting relations (2.3-47) to (2.3-50) into (2.3-46) we obtain expressions (2.3-51) and (2.3-52) for the distribution functions for electrons and holes, respectively.

$$f_n \cong f_{no} - \tau_n \cdot f_{no} \cdot (1 - f_{no}) \cdot \frac{\vec{u}_n}{k \cdot T} \cdot \text{grad}_x\, E_{fn} \tag{2.3-51}$$

$$f_p \cong f_{po} + \tau_p \cdot f_{po} \cdot (1 - f_{po}) \cdot \frac{\vec{u}_p}{k \cdot T} \cdot \text{grad}_x\, E_{fp} \tag{2.3-52}$$

The current densities can now be evaluated as the integrals of the product of group velocity and distribution function over momentum space by further assuming $f_{\nu o} \ll 1$.

$$\vec{J}_n = \frac{-q}{4 \cdot \pi^3} \cdot \int\limits_{Vk} \vec{u}_n \cdot f_n \cdot d\vec{k} = -q \cdot \mu_n \cdot n \cdot \text{grad}\, \varphi_n \tag{2.3-53}$$

$$\vec{J}_p = \frac{q}{4 \cdot \pi^3} \cdot \int\limits_{Vk} \vec{u}_p \cdot f_p \cdot d\vec{k} = -q \cdot \mu_p \cdot p \cdot \text{grad}\, \varphi_p \tag{2.3-54}$$

φ_n and φ_p denote the quasi-Fermi potentials for electrons and holes which are related to the Fermi energies for small external forces by:

$$\text{grad}\, \varphi_\nu = -q \cdot \text{grad}_x\, E_{f\nu} \tag{2.3-55}$$

The simplifying assumptions necessary to derive (2.3-53) and (2.3-54) are quite similar to those required for the derivation of the drift-diffusion current relations (2.3-36) and (2.3-37). However, if we introduce an effective intrinsic concentration to fit moderate heavy doping effects (cf. Section 2.4)

$$n = n_{ie} \cdot \exp\left(\frac{q \cdot (\psi - \varphi_n)}{k \cdot T}\right) \tag{2.3-56}$$

$$p = n_{ie} \cdot \exp\left(\frac{q \cdot (\varphi_p - \psi)}{k \cdot T}\right) \tag{2.3-57}$$

in (2.3-53), (2.3-54), we obtain directly the drift-diffusion formulation of the current densities. We get after straightforward calculation using (2.3-56) and (2.3-57) the following expressions for the quasi-Fermi potentials φ_ν:

$$\varphi_n = \psi - \frac{k \cdot T}{q} \cdot \ln\left(\frac{n}{n_{ie}}\right) \tag{2.3-58}$$

$$\varphi_p = \psi + \frac{k \cdot T}{q} \cdot \ln\left(\frac{p}{n_{ie}}\right) \tag{2.3-59}$$

We substitute these expressions into (2.3-53) and (2.3-54).

$$\vec{J}_n = -q \cdot \mu_n \cdot n \cdot \text{grad}\left(\psi - \frac{k \cdot T}{q} \cdot \ln\left(\frac{n}{n_{ie}}\right)\right) \tag{2.3-60}$$

$$\vec{J}_p = -q \cdot \mu_p \cdot p \cdot \text{grad}\left(\psi + \frac{k \cdot T}{q} \cdot \ln\left(\frac{p}{n_{ie}}\right)\right) \tag{2.3-61}$$

Then we evaluate the "grad" operator and obtain relations (2.3-62) and (2.3-63) for the electron and hole current, respectively.

$$\vec{J}_n = q \cdot n \cdot \mu_n \cdot \vec{E} + q \cdot D_n \cdot \text{grad}\, n - q \cdot \mu_n \cdot n \cdot \left(\frac{k \cdot T}{q} \cdot \text{grad} \ln(n_{ie})\right) \tag{2.3-62}$$

$$\vec{J} = q \cdot p \cdot \mu_p \cdot \vec{E} - q \cdot D_p \cdot \text{grad}\, p + q \cdot \mu_p \cdot p \cdot \left(\frac{k \cdot T}{q} \cdot \text{grad} \ln(n_{ie})\right) \tag{2.3-63}$$

The last term in these relations represents a current which is caused by a possible dependence on position of the intrinsic concentration. It thus accounts for variations in the bandgap, and it will describe the bandgap narrowing effect observed in heavily doped semiconductors (see also [2.91]). If we assume a constant intrinsic concentration we obviously do not have a gradient of the intrinsic concentration, and then relations (2.3-62) and (2.3-63) are indeed identical to (2.3-36) and (2.3-37). For practical purpose it is often useful to define effective fields for the drift current components of the electron and hole current density.

$$\vec{E}_n = \vec{E} - \frac{k \cdot T}{q} \cdot \text{grad} \ln(n_{ie}) \tag{2.3-64}$$

$$\vec{E}_p = \vec{E} + \frac{k \cdot T}{q} \cdot \text{grad} \ln(n_{ie}) \tag{2.3-65}$$

By rewriting (2.3-62) and (2.3-63) we obtain a form of the current relations which is very similar to the classical drift-diffusion approximations but which can, as in our example, take into account positional variations of the band gap.

$$\vec{J}_n = q \cdot n \cdot \mu_n \cdot \vec{E}_n + q \cdot D_n \cdot \text{grad}\, n \tag{2.3-66}$$

$$\vec{J}_p = q \cdot p \cdot \mu_p \cdot \vec{E}_p - q \cdot D_p \cdot \text{grad}\, p \tag{2.3-67}$$

It is worth noting that Boltzmann statistics for the carrier concentrations have been used in (2.3-66) and (2.3-67). With Fermi statistics for the carrier concentrations an equally simple form of the current relations compared to the classical drift-diffusion approximations is not derived so easily.

In the following I would like to summarize some of the simplifying assumptions which became clear in the derivation of (2.3-51) and (2.3-52), and which, additionally have been implicitly used for the derivation of the drift diffusion relations (2.3-34) and (2.3-35) (see also [2.10]).

- Higher order derivatives of the quasi-Fermi potentials have been neglected (cf. (2.3-46)). This means that we transform a non-local solution of the Boltzmann equation into an approximate one depending only upon the local gradient of the quasi-Fermi potential.
- The dependence of the distribution function upon the gradient of the quasi-Fermi potential has been linearized. That means that the scale of length over which the

quasi-Fermi potential varies by $k \cdot T/q$ must be large compared to the carrier mean free path.

- Only to first order is the carrier transport driving force the gradient of the quasi-Fermi potential. Away from equilibrium the electric field vector will become important.

Various authors have published approaches for a more sophisticated treatment of carrier transport in semiconductors. Froelich and Blakey, for instance, have carried out a one dimensional simulation using energy and momentum conservation laws [2.31]. They use for the description of electron transport:

$$\frac{\partial v}{\partial t} = -v \cdot \frac{\partial v}{\partial x} + \frac{q \cdot E}{m^*} - \frac{2}{3 \cdot m^* \cdot n} \cdot \frac{\partial}{\partial x}\left(n \cdot \left(w - \frac{m^* \cdot v^2}{2}\right)\right) - \frac{v}{\tau_v} \tag{2.3-68}$$

$$\frac{\partial w}{\partial t} = -v \cdot \frac{\partial w}{\partial x} + q \cdot E \cdot v - \frac{2}{3 \cdot n} \cdot \frac{\partial}{\partial x}\left(n \cdot v \cdot \left(w - \frac{m^* \cdot v^2}{2}\right)\right) - \frac{w - w_0}{\tau_w} \tag{2.3-69}$$

v denotes the electron drift velocity, τ_v is the momentum relaxation time, w is the electron energy and τ_w represents the energy relaxation time. It can be seen that (2.3-68) is almost equivalent to (2.3-21) which is an intermediate result we have obtained during the derivation of the drift-diffusion relations. The total electron energy w has been assumed to be:

$$w = \frac{3 \cdot k \cdot T_n}{2} + \frac{m^* \cdot v^2}{2} \tag{2.3-70}$$

Equation (2.3-69) is obtained by multiplying the Boltzmann equation with the energy E and then integrating the result over $\vec{k}$ space [2.14], fairly similar to the procedure we had to go through to derive (2.3-21). However, in order to avoid terms of the order $\vec{u} \cdot (\vec{u} \cdot \vec{u})$ one has to specify the form of the distribution function to be a "displaced" Maxwellian [2.42].

$$f_v(\vec{x}, \vec{k}) \sim \exp\left(-\frac{E(\vec{x}, \vec{k} - \langle\vec{k}\rangle)}{k \cdot T_v}\right) \tag{2.3-71}$$

If one uses such a model based on energy and momentum conservation, one circumvents the assumption that the carrier energy and momentum distribution are always in equilibrium with the local electric field. However, the assumptions required to derive (2.3-69), e.g. a displaced Maxwellian distribution function (cf. [2.42], [2.95]), and the additional parameters to model, e.g. energy relaxation time, lead to many open questions which will have to be discussed quite thoroughly until these types of models become suitable for engineering applications. A comparison of such an electron temperature model, a classical drift-diffusion model and first principles particle model has been published by, e.g., Curtice [2.23], the summary of which is that simulation results differ quite significantly when using a model based upon classical current relations, an electron temperature model or a particle model. A two dimensional simulation with this type of equations has been presented in [2.16], [2.22].

Thornber [2.88] has published a generalized current equation for the simulation of submicron devices by supplementing the drift and diffusion current components with so-called gradient, rate and relaxation current components in order to include the most important features of velocity overshoot.

$$J_n = q \cdot n \cdot \left(v(E) + W(E) \cdot \frac{\partial E}{\partial x} + B(E) \cdot \frac{\partial E}{\partial t} \right) - q \cdot D(E) \cdot \frac{\partial n}{\partial x} - q \cdot A(E) \cdot \frac{\partial n}{\partial t} \tag{2.3-72}$$

Graphs of the gradient coefficient $W(E)$, the rate coefficient $B(E)$ and the relaxation coefficient $A(E)$ have been presented by Thornber for electrons in silicon at room temperature. $v(E)$ and $D(E)$ are the well known terms for the drift velocity and the diffusion coefficient. In the classical drift-diffusion current approximations all terms except those two are assumed to be negligibly small. Thornber stated in his article [2.88] that relations of the form (2.3-72) are adequate to represent current densities whenever the characteristic distances over which the particle concentration or electric field changes exceed 20 nm in silicon (200 nm in GaAs) and the characteristic time intervals of such changes exceed about 0.4 picoseconds in silicon (2 picoseconds in GaAs). However, as far as I know, relations of the form (2.3-72) – a generalization of which to higher dimensional form is supposed to be straightforward – have not been tested for practical applicability, although it can be speculated that the range of validity of such treatments is greatly extended.

In recent work a new concept of device operation has been brought about, namely ballistic transport. It was argued that by properly selecting the material, temperature, geometry and bias, a device could be built which is much smaller than the mean free path between scattering events [2.41], [2.78]. This question, however, is still open, although quite a few activities in that field can be observed [2.3], [2.6], [2.7], [2.37], [2.45], [2.74], [2.78], [2.87], [2.93].

As review papers on which kind of problems have to be faced especially for the simulation of miniaturized devices, references [2.10], [2.29], [2.30] can be suggested. Investigations on how the carrier transport equations, i.e. current relations, are changed, particularly by heavy doping effects, are presented in [2.59], [2.63], [2.70], [2.73], [2.90], [2.91]. In the next section we shall address these problems with emphasis on adequate models for the carrier concentrations. However, considering the current relations, throughout this text we shall favour current relations which have a structural appearing like (2.3-66) and (2.3-67). These formulations will allow us to best characterize, in a more mathematical sense, the problem of carrier transport. From a pragmatic physical point of view equations (2.3-66) and (2.3-67) offer, considering the state of the art in understanding their background, a sufficiently large set of parameters (effective mobilities μ_ν, effective fields $\vec{E}_\nu$, effective diffusivities D_ν) to be invoked in order to reach a specific goal for agreement between results obtained by simulation and measurement. One must keep in mind that all these equations are just models in any case, which only imitatively simulate a real process, more or less accurately, in a qualitative and quantitative sense.

2.4 Carrier Concentrations

Accurate models for the carrier concentrations in semiconductor devices are a necessity if qualitatively and quantitatively correct simulation results are to be obtained. I first shall review the "classical" approaches of modeling the carrier concentrations, which give fairly simple, closed form algebraic results. These approaches are certainly well documented in many books on semiconductor physics. However, I shall place particular emphasis on properly pointing out assumptions which are very possibly going to be violated when device sophistication keeps pace with the current trends.

Assuming a parabolic and isotropic band structure the density of possible states for the conduction band (2.4-1) and the valence band (2.4-2) as a function of energy E are given in the well known form with properly averaged effective masses [2.42]:

$$\rho_c(E) = \frac{4 \cdot \pi \cdot (2 \cdot m_n^*)^{3/2}}{h^3} \cdot \sqrt{E - E_c} \tag{2.4-1}$$

$$\rho_v(E) = \frac{4 \cdot \pi \cdot (2 \cdot m_p^*)^{3/2}}{h^3} \cdot \sqrt{E_v - E} \tag{2.4-2}$$

I have avoided in (2.4-1), (2.4-2) – as I shall do for all other formulae in this section – introducing a reference energy with a specific absolute value, because only energy differences are of relevance. One should, however, be quite careful in reading the literature because that problem is absolutely not treated in a unique manner. It seems to be very convenient, a matter of taste, to some people to introduce an arbitrary reference point with zero energy on the energy scale.

E_c and E_v are the so-called band edges. Their difference E_g denotes the band gap, i.e. the width of the forbidden band between conduction band and valence band.

$$E_g = E_c - E_v \tag{2.4-3}$$

Numerical values of the band gap and its linear temperature coefficient for the most frequently used materials semiconductor devices are made of are summarized in Table 2.4-1.

Table 2.4-1. *Band gaps in undoped material at $T = 300\,K$*

material	E_g [eV]	dE_g/dT [eV K^{-1}]
Si	1.12	$-2.7 \cdot 10^{-4}$
GaAs	1.35	$-5.0 \cdot 10^{-4}$
Ge	0.67	$-3.7 \cdot 10^{-4}$

For silicon the temperature dependence of the band gap can be modeled more accurately with (2.4-4) or (2.4-5) [2.32], [2.34].

$$E_g = 1.17\,\text{eV} - \frac{7.02 \cdot 10^{-4}\,\text{eV} \cdot \left(\frac{T}{K}\right)^2}{1108 + \left(\frac{T}{K}\right)} \tag{2.4-4}$$

$$E_g = 1.1785\,\text{eV} - 9.025 \cdot 10^{-5}\,\text{eV} \cdot \left(\frac{T}{K}\right) - 3.05 \cdot 10^{-7}\,\text{eV} \cdot \left(\frac{T}{K}\right)^2 \tag{2.4-5}$$

(2.4-4) is the older formula; it has also been published with slightly different constants, e.g. [2.5], [2.33].
The temperature dependence of the effective masses m_n^* and m_p^* of electrons and holes for the density of states in silicon can be modeled with polynomials which are fitted to experimental data [2.32], [2.34].

$$m_n^* = m_o \cdot \left(1.045 + 4.5 \cdot 10^{-4} \cdot \left(\frac{T}{K}\right)\right) \tag{2.4-6}$$

$$m_p^* = m_o \cdot \left(0.523 + 1.4 \cdot 10^{-3} \cdot \left(\frac{T}{K}\right) - 1.48 \cdot 10^{-6} \cdot \left(\frac{T}{K}\right)^2\right) \tag{2.4-7}$$

Values for the effective masses at room temperature are summarized for some of the relevant semiconductor materials in Table 2.4-2.

Table 2.4-2. *Effective mass ratios at $T = 300\,K$*

material	m_n^*/m_0 [ ]	m_p^*/m_0 [ ]
Si	1.18	0.5
GaAs	0.068	0.5
Ge	0.55	0.3

In order to obtain expressions for the carrier concentrations we have to integrate the density of states function multiplied with the corresponding carrier distribution function over the energy space.

$$n = \int_{E_c}^{\infty} \rho_c(E) \cdot f_n(E) \cdot dE \tag{2.4-8}$$

$$p = \int_{-\infty}^{E_v} \rho_v(E) \cdot f_p(E) \cdot dE \tag{2.4-9}$$

The lower integration bound in (2.4-8) is E_c because no possible states for electrons do exist for energies below the conduction band edge. For a similar reason the upper bound in (2.4-9) is E_v. The distribution functions $f_n(E)$ and $f_p(E)$ are Fermi functions.

$$f_n(E) = \frac{1}{1 + \exp\left(\dfrac{E - E_{fn}}{k \cdot T}\right)} \tag{2.4-10}$$

$$f_p(E) = \frac{1}{1 + \exp\left(\dfrac{E_{fp} - E}{k \cdot T}\right)} \tag{2.4-11}$$

E_{fn} and E_{fp} denote the Fermi energies for electrons and holes, respectively. Their exact meaning will be discussed later. The integrals in (2.4-8) and (2.4-9) can be evaluated to

$$n = N_c \cdot \frac{2}{\sqrt{\pi}} \cdot F_{1/2}\left(\frac{E_{fn} - E_c}{k \cdot T}\right) \tag{2.4-12}$$

$$p = N_v \cdot \frac{2}{\sqrt{\pi}} \cdot F_{1/2}\left(\frac{E_v - E_{fp}}{k \cdot T}\right) \tag{2.4-13}$$

where N_c and N_v denote the effective density of states in the conduction band and in the valence band, respectively.

$$N_c = 2 \cdot \left(\frac{2 \cdot \pi \cdot k \cdot T \cdot m_n^*}{h^2}\right)^{3/2} \tag{2.4-14}$$

$$N_v = 2 \cdot \left(\frac{2 \cdot \pi \cdot k \cdot T \cdot m_p^*}{h^2}\right)^{3/2} \tag{2.4-15}$$

$F_{1/2}(x)$ is the Fermi integral of order 1/2 which, unfortunately, does not have a closed form solution.

$$F_{1/2}(x) = \int_0^\infty \frac{\sqrt{y}}{1 + e^{y-x}} \cdot dy \tag{2.4-16}$$

The asymptotic behavior of $F_{1/2}(x)$ for large negative and large positive argument, however, is analytic.

$$F_{1/2}(x) \cong \frac{\sqrt{\pi}}{2} \cdot e^x, \quad x \ll -1 \tag{2.4-17}$$

$$F_{1/2}(x) \cong \frac{2}{3} \cdot x^{3/2}, \quad x \gg 1 \tag{2.4-18}$$

The qualitative behavior of $F_{1/2}(x)$ and its asymptotic expansions are shown in Fig. 2.4-1. For arguments close to zero it has been shown that $F_{1/2}(x)$ can be approximated with an expression of the following type:

$$F_{1/2}(x) \cong \frac{\frac{\sqrt{\pi}}{2}}{c(x) + e^{-x}} \tag{2.4-19}$$

Quite a few suggestions have been made in the literature for $c(x)$. A most simple but very crude approximation reads:

$$c(x) = 1/4, \quad -1 < x < 2.5 \tag{2.4-20}$$

However, due to this simplicity it is possible to perform analytical investigations on expressions where $F_{1/2}(x)$ is involved [2.61], [2.82]. Another formula for $c(x)$ has been proposed and used in [2.58], [2.77].

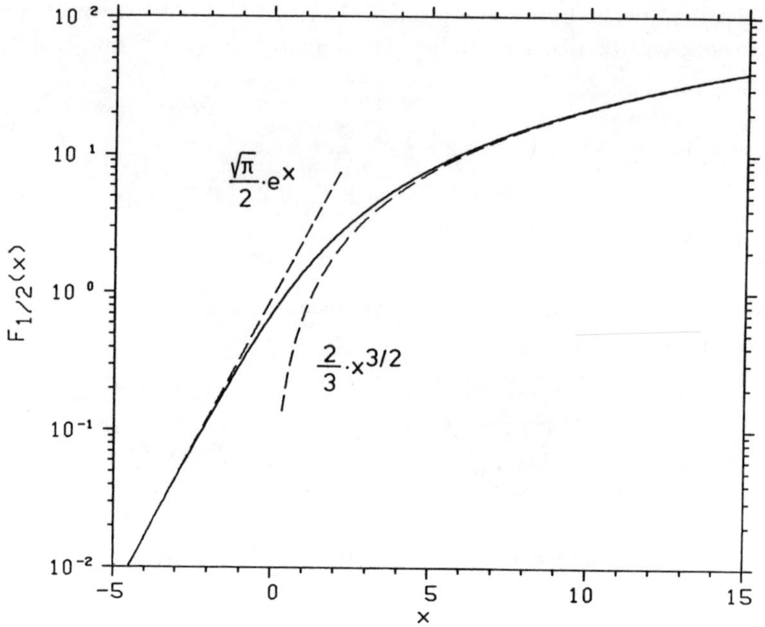

Fig. 2.4-1. Fermi-integral of order 1/2 and its asymptotic expansions

$$c(x) = 0.31 - 0.044 \cdot x, \quad x < 2$$
$$c(x) = \exp(-0.88 - 0.32 \cdot x + 0.0086 \cdot x^2), \quad 2 < x < 12 \tag{2.4-21}$$

The error associated with (2.4-21) is always smaller than four percent in the specified range for the argument. There is obviously a need for approximations which are valid over the entire range of possible arguments $[-\infty, \infty]$. Two formulae with this property are:

$$c(x) = \frac{\frac{3}{4} \cdot \sqrt{\pi}}{(50 + x^4 + 33.6 \cdot x \cdot (1 - 0.68 \cdot \exp(-0.17 \cdot (x+1)^2))^{3/8}} \tag{2.4-22}$$

$$c(x) = \frac{3 \cdot \sqrt{\frac{\pi}{2}}}{(2.13 + x + (|x - 2.13|^{12/5} + 9.6)^{5/12})^{3/2}} \tag{2.4-23}$$

(2.4-22) has been pointed out to be useful in [2.12], and (2.4-23) has been presented in [2.8]. The maximum error associated with both expressions is in the order of 0.5 percent only. Both formulae nicely accommodate the asymptotic behavior (2.4-17), (2.4-18).

Another approach to approximating $F_{1/2}(x)$ has been suggested in [2.12]. The method of least squares has been used to calculate coefficients of polynomials the exponential of which represents the fitting function.

$$F_{1/2}(x) \cong e^{p(x)} \tag{2.4-24}$$

However, this approach will only deliver formulae which are valid for a restricted range of arguments. A review on approximations for Fermi integrals and their inverse function is presented in [2.12]. For the purpose of implementation of formulae with high accuracy on large computers it is better to use rational Chebyshev approximations as demonstrated in [2.19].
To come back to the carrier concentrations, we can use the asymptotic expansion (2.4-17) for the Fermi integral in the expressions (2.4-12), (2.4-13) if

$$\frac{E_{fn}-E_c}{k \cdot T} \ll -1 \tag{2.4-25}$$

$$\frac{E_v-E_{fp}}{k \cdot T} \ll -1 \tag{2.4-26}$$

holds. The validity of these assumptions thus requires that the Fermi energy for electrons is sufficiently smaller than the conduction band edge and that the Fermi energy for holes is sufficiently larger than the valence band edge. These assumptions are equivalent to the use of Boltzmann statistics for the carrier concentrations which will then simplify to

$$n = N_c \cdot \exp\left(\frac{E_{fn}-E_c}{k \cdot T}\right) \tag{2.4-27}$$

$$p = N_v \cdot \exp\left(\frac{E_v-E_{fp}}{k \cdot T}\right) \tag{2.4-28}$$

In order to be able to investigate the expressions for the carrier concentrations more thoroughly (with Fermi statistics (2.4-12), (2.4-13) and with Boltzmann statistics (2.4-27), (2.4-28)) we have to define precisely the meaning of the band edges E_c, E_v and the Fermi energies E_{fn}, E_{fp}. The band edges can be split into essentially three parts:

$$E_c = E_{co} - \delta E_c - q \cdot \psi \tag{2.4-29}$$

$$E_v = E_{vo} + \delta E_v - q \cdot \psi \tag{2.4-30}$$

E_{co}, E_{vo} denote the band edges for pure material, i.e. an intrinsic semiconductor. δE_c and δE_v describe shifts of the band edges caused by a non-uniform composition of the semiconductor under consideration such as dopants. These quantities have to be assumed to be functions of position, whereas E_{co}, E_{vo} are usually not position dependent, except for devices which include heterojunctions [2.60]. However, for this type of devices it is also possible to introduce artificial band edges E_{co}, E_{vo} which are not position dependent and to account for rigid bands with shift energies δE_c, δE_v [2.91]. ψ represents the electrostatic potential as already used in the previous sections.
The Fermi energies E_{fn}, E_{fp} are usually separated into two parts:

$$E_{fn} = E_i - q \cdot \varphi_n \tag{2.4-31}$$

$$E_{fp} = E_i - q \cdot \varphi_p \tag{2.4-32}$$

φ_n and φ_p are the quasi-Fermi potentials of electrons and holes. They describe deviations of the corresponding distribution function from the equilibrium state. The gradients of the quasi-Fermi potentials are first order approximations to the driving forces of the current densities (cf. Section 2.3). Note that only differences between the quasi-Fermi potentials and the electrostatic potential are relevant for the carrier concentrations. One may again define a reference point arbitrarily. It is sound, although arbitrary, to define the quasi-Fermi potentials to be zero if the distribution functions are in equilibrium, which is the case for a structure to which no external forces are applied. We define the electrostatic potential to be zero for the intrinsic semiconductor when the distribution functions are in equilibrium. Thus, E_i denotes the Fermi energy for the intrinsic semiconductor. It will be calculated in the following. For that purpose we rewrite the carrier concentrations using the above given nomenclature.

$$n = N_c \cdot \exp\left(\frac{E_i - E_{co} + \delta E_{co}}{k \cdot T}\right) \cdot \exp\left(\frac{q \cdot (\psi - \varphi_n)}{k \cdot T}\right) \qquad (2.4\text{-}33)$$

$$p = N_v \cdot \exp\left(\frac{E_{vo} + \delta E_{vo} - E_i}{k \cdot T}\right) \cdot \exp\left(\frac{q \cdot (\varphi_p - \psi)}{k \cdot T}\right) \qquad (2.4\text{-}34)$$

Then we recall Poisson's equation from Section 2.1.

$$\operatorname{div} \operatorname{grad} \psi = \frac{q}{\varepsilon} \cdot (n - p - C) \qquad (2.4\text{-}35)$$

In the absence of dopants and external forces Poisson's equation reduces to the trivial form:

$$p(\psi = \varphi_n = 0) - n(\psi = \varphi_p = 0) = 0 \qquad (2.4\text{-}36)$$

Using the expressions just derived for the carrier concentrations we obtain:

$$N_c \cdot \exp\left(\frac{E_i - E_c}{k \cdot T}\right) = N_v \cdot \exp\left(\frac{E_v - E_i}{k \cdot T}\right) \qquad (2.4\text{-}37)$$

from where we can calculate the intrinsic Fermi energy E_i with a small amount of algebra. Note, we sloppily say that we calculate the intrinsic Fermi energy, but clearly we can only calculate the difference of the intrinsic Fermi energy to one of the band edges.

$$E_i - E_c = \frac{k \cdot T}{2} \cdot \ln\left(\frac{N_v}{N_c}\right) - \frac{E_g}{2} \qquad (2.4\text{-}38)$$

In case that the assumptions (2.4-25), (2.4-26) are violated we need to solve (2.4-39) instead of (2.4-38) for E_i.

$$N_c \cdot F_{1/2}\left(\frac{E_i - E_c}{k \cdot T}\right) = N_v \cdot F_{1/2}\left(\frac{E_v - E_i}{k \cdot T}\right) \qquad (2.4\text{-}39)$$

E_i can only be found from equation (2.4-39) by means of numerical methods. However, in almost all semiconductors E_i lies in about the middle of the forbidden

band and it is, therefore, well separated from both band edges. Thus Boltzmann statistics for intrinsic semiconductors in equilibrium are usually valid.
For many applications it is convenient to define a so-called intrinsic concentration n_i as the geometric average of the carrier concentrations in a semiconductor in equilibrium.

$$n_i=\sqrt{n \cdot p} \tag{2.4-40}$$

The existence of dopants is allowed in (2.4-40). If Boltzmann statistics are valid for describing the carrier concentrations, n_i is evaluated with small algebraic effort:

$$n_i=\sqrt{N_c \cdot N_v} \cdot \exp\left(-\frac{E_g}{2 \cdot k \cdot T}\right) \tag{2.4-41}$$

We see that n_i is position dependent if the band gap E_g is position dependent. The carrier concentrations can now be rewritten into the well known form with five parameters: intrinsic concentration n_i, electrostatic potential ψ, quasi-Fermi potentials φ_n, φ_p and temperature T.

$$n=n_i \cdot \exp\left(\frac{q \cdot (\psi-\varphi_n)}{k \cdot T}\right) \tag{2.4-42}$$

$$p=n_i \cdot \exp\left(\frac{q \cdot (\varphi_p-\psi)}{k \cdot T}\right) \tag{2.4-43}$$

In the presence of dopants we can again calculate from Poisson's equation the so-called "built-in" potential ψ_b which will shift the Fermi energies E_{fn}, E_{fp}, depending on the sign of the fixed charges C. In many textbooks one reads that in the presence of dopants the Fermi energies are shifted towards one of the band edges. This is simply wrong; it could be accepted in viewing the relative change of position, but such an interpretation should be strictly avoided for didactical reasons.
For the calculation of the built-in potential we have to assume a homogeneously doped semiconductor and no external forces. Then the Laplacian of the electrostatic potential is identically zero and Poisson's equation reduces to (2.4-44), still assuming the validity of Boltzmann statistics.

$$n_i \cdot \exp\left(-\frac{q \cdot \psi_b}{k \cdot T}\right)-n_i \cdot \exp\left(\frac{q \cdot \psi_b}{k \cdot T}\right)+N_D^+-N_A^-=0 \tag{2.4-44}$$

For the sake of clarity we assume that the fixed charges C are composed only of singly ionized donors N_D^+ and singly ionized acceptors N_A^-.

$$C=N_D^+-N_A^- \tag{2.4-45}$$

From (2.4-44) the built-in potential is evaluated to:

$$\psi_b=\frac{k \cdot T}{q} \cdot \operatorname{arsinh}\left(\frac{N_D^+-N_A^-}{2 \cdot n_i}\right) \tag{2.4-46}$$

If we have one type of dopants dominating the other type, (2.4-46) can be simplified in order to obtain an even simpler result.

$$N_D^+ \gg N_A^- \quad \rightarrow \quad \psi_b \cong \frac{k \cdot T}{q} \cdot \ln\left(\frac{N_D^+}{n_i}\right) \tag{2.4-47}$$

$$N_A^- \gg N_D^+ \quad \rightarrow \quad \psi_b \cong -\frac{k \cdot T}{q} \cdot \ln\left(\frac{N_A^-}{n_i}\right) \tag{2.4-48}$$

However, it is to note that the validity of Boltzmann statistics becomes a very poor assumption for high doping concentrations, because, as already mentioned, the Fermi energies E_{fn}, E_{fp} are shifted towards one of the band edges. If the error introduced by the assumption of Boltzmann statistics is not acceptable, one has to solve (2.4-49) for the built-in potential.

$$N_v \cdot \frac{2}{\sqrt{\pi}} \cdot F_{1/2}\left(\frac{E_v - E_{fp}}{k \cdot T}\right) - N_c \cdot \frac{2}{\sqrt{\pi}} \cdot F_{1/2}\left(\frac{E_{fn} - E_c}{k \cdot T}\right) + N_D^+ - N_A^- = 0 \tag{2.4-49}$$

Again, this can only be done with numerical methods. It is obvious that the sum of the intrinsic Fermi energy E_i and built-in potential ψ_b, which is often termed the extrinsic Fermi energy, can be calculated simultaneously from (2.4-49).

Most semiconductor devices contain regions with doping levels above $10^{18}\,\mathrm{cm}^{-3}$ and the transport of carriers through these heavily doped parts can play an essential role in determining device behavior and performance. Therefore, the models for the carrier concentrations have to properly reflect the underlying physics of heavy doping effects. In the preceding considerations we have only addressed the problem of carrier statistics in this context. All of the possible problems associated with the density of states functions have been ignored, except that shift energies for the conduction and the valence band, which have been assumed to be parabolic, have been allowed. In the following we shall examine more in depth why and how the band structure is changed in heavily doped semiconductors. However, the statements we shall make have to some extent a speculative character, because, as it has to be said, our understanding of the physics of heavily doped semiconductors is fairly limited.

The density of states function for electrons and holes is influenced by essentially two categories of phenomena [2.62]. The first category consists of interactions between carriers and between carriers and ionized impurity atoms. The second category comprises the effects of electrostatic potential fluctuations which account for the random distribution of impurities together with the overlap of the electron wave functions at the impurity states causing bandtails [2.48] and impurity bands [2.66]. While the second category of heavy doping phenomena alters the shape of the density of states functions for electrons and holes, the first category produces only rigid shifts of both the conduction and the valence bands towards each other. To give an example we shall discuss the possible carrier interaction phenomena in *n*-type silicon. For *p*-type material the facts are analogous. In *n*-type semiconductors three phenomena become apparent: electron-donor interaction, electron-electron interaction and electron-hole interaction. The electron-donor interaction does not yield changes in the band edges, but the number of electrons in the semiconductor becomes so large that they screen the donor ions, which effectively reduces the impurity ionization energy so that the donor levels ultimately disappear into the conduction band (see also [2.62]). Electron-electron interaction yields a rigid shift

δE_c of the conduction band towards the valence band. Electron-hole interaction causes a shift δE_v of the valence band towards the conduction band, because the majority electrons also screen the mobile minority holes in addition to the immobile donor ions. As already said, completely analogous statements hold for the description of the interaction phenomena in p-type material. An excellent review on these subjects can be found in [2.57]. We shall primarily concentrate in the following on the results which are established without going very much into details of their derivation.

Kane [2.48] has derived approximations for the density of states function for the conduction and the valence bands in heavily doped semiconductors by assuming that the local potential fluctuations are sufficiently slow that a local density of states function can be defined as if the local potential were constant. The macroscopic density of states functions which are the statistical average of the local density of states functions over the lattice can then be expressed as:

$$\rho_c(E) = \frac{4 \cdot \pi \cdot (2 \cdot m_n^*)^{3/2}}{h^3} \cdot \sqrt{\sigma_{cv}} \cdot y\left(\frac{E - E_c}{\sigma_{cv}}\right) \tag{2.4-50}$$

$$\rho_v(E) = \frac{4 \cdot \pi \cdot (2 \cdot m_p^*)^{3/2}}{h^3} \cdot \sqrt{\sigma_{cv}} \cdot y\left(\frac{E_v - E}{\sigma_{cv}}\right) \tag{2.4-51}$$

with:

$$y(x) = \frac{1}{\sqrt{\pi}} \cdot \int_{-\infty}^{x} \sqrt{x-u} \cdot \exp(-u^2) \cdot du \tag{2.4-52}$$

A simple approximation for the unwieldy equation (2.4-52) has been suggested by Slotboom [2.81].

$$y(x) \cong \begin{cases} x \leq 0.601 & \dfrac{1}{2 \cdot \sqrt{\pi}} \cdot e^{-x^2} \cdot (1.225 - 0.906 \cdot (1 - e^{2x})) \\ x \geq 0.601 & \sqrt{x} \cdot \left(1 - \dfrac{1}{16 \cdot x^2}\right) \end{cases} \tag{2.4-53}$$

Results which are fairly similar to (2.4-50), (2.4-51) have been presented at almost the same time by Bonch-Bruevich [2.15]. These density of states functions include infinite tails for the conduction and the valence bands. That means the density of states functions are principally different from zero everywhere in the forbidden band, but they fall off rapidly with increasing distance from the corresponding band edge. As expected (2.4-50) and (2.4-51) are for small doping concentrations asymptotically equivalent to the parabolic density of states functions (2.4-1) and (2.4-2), respectively. A more rigorous approach to the derivation of density of states functions has been carried out by Halperin and Lax [2.38]. However, their results are remarkably more complex. For strongly compensated, heavily doped semiconductors $|N_D^+| \cong |N_A^-| \gg 0$ only, the Halperin and Lax theory is expected to be superior to Kane's method (cf. [2.72]).

σ_{cv} is the characteristic standard deviation of the Gaussian tails of the density of states functions (2.4-50), (2.4-51). The best established model for σ_{cv} has been published by Morgan [2.66].

$$\sigma_{cv}=\frac{q^2}{\varepsilon}\cdot\sqrt{\frac{(N_D^+ + N_A^-)\cdot\lambda}{4\cdot\pi}}\cdot\exp\left(-\frac{a}{2\cdot\lambda}\right) \quad (2.4\text{-}54)$$

λ denotes the screening length, and a is the crystal lattice constant, numerical values of which are summarized in Table 2.4-3.

Table 2.4-3. *Crystal lattice constants*

material	a [10^{-9} m]
Si	0.543072
GaAs	0.565315
Ge	0.565754

Kane [2.48] as well as Morgan [2.66] in their original work have used a so-called cutoff radius instead of $a/2$ in the exponential term of (2.4-54). However, there is evidence to relate this quantity to the lattice constant [2.66]. VanOverstraeten et al. [2.90] and Slotboom [2.81] have in their investigations fully neglected the exponential factor of (2.4-54). For the screening length λ two models are most frequently in use. The first one has been proposed by Stern [2.84].

$$\lambda=\frac{1}{q}\cdot\sqrt{\frac{\varepsilon}{\left|\frac{\partial n}{\partial E_{fn}}\right|+\left|\frac{\partial p}{\partial E_{fp}}\right|+\frac{N_D^+ + N_A^-}{k\cdot T_{ion}}}} \quad (2.4\text{-}55)$$

For non-degenerate material when Boltzmann statistics can be applied this formula reduces to the well known Debye length.

$$\lambda=\frac{1}{q}\cdot\sqrt{\frac{\varepsilon\cdot k\cdot T}{n+p}} \quad (2.4\text{-}56)$$

T_{ion} in (2.4-55) represents an effective temperature for ion screening. In the original paper of Stern [2.84] this quantity is treated as adjustable parameter in order to fit experimental data. Stern has speculated that at room temperature T_{ion} should be in the range from about 7000 K to 9000 K. Mock [2.63] and Polsky et al. [2.73] have used 9000 K in their work; Nakagawa [2.70] has claimed that 6000 K is more appropriate to obtain quantitatively correct results; and Slotboom [2.81] has assumed T_{ion} to be infinite so that the last term in the denominator of (2.4-55) vanishes. In [2.51] and [2.90] a different model for the screening length which has also been suggested by Stern [2.83] in an early work has been used.

$$\lambda=\frac{1}{2\cdot\frac{q}{h}\cdot\sqrt{\frac{m^*}{\varepsilon}}\cdot\left(\frac{3\cdot(N_D^+ - N_A^-)}{\pi}\right)^{1/6}} \quad (2.4\text{-}57)$$

For the derivation of this formula it has been assumed that the conduction and valence band structures are parabolic so that (2.4-57) seems to be inconsistent with (2.4-50), (2.4-51). Fermi statistics for the carrier distribution functions have been accounted for in the calculation of (2.4-57) which can also be identified as the screening length in metals [2.50]. A requirement for the applicability of (2.4-57) is that the Fermi energy lies in one of the carrier bands, and not as usual in the forbidden band. This does not happen unless the doping concentration is extraordinarily high which should lead to the conclusion that (2.4-57) is inappropriate for semiconductor device modeling.

In Fig. 2.4-2 a comparison of the models for the screening length in *n*-type silicon is given. The solid line corresponds to (2.4-55) with T_{ion} equal to 9000 K (the model of Mock, Polsky et al.); the dashed line is also (2.4-55) but with T_{ion} assumed to be infinite (the model of Slotboom); and the dot-dashed line corresponds to (2.4-57) (the model of VanOverstraeten et al.). The dotted line denotes the classical Debye length (2.4-56) as a reference.

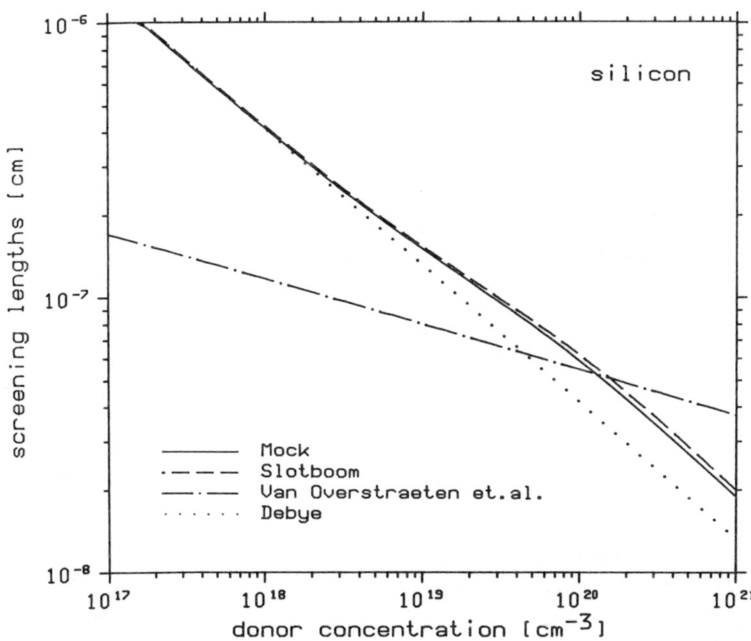

Fig. 2.4-2. Screening length versus donor concentration in silicon at 300 K temperature

When the doping concentration is large, the impurity energy level cannot be described by a delta function as it is the case in simple theory. The wave function of the electrons of the impurity atoms overlap, thus causing the formation of an impurity band. Morgan [2.66] has developed a theory which predicts a Gaussian shape for the impurity band.

$$\rho_D = \frac{2 \cdot N_D^+}{\sqrt{\pi} \cdot \sigma_{DA}} \cdot \exp\left(-\frac{(E-E_D)^2}{\sigma_{DA}^2}\right) \tag{2.4-58}$$

$$\rho_A = \frac{2 \cdot N_A^-}{\sqrt{\pi} \cdot \sigma_{DA}} \cdot \exp\left(-\frac{(E-E_A)^2}{\sigma_{DA}^2}\right) \tag{2.4-59}$$

E_D and E_A are the activation energies for specific donor and acceptor atoms, respectively. Numerical values for E_D and E_A are well documented in the literature, e.g. [2.86]. The expression for σ_{DA} like equation (2.4-54) for σ_{cv} has also been proposed by Morgan [2.66].

$$\sigma_{DA} = \frac{q^2}{\varepsilon} \cdot \sqrt{\frac{(N_D^+ + N_A^-) \cdot \lambda}{4 \cdot \pi}} \cdot 1.0344 \cdot \exp\left(-\frac{1}{\sqrt{11.3206 \cdot \pi \cdot (N_D^+ + N_A^-) \cdot \lambda^3}}\right) \tag{2.4-60}$$

Some discussion about the models for σ_{DA} can be found in, e.g., [2.39], [2.72].
In order to obtain a density of states function for electrons and holes, the density of states functions of the conduction band (2.4-50), the valence band (2.4-51) and the impurity bands (2.4-58), (2.4-59) have to be combined. Kleppinger and Lindholm [2.51] have simply added up the corresponding functions for that purpose. VanOverstraeten et al. [2.30], however, have assumed that the total density of states function of the mobile carriers is composed of the envelope of the conduction (valence) density of states and the corresponding impurity band density of states function. This approach is physically much more sound since adding up the density of states functions implies that a substitute impurity atom and a silicon atom that were at that same lattice site both contribute to the density of states (cf. [2.72]).
The concentration of electrons and holes can now, finally, be calculated by:

$$n = \int_{-\infty}^{\infty} \max\left(\rho_c(E), \rho_D(E)\right) \cdot f_n(E) \cdot dE \tag{2.4-61}$$

$$p = \int_{-\infty}^{\infty} \max\left(\rho_v(E), \rho_A(E)\right) \cdot f_p(E) \cdot dE \tag{2.4-62}$$

The integration bounds are now $-\infty$ and ∞ in contrast to (2.4-8) and (2.4-9) because of the infinite tails of the density of states functions. It is obvious that the integrals (2.4-61), (2.4-62) do not have a closed form algebraic solution; they have to be solved with numerical methods. Details on how to design efficient algorithms for the self-consistent solution of the carrier concentrations and the built-in potential are given in, e.g., [2.46].
Fig. 2.4-3, Fig. 2.4-4 and Fig. 2.4-5 summarize the results we have obtained in a graphical way. They show the density of states function for electrons $\max(\rho_c(E), \rho_D(E))$ and the density of states function for holes $\max(\rho_v(E), \rho_A(E))$. The dashed line in the conduction band denotes the distribution function of electrons, i.e. the integrand of (2.4-61). Fig. 2.4-3 corresponds to a doping of $N_D^+ = 10^{16}\,\mathrm{cm}^{-3}$, $N_A^- = 0$, i.e. fairly low doping concentration. Fig. 2.4-4 has been

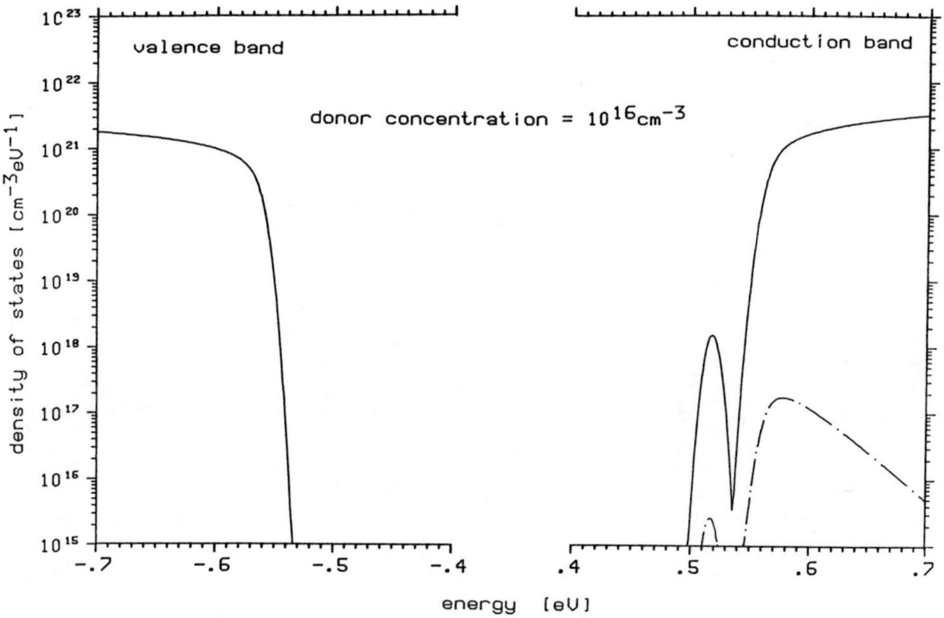

Fig. 2.4-3. Band structure for $N_D^+ = 10^{16}\,\mathrm{cm}^{-3}$, $N_A^- = 0$ in silicon at 300 K temperature

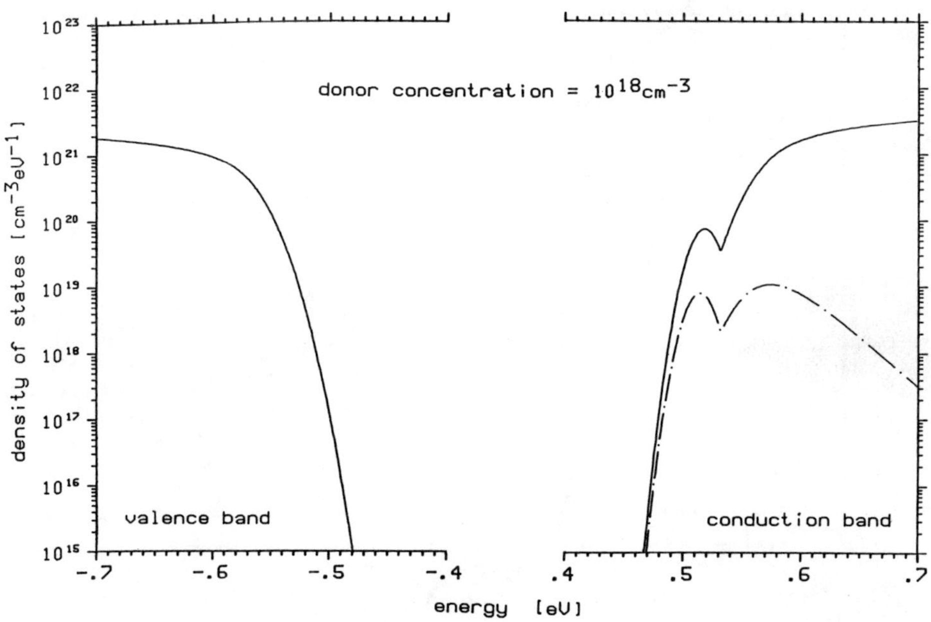

Fig. 2.4-4. Band structure for $N_D^+ = 10^{18}\,\mathrm{cm}^{-3}$, $N_A^- = 0$ in silicon at 300 K temperature

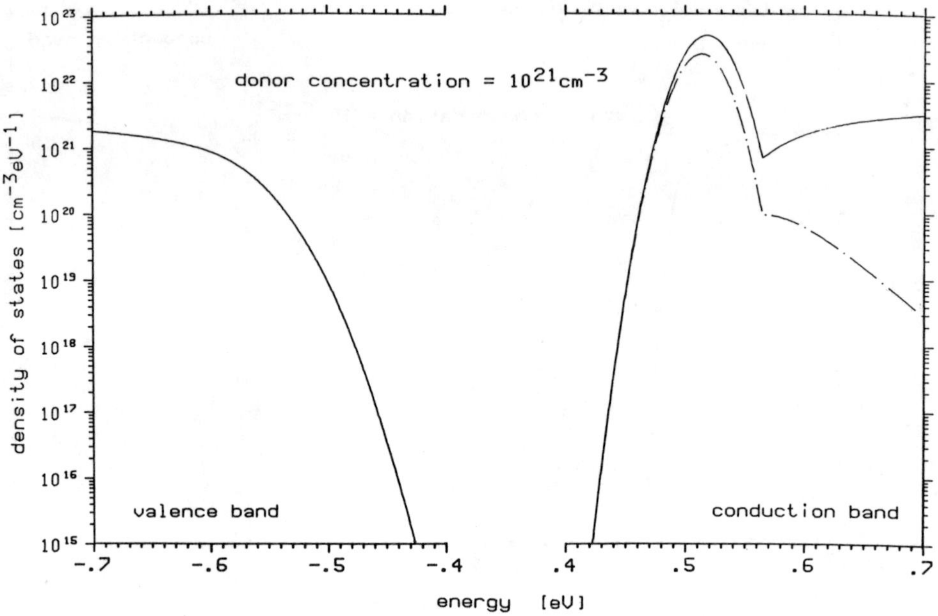

Fig. 2.4-5. Band structure for $N_D^+ = 10^{21}\,\mathrm{cm}^{-3}$, $N_A^- = 0$ in silicon at 300 K temperature

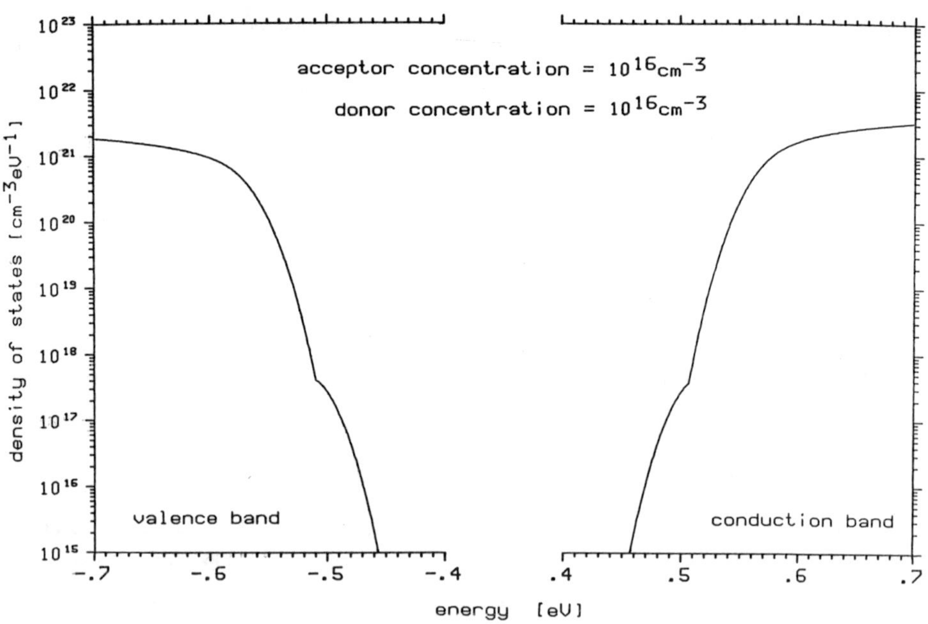

Fig. 2.4-6. Band structure for $N_D^+ = 10^{16}\,\mathrm{cm}^{-3}$, $N_A^- = 10^{16}\,\mathrm{cm}^{-3}$ in silicon at 300 K temperature

calculated for medium large doping $N_D^+ = 10^{18}\,\text{cm}^{-3}$, $N_A^- = 0$ and Fig. 2.4-5 corresponds to extremely heavy doping, i.e. $N_D^+ = 10^{21}\,\text{cm}^{-3}$, $N_A^- = 0$. One can see quite nicely how the impurity band is formed for increasing doping and how it even dominates the conduction band in Fig. 2.4-5. The sharp kink at the transition from the donor band to the conduction band is caused by taking the envelope of both individual bands for the overall density of states function. This kink is indeed unrealistic but I am not aware of any more realistic model for the transition. It should be noted that the extraordinarily heavy doping in Fig. 2.4-5 is almost unobtainable in real processing. The distribution function of holes is too small to have been drawn within the same scales.

Fig. 2.4-6 shows the same quantities as the previous figures for strongly compensated material $N_D^+ = N_A^- = 10^{16}\,\text{cm}^{-3}$. Obviously, we can now see two impurity bands which remarkably narrow the effective band gap. The effect of band gap narrowing is much more pronounced in Fig. 2.4-6 compared to Fig. 2.4-3, although the total doping concentration is only higher by a factor of two. In compensated material screening is weaker and the screening length is larger which leads to a stronger influence of the potential fluctuations on the band tails. The distribution functions of electrons and holes are too small to be drawn within the scale of Fig. 2.4-6.

The models for the carrier concentrations which account for Fermi statistics, deformation of the conduction and valence band and the formation of impurity bands are, unfortunately, so complex, in a mathematical sense, that it is very cumbersome to implement them in device modeling programs, although it can be done in principle, e.g. [2.70], [2.73]. It is instead most attractive to use a so-called effective intrinsic concentration and an exponential dependence upon the potentials for the carrier concentrations.

$$n = n_{ie} \cdot \exp\left(\frac{q \cdot (\psi - \varphi_n)}{k \cdot T}\right) \tag{2.4-63}$$

$$p = n_{ie} \cdot \exp\left(\frac{q \cdot (\varphi_p - \psi)}{k \cdot T}\right) \tag{2.4-64}$$

These expressions are quite similar to (2.4-42), (2.4-43) which have been derived based upon the assumptions of Boltzmann statistics and parabolic band structure. Probably the first empirical formula for n_{ie} has been proposed by Slotboom [2.79], [2.80], [2.81].

$$n_{ie} = n_i(T) \cdot \exp \frac{q \cdot V_1 \cdot \left(\ln\left(\frac{N_D^+ + N_A^-}{N_o}\right) + \sqrt{\ln^2\left(\frac{N_D^+ + N_A^-}{N_o}\right) + C}\right)}{2 \cdot k \cdot T} \tag{2.4-65}$$

with:

$$V_1 = 9 \cdot 10^{-3}\ \text{V}, \quad N_o = 10^{17}\ \text{cm}^{-3}, \quad C = 0.5$$

This formula has been derived by a fit to experimental values of the intrinsic concentration obtained from measurements of bipolar transistors. In [2.26] the structure of Slotboom's formula has been made plausible by theoretical investigations. (2.4-65) and formulae which are very similar, e.g. [2.61], have proved by many authors to be extraordinarily valuable in the simulation of bipolar devices, e.g. [2.2], [2.24], [2.28]. In [2.46] the range of validity of an approach with (2.4-63), (2.4-64) has been investigated very thoroughly. One can state that the concept of an effective intrinsic concentration is valid (error $< 10\%$) for total impurity concentrations smaller than about $8 \cdot 10^{19}\,\mathrm{cm}^{-3}$ when the amount of compensation is not too large ($< 10\%$). For heavily doped and compensated material any approach with an effective intrinsic concentration fails dramatically. It is to note that for degenerate material the product $n \cdot p$ of electrons and holes away from equilibrium is always smaller as it would be predicted by (2.4-63), (2.4-64). The reason can be found in the strong bending of the Fermi integral (cf. Fig. 2.4-1, [2.40]) compared to the exponential function.

Another formula for the purpose of fitting an effective intrinsic concentration has been derived by Lanyon and Tuft [2.55], [2.56].

$$n_{ie} = n_i(T) \cdot \exp\left(\frac{3 \cdot q^3}{32 \cdot \pi\,(\varepsilon \cdot k \cdot T)^{3/2}} \cdot \sqrt{N_D^+ + N_A^-}\right) \tag{2.4-66}$$

This expression is claimed to be in excellent agreement with experimental data up to doping concentrations of $3.3 \cdot 10^{19}\,\mathrm{cm}^{-3}$. For larger concentrations the theoretical constants in (2.4-66) have been replaced by empirical values to fit experimental results [2.56], [2.75]. A similar expression has been presented in [2.11].

A point of uncertainty has been in recent years that optically measured band gaps and electrically measured band gaps differ quite significantly for heavily doped material. The reason for that fact is that by optical measurement only the rigid shifts of the band edges can be detected [2.25], whereas all effects due to lattice disorder, e.g. band tails, cannot be found. By electrical measurements both contributions to the band gap narrowing are seen (cf. [2.62]). Therefore, the effective electrical band gap is always smaller than the one predicted by optical investigations.

Formulae which fit the effective intrinsic concentration to the complex models of Mock [2.63], Slotboom [2.81] and VanOverstraeten et al. [2.90] which have been discussed above, have been developed in [2.46]. The following structure for the empirical expression has been used.

$$n_{ie}(T, N_D^+) = \exp\left(a1(T) + a2(T) \cdot \left(\frac{N_D^+}{10^{17}\,\mathrm{cm}^{-3}}\right)^{a3(T)}\right)\,\mathrm{cm}^{-3} \tag{2.4-67}$$

with:

$$a1(T) = -1.99765 \cdot 10^{+1} + 2.01814 \cdot 10^{-1} \cdot T - 1.97040 \cdot 10^{-4} \cdot T^2 \tag{2.4-68}$$

The coefficients $a2(T)$ and $a3(T)$ which fit the empirical formulae best to Mock's model in the doping range $[10^{12}, 10^{20}]\,\mathrm{cm}^{-3}$ are:

$$\begin{aligned} a2(T) &= 9.60563 \cdot 10^{-1} - 3.94127 \cdot 10^{-3} \cdot T + 4.41488 \cdot 10^{-6} \cdot T^2 \\ a3(T) &= 1.29363 \cdot 10^{-1} + 1.10709 \cdot 10^{-3} \cdot T - 9.56981 \cdot 10^{-7} \cdot T^2 \end{aligned} \tag{2.4-69}$$

whereas for Slotboom's model in the doping range $[10^{12}, 3 \cdot 10^{20}]\,\mathrm{cm}^{-3}$ they read:

$$a2(T) = 7.95811 \cdot 10^{-1} - 3.20439 \cdot 10^{-3} \cdot T + 3.54153 \cdot 10^{-6} \cdot T^2$$
$$a3(T) = 2.97104 \cdot 10^{-1} + 6.75707 \cdot 10^{-4} \cdot T - 4.90892 \cdot 10^{-7} \cdot T^2 \quad (2.4\text{-}70)$$

and for VanOverstraeten's et al. model in the doping range $[10^{17}, 10^{21}]\,\mathrm{cm}^{-3}$ they evaluate to:

$$a2(T) = 2.38838 \cdot 10^{-1} - 9.57814 \cdot 10^{-4} \cdot T + 1.07551 \cdot 10^{-6} \cdot T^2$$
$$a3(T) = 5.10190 \cdot 10^{-1} + 5.75190 \cdot 10^{-4} \cdot T - 7.01029 \cdot 10^{-7} \cdot T^2 \quad (2.4\text{-}71)$$

The temperature T has to be given in Kelvin in (2.4-68) to (2.4-71). The maximum relative difference of formula (2.4-67) with the above given coefficients and the exactly evaluated models is always smaller than ten percent (cf. [2.46], [2.47]) in the temperature range [250, 400] K. Formulae for an effective intrinsic concentration for compensated material are also given in [2.46], however, they are much more complicated.

In Fig. 2.4-7 the effective intrinsic concentration for Mock's model (solid line), Slotboom's model (dashed line) and VanOverstraeten's et al. model (dot dashed line) are shown in conjunction with the experimental values of Mertens et al. [2.61], Slotboom [2.81], Wieder [2.92] and Wulms [2.94]. Although the agreement between the models and the experimental data is not overwhelming, it can be considered pragmatically to be quite good, because of the fairly pronounced scatter of the measured data. However, a judgment as to which of the models is to be prefered can not, therefore, be given.

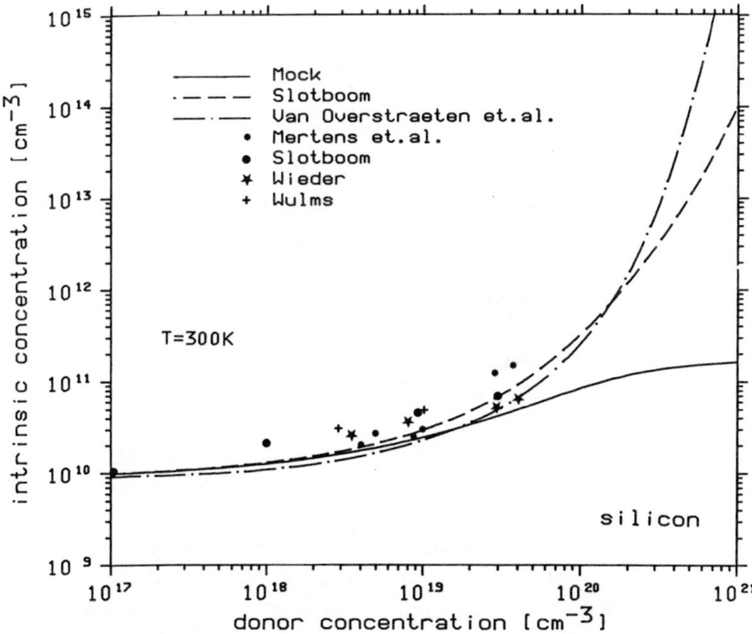

Fig. 2.4-7. Intrinsic concentration versus donor concentration in silicon at 300 K temperature

2.5 Heat Flow Equation

For the design of power devices it is often desired to simulate interaction of electrothermal phenomena. Changes in the temperature and its distribution in the interior of a device can influence significantly the electrical device behavior. Particularly, two effects usually have to be considered. Thermal runaway is one, a rather common mechanism where the electrical energy dissipated causes a temperature rise over an extended area of a device resulting in increased power dissipation. The device temperature increases which leads to an irrecoverable device failure (burn out), unless an equilibrium situation can occur with a heat sink removing all of the energy dissipated. The existence of such an equilibrium situation is the second question which is sometimes quite difficult to answer [2.54]. In order to account for thermal effects in semiconductor devices the heat flow equation (2.5-1) has to be solved.

$$\rho \cdot c \cdot \frac{\partial T}{\partial t} - H = \operatorname{div} k(T) \cdot \operatorname{grad} T \tag{2.5-1}$$

ρ and c are the specific mass density and specific heat of the material. Numerical values for ρ and c at room temperature are summarized in Table 2.5-1 for the most frequently used materials in device processing.

Table 2.5-1. *Specific heat and density constants at $T = 300\,K$*

material	c [$m^2 s^{-2} K^{-1}$]	ρ [$VA s^3 m^{-5}$]	
Si	703	2328	
SiO_2	782	2650	typical
Si_3N_4	787	3440	typical
GaAs	351	5316	
Ge	322	5323	

The temperature dependence of ρ and c can be assumed to be negligibly small in consideration of practical device applications [2.50]. If one is not interested in thermal transients one can assume for the simulation that the partial derivative of the temperature with respect to time vanishes, which eases the problem of solving the heat flow equation by one dimension. However, one is absolutely incorrect in using this assumption in a simulation for which an equilibrium condition does not exist. The simulation program will "blow up" in a manner analogous to the real device.

$k(T)$ and H denote the thermal conductivity and the locally generated heat. Models for these quantities will be given and discussed in Section 4.3 and Section 4.4, respectively.

To just calculate the temperature distribution and the associated thermal power dissipation without taking into account the current induced by gradients of the temperature is a fairly crude approach which is only appropriate for limited application [2.35]. In a more rigorous approach the current density equations have to be supplemented by additional terms.

$$\vec{J}_n = q \cdot n \cdot \mu_n \cdot \vec{E} + q \cdot D_n \cdot \operatorname{grad} n + q \cdot n \cdot D_n^T \cdot \operatorname{grad} T \qquad (2.5\text{-}2)$$

$$\vec{J}_p = q \cdot p \cdot \mu_p \cdot \vec{E} - q \cdot D_p \cdot \operatorname{grad} p - q \cdot p \cdot D_p^T \cdot \operatorname{grad} T \qquad (2.5\text{-}3)$$

The last expression in (2.5-2), (2.5-3) represents a drift current component with the temperature field as the driving force. In Section 2.3 we did assume temperature being constant for the derivation of the classical drift-diffusion relations (cf. (2.3-33)). As can be proved with minor algebraic effort, by assuming non-constant temperature in (2.3-29), (2.3-30) we obtain equations (2.5-2), (2.5-3). Stratton [2.85] has verified these relations with a much more rigorous approach, from a perturbation solution of the Boltzmann equation. He also derived in his paper approximations for the thermal diffusion coefficients D_ν^T.

$$D_n^T \cong \frac{D_n}{2 \cdot T} \qquad (2.5\text{-}4)$$

$$D_p^T \cong \frac{D_p}{2 \cdot T} \qquad (2.5\text{-}5)$$

These coefficients are smaller by a factor of two compared to those we obtain with the procedure just sketched above. However, as pointed out in [2.85] a more exact result owing to the complexity of the problem is cumbersome, if at all possible, to obtain and discrepancies of that order are not at all surprising. Dorkel [2.27] demonstrated that Stratton's result is applicable for intrinsic semiconductors; in the presence of dopands the thermal diffusion coefficient is underestimated by at most a factor of five. Some more considerations on this subject can be found in, e.g. [2.14], [2.76]. However, one need not worry as all publications on non-isothermal effects in the context of semiconductor device modeling certify more or less the applicability of these relatively rough models for describing the feedback of temperature gradients on the current densities, e.g. [2.1], [2.18].

2.6 The Basic Semiconductor Equations

We shall now summarize the results which we have obtained in the previous sections in order to be able to write down a set of equations, the "basic" semiconductor equations, which we shall use in all further investigations. It is obvious that for the sake of transparency and efficiency, we shall perform a trade-off between accuracy and complexity of our model. The equations we shall concentrate on are valid for the major number of engineering applications, particularly for silicon devices. Certainly, conditions do exist for which their validity is not guaranteed, or at least in doubt. However, as I tried to express in the previous sections, the more sophisticated results in semiconductor physics are too complex to give a rigorous, generally applicable and still sufficiently simple model for the purpose of device simulation.

The basic semiconductor equations consist of Poisson's equation (2.6-1), the continuity equations for electrons (2.6-2) and holes (2.6-3) and the current relations for electrons (2.6-4) and holes (2.6-5). For some applications it is desired to add to this set the heat flow equations (2.6-6).

$$\operatorname{div} \operatorname{grad} \psi = \frac{q}{\varepsilon} \cdot (n - p - C) \tag{2.6-1}$$

$$\operatorname{div} \vec{J}_n - q \cdot \frac{\partial n}{\partial t} = q \cdot R \tag{2.6-2}$$

$$\operatorname{div} \vec{J}_p + q \cdot \frac{\partial p}{\partial t} = -q \cdot R \tag{2.6-3}$$

$$\vec{J}_n = q \cdot n \cdot \mu_n \cdot \vec{E}_n + q \cdot D_n \cdot \operatorname{grad} n \tag{2.6-4}$$

$$\vec{J}_p = q \cdot p \cdot \mu_p \cdot \vec{E}_p - q \cdot D_p \cdot \operatorname{grad} p \tag{2.6-5}$$

$$\rho \cdot c \cdot \frac{\partial T}{\partial t} - H = \operatorname{div} k(T) \cdot \operatorname{grad} T \tag{2.6-6}$$

To almost this level of completeness, these equations were first presented by VanRoosbroeck [2.89].

Models for C, the net doping concentration, for R, the net generation/recombination, for μ_n, μ_p, the carrier mobilities, for H, the thermal generation and for $k(T)$, the thermal conductivity will be discussed in the following chapters. $\vec{E}_n$ and $\vec{E}_p$, the effective fields in the current relations are to first order the electric field, however, we may use supplementary correction terms to account for heavy doping (cf. Section 2.3, Section 2.4) or thermally induced currents (cf. Section 2.5). For such mathematical investigations, relatively slight perturbations are of only secondary importance. Hence, for most applications, accounting for some specific effect is possible by properly modeling the parameters in the basic equations.

2.7 References

[2.1] Adler, M. S.: Accurate Calculations of the Forward Drop and Power Dissipation in Thyristors. IEEE Trans. Electron Devices *ED-25,* No. 1, 16–22 (1978).

[2.2] Adler, M. S.: An Operational Method to Model Carrier Degeneracy and Band Gap Narrowing. Solid-State Electron. *26,* No. 5, 387–396 (1983).

[2.3] Ankri, D., Eastman, L. F.: GaAlAs-GaAs Ballistic Hetero-Junction Bipolar Transistor. Electronics Lett. *18,* No. 17, 750–751 (1982).

[2.4] Anselm, A. I.: Einführung in die Halbleitertheorie. Berlin: Akademie-Verlag 1964.

[2.5] Antoniadis, D. A., Dutton, R. W.: Models for Computer Simulation of Complete IC Fabrication Process. IEEE J. Solid State Circuits *SC-14,* No. 2, 412–422 (1979).

[2.6] Awano, Y., Tomizawa, K., Hashizume, N., Kawashima, M.: Monte Carlo Particle Simulation of a GaAs Submicron n+-i-n+ Diode. Electronics Lett. *18,* No. 3, 133–135 (1982).

[2.7] Awano, Y., Tomizawa, K., Hashizume, N., Kawashima, M.: Monte Carlo Particle Simulation of a GaAs Short-Channel MESFET. Electronics Lett. *19,* No. 1, 20–21 (1983).

[2.8] Aymerich-Humett, X., Serra-Mestres, F., Millan, J.: An Analytical Approximation for the Fermi-Dirac Integral F 3/2(x). Solid-State Electron. *24,* No. 10, 981–982 (1981).

[2.9] Baccarani, G., Mazzone, A. M.: On the Diffusion Current in Heavily Doped Silicon. Solid-State Electron. *18,* 469–470 (1975).

[2.10] Baccarani, G.: Physics of Submicron Devices. Proc. VLSI Process and Device Modeling, pp. 1–23. Katholieke Universiteit Leuven, 1983.

[2.11] Bennett, H. S.: Improved Concepts for Predicting the Electrical Behavior of Bipolar Structures in Silicon. IEEE Trans. Electron Devices *ED-30,* No. 8, 920–927 (1983).

[2.12] Blakemore, J. S.: Approximations for Fermi-Dirac Integrals, especially the Function F1/2(x) Used to Describe Electron Density in a Semiconductor. Solid-State Electron. *25,* No. 11, 1067 – 1076 (1982).

[2.13] Blatt, F. J.: Physics of Electronic Conduction in Solids. New York: McGraw-Hill 1968.

[2.14] Bløtekjær, K.: Transport Equations for Electrons in Two-Valley Semiconductors. IEEE Trans. Electron Devices *ED-17,* No. 1, 38 – 47 (1970).

[2.15] Bonch-Bruevich, V. L.: On the Theory of Heavily Doped Semiconductors. Soviet Physics Solid State *4,* No. 10, 1953 – 1962 (1963).

[2.16] Buot, F. A., Frey, J.: Effects of Velocity Overshoot on Performance of GaAs Devices, with Design Information. Solid-State Electron. *26,* No. 7, 617 – 632 (1983).

[2.17] Capasso, F., Pearsall, T. P., Thornber, K. K.: The Effect of Collisional Broadening on Monte Carlo Simulations of High-Field Transport in Semiconductor Devices. IEEE Electron Dev. Lett. *EDL-2,* 295 (1981).

[2.18] Chryssafis, A., Love, W.: A Computer-Aided Analysis of One-Dimensional Thermal Transients in n-p-n Power Transistors. Solid-State Electron. *22,* 249 – 256 (1979).

[2.19] Cody, W. J., Thacher, H. C.: Rational Chebyshev Approximations for Fermi-Dirac Integrals of Orders – 1/2, 1/2 and 3/2. Math. Comp. *21,* 30 – 40 (1967).

[2.20] Cohen, M. L., Bergstresser, T. K.: Band Structures and Pseudopotential Form Factors for Fourteen Semiconductors of the Diamond and Zinc-blende Structures. Physical Review *141,* 789 – 796 (1966).

[2.21] Conwell, E. M.: High Field Transport in Semiconductors. New York: Academic Press 1967.

[2.22] Cook, R. K., Frey, J.: Two-Dimensional Numerical Simulation of Energy Transport Effects in Si and GaAs MESFET's. IEEE Trans. Electron Devices *ED-29,* No. 6, 970 – 977 (1982).

[2.23] Curtice, W. R.: Direct Comparison of the Electron Temperature Model with the Particle Mesh (Monte-Carlo) Model for the GaAs MESFET. IEEE Trans. Electron Devices *ED-29,* No. 12, 1942 – 1943 (1982).

[2.24] DeMan, H. J. J.: The Influence of Heavy Doping on the Emitter Efficiency of a Bipolar Transistor. IEEE Trans. Electron Devices *ED-18,* No. 10, 833 – 835 (1971).

[2.25] DelAlamo, J. A., Swanson, R. M., Lietoila, A.: Photovoltaic Measurement of Bandgap Narrowing in Moderately Doped Silicon. Solid-State Electron. *26,* No. 5, 483 – 489 (1983).

[2.26] Dhariwal, S. R., Ojha, V. N.: Band Gap Narrowing in Heavily Doped Silicon. Solid-State Electron. *25,* No. 9, 909 – 911 (1982).

[2.27] Dorkel, J. M.: On Electrical Transport in Non-Isothermal Semiconductors. Solid-State Electron. *26,* No. 8, 819 – 821 (1983).

[2.28] Engl, W. L., Dirks, H. K., Meinerzhagen, B.: Device Modeling. Proc. IEEE *71,* No. 1, 10 – 33 (1983).

[2.29] Frey, J.: Transport Physics for VLSI. In: Introduction to the Numerical Analysis of Semiconductor Devices and Integrated Circuits, pp. 51 – 57. Dublin: Boole Press 1981.

[2.30] Frey, J.: Physics Problems in VLSI Devices. In: Introduction to the Numerical Analysis of Semiconductor Devices and Integrated Circuits, pp. 47 – 50. Dublin: Boole Press 1981.

[2.31] Froelich, R. K., Blakey, P. A.: Energy and Momentum Conserving Simulation of Millimeter Wave Impatt Diodes. Proc. NASECODE II Conf., pp. 208 – 212. Dublin: Boole Press 1981.

[2.32] Gaensslen, F. H., Jaeger, R. C., Walker, J. J.: Low-Temperature Threshold Behavior of Depletion Mode Devices – Characterization and Simulation. Proc. Int. Electron Devices Meeting, pp. 520 – 524 (1976).

[2.33] Gaensslen, F. H., Rideout, V. L., Walker, E. J., Walker, J. J.: Very Small MOSFET's for Low Temperature Operation. IEEE Trans. Electron Devices *ED-24,* No. 3, 218 – 229 (1977).

[2.34] Gaensslen, F. H., Jaeger, R. C.: Temperature Dependent Threshold Behaviour of Depletion Mode MOSFET's. Solid-State Electron. *22,* 423 – 430 (1979).

[2.35] Gaur, S. P., Navon, D. H.: Two-Dimensional Carrier Flow in a Transistor Structure under Nonisothermal Conditions. IEEE Trans. Electron Devices *ED-23,* 50 – 57 (1976).

[2.36] Gnädinger, A. P., Talley, H. E.: Quantum Mechanical Calculations of the Carrier Distribution and the Thickness of the Inversion Layer of a MOS Field-Effect Transistor. Solid-State Electron. *13,* 1301 – 1309 (1970).

[2.37] Grondin, R. O., Lugli, P., Ferry, D. K.: Ballistic Transport in Semiconductors. IEEE Electron Device Lett. *EDL-3,* No. 12, 373 – 375 (1982).

[2.38] Halperin, B. I., Lax, M.: Impurity Band Tails in the High Density Limit. I. Minimum Counting Method. Physical Review *148,* 722–740 (1966).

[2.39] Heasell, E. L.: A Self-Consistent Calculation of Effective Intrinsic Concentration in Heavily Doped Silicon. Int. J. Electronics *38,* No. 1, 127–135 (1975).

[2.40] Heasell, E. L.: On the Role of Degeneracy in the "Heavy Doping" Phenomenon. Solid-State Electron. *23,* 183 (1980).

[2.41] Hess, K.: Ballistic Electron Transport in Semiconductors. IEEE Trans. Electron Devices *ED-28,* 937–940 (1981).

[2.42] Heywang, W., Pötzl, H. W.: Bandstruktur und Stromtransport. Berlin-Heidelberg-New York: Springer 1976.

[2.43] Hillbrand, H.: Untersuchungen des Transportverhaltens von III-V Halbleitern bei hohen elektrischen Feldern mit Monte-Carlo-Methoden. Dissertation, Technische Hochschule Wien, 1974.

[2.44] Hofmann, H.: Das elektromagnetische Feld, 2. Aufl. Wien-New York: Springer 1982.

[2.45] Holden, A. J., Debney, B. T.: Improved Theory of Ballistic Transport in One Dimension. Electronics Lett. *18,* No. 13, 558–559 (1982).

[2.46] Jüngling, W.: Hochdotierungseffekte in Silizium. Diplomarbeit, Technische Universität Wien, 1983.

[2.47] Jüngling, W., Guerrero, E., Selberherr, S.: On Modeling the Intrinsic Number and Fermi Levels for Device and Process Simulation. Proc. NASECODE III Conf., pp. 144–149. Dublin: Boole Press 1983.

[2.48] Kane, E. O.: Thomas Fermi Approach to Impure Semiconductor Band Structure. Physical Review *131,* No. 1, 79–88 (1963).

[2.49] Kireev, P. S.: Semiconductor Physics. Moscow: MIR Publishers 1978.

[2.50] Kittel, C.: Introduction to Solid-State Physics. New York: Wiley 1967.

[2.51] Kleppinger, D. D., Lindholm, F. A.: Impurity Concentration Dependent Density of States and Resulting Fermi Level for Silicon. Solid-State Electron. *14,* 407–416 (1971).

[2.52] Landsberg, P. T., Hope, S. A.: Diffusion Currents in Semiconductors. Solid-State Electron. *19,* 173–174 (1976).

[2.53] Landsberg, P. T., Hope, S. A.: Two Formulations of Semiconductor Transport Equations. Solid-State Electron. *20,* 421–429 (1977).

[2.54] Langer, E.: Numerische Simulation der Halbleiterdiode. Diplomarbeit, Technische Universität Wien, 1980.

[2.55] Lanyon, H. P. D., Tuft, R. A.: Bandgap Narrowing in Heavily Doped Silicon. Proc. International Electron Device Meeting, pp. 316–319 (1978).

[2.56] Lanyon, H. P. D., Tuft, R. A.: Bandgap Narrowing in Moderately to Heavily Doped Silicon. IEEE Trans. Electron Devices *ED-26,* No. 7, 1014–1018 (1979).

[2.57] Lee, D. S., Fossum, J. G.: Energy Band Distortion in Highly Doped Silicon. IEEE Trans. Electron Devices *ED-30,* No. 6, 626–634 (1983).

[2.58] Lue, J. T.: Theory of Schottky Barrier Heights of Amorphous MIS Solar Cells. Solid-State Electron. *25,* No. 9, 869–874 (1982).

[2.59] Lundstrom, M. S., Schwartz, R. J., Gray, J. L.: Transport Equations for the Analysis of Heavily Doped Semiconductor Devices. Solid-State Electron. *24,* 195–202 (1981).

[2.60] Lundstrom, M. S., Schuelke, R. J.: Modeling Semiconductor Heterojunctions in Equilibrium. Solid-State Electron. *25,* No. 8, 683–691 (1982).

[2.61] Mertens, R. P., VanMeerenbergen, J. L., Nijs, J. F., VanOverstraeten, R. J.: Measurement of the Minority-Carrier Transport Parameters in Heavily Doped Silicon. IEEE Trans. Electron Devices *ED-27,* 949–955 (1980).

[2.62] Mertens, R. P.: Modeling of Heavy Doping Effects. Proc. VLSI Process and Device Modeling, pp. 1–29. Katholieke Universiteit Leuven, 1983.

[2.63] Mock, M. S.: Transport Equations in Heavily Doped Silicon, and the Current Gain of a Bipolar Transistor. Solid-State Electron. *16,* 1251–1259 (1973).

[2.64] Moglestue, C., Beard, S. J.: A Particle Model Simulation of Field Effect Transistors. Proc. NASECODE I Conf., pp. 232–236. Dublin: Boole Press 1979.

[2.65] Moglestue, C.: A Monte-Carlo Particle Model Study of the Influence of the Doping Profiles on the Characteristics of Field-Effect Transistors. Proc. NASECODE II Conf., pp. 244–249. Dublin: Boole Press 1981.

[2.66] Morgan, T. N.: Broadening of Impurity Bands in Heavily Doped Semiconductors. Physical Review. *139,* No. 1A, A343 – A348 (1965).

[2.67] Nag, B. R.: Diffusion Equation for Hot Electrons. Phys. Rev. *B11,* No. 8, 3031 – 3036 (1974).

[2.68] Nag, B. R., Chakravarti, A. N.: Comments on the Generalized Einstein Relation for Semiconductors. Solid-State Electron. *18,* 109 – 110 (1975).

[2.69] Nag, B. R.: Parallel Diffusion Constant of Hot Electrons in Silicon. Appl. Phys. Lett. *28,* No. 9, 550 – 551 (1976).

[2.70] Nakagawa, A.: One-Dimensional Device Model of the npn Bipolar Transistor Including Heavy Doping Effects under Fermi Statistics. Solid-State Electron. *22,* 943 – 949 (1979).

[2.71] Paul, R.: Halbleiterphysik. Heidelberg: Hüthig-Verlag 1975.

[2.72] Phillips, A.: On Modeling the Bipolar Transistor with Realistic Impurity Distributions, Heavy Doping Physics. Technical Report TR22.2045, IBM (1976).

[2.73] Polsky, B. S., Rimshans, J. S.: Two-Dimensional Numerical Simulation of Bipolar Semiconductor Devices Taking into Account Heavy Doping Effects and Fermi Statistics. Solid-State Electron. *26,* No. 4, 275 – 279 (1983).

[2.74] Raychaudhuri, D., Chattopadhyay, D.: Harmonic Generation Due to Ballistic Electron Transport in GaAs. Proc. IEEE *71,* No. 3, 440 – 441 (1983).

[2.75] Robinson, J. E., Rodriguez, S.: Ionized Impurity Scattering in Degenerate Many-Valley Semiconductors. Physical Review *135,* No. 3A, A779 – A784 (1964).

[2.76] Seeger, K.: Semiconductor Physics. Wien-New York: Springer 1973.

[2.77] Shibib, M. A.: Inclusion of Degeneracy in the Analysis of Heavily Doped Regions in Silicon Solar Cells and Other Semiconductor Devices. Solar Cells *3,* 81 – 85 (1981).

[2.78] Shur, M. S., Eastman, L. F.: Ballistic Transport in Semiconductor at Low Temperature for Low-Power High-Speed Logic. IEEE Trans. Electron Devices *ED-26,* No. 11, 1677 – 1683 (1979).

[2.79] Slotboom, J. W., DeGraaff, H. C.: Measurements of Bandgap Narrowing in Si Bipolar Transistors. Solid-State Electron. *19,* 857 – 862 (1976).

[2.80] Slotboom, J. W., DeGraaff, H. C.: Bandgap Narrowing in Silicon Bipolar Transistors. IEEE Trans. Electron Devices *ED-24,* No. 8, 1123 – 1125 (1977).

[2.81] Slotboom, J. W.: The pn-Product in Silicon. Solid-State Electron. *20,* 279 – 283 (1977).

[2.82] Smith, R. A.: Semiconductors. Cambridge: Cambridge University Press 1978.

[2.83] Stern, F.: Effect of Band Tails on Stimulated Emission of Light in Semiconductors. Physical Review *148,* No. 1, 186 – 193 (1966).

[2.84] Stern, F.: Optical Absorption Edge of Compensated Germanium. Physical Review *B3,* No. 10, 3559 – 3560 (1971).

[2.85] Stratton, R.: Semiconductor Current-Flow Equations (Diffusion and Degeneracy). IEEE Trans. Electron Devices *ED-19,* No. 12, 1288 – 1292 (1972).

[2.86] Sze, S. M.: Physics of Semiconductor Devices. New York: Wiley 1969.

[2.87] Teitel, S. L., Wilkins, J. W.: Ballistic Transport and Velocity Overshoot in Semiconductors: Part I – Uniform Field Effects. IEEE Trans. Electron Devices *ED-30,* No. 2, 150 – 153 (1983).

[2.88] Thornber, K. K.: Current Equations for Velocity Overshoot. IEEE Electron Device Lett. *EDL-3,* No. 3, 69 – 70 (1982).

[2.89] Tihanyi, J.: Integrated Power Devices. Proc. International Electron Devices Meeting, pp. 6 – 10 (1982).

[2.90] VanOverstraeten, R. J., DeMan, H. J., Mertens, R. P.: Transport Equations in Heavy Doped Silicon. IEEE Trans. Electron Devices *ED-20,* 290 – 298 (1973).

[2.91] VanVliet, K. M.: The Shockley-Like Equations for the Carrier Densities and the Current Flows in Materials with a Nonuniform Composition. Solid-State Electron. *23,* 49 – 53 (1980).

[2.92] Wieder, A. W.: Emitter Effects in Shallow Bipolar Devices: Measurements and Consequences. IEEE Trans. Electron Devices *ED-27,* No. 8, 1402 – 1417 (1980).

[2.93] Williams, C. K., Glisson, T. H., Littlejohn, M. A., Hauser, J. R.: Ballistic Transport in GaAs. IEEE Electron Device Lett. *EDL-4,* No. 6, 161 – 163 (1983).

[2.94] Wulms, H. E. J.: Base Current of I2L Transistors. IEEE J. Solid-State Circuits *SC-12,* No. 2, 143 – 150 (1977).

[2.95] Ziman, J. M.: Electrons and Phonons. London: Clarendon Press 1963.

[2.96] Zimmerl, O.: Iterative Lösung der Boltzmanngleichung für heiße Elektronen in InSb. Dissertation, Technische Hochschule Wien, 1972.

3 Process Modeling

To enable the simulation of the electric behavior of a device the configuration of the device (i. e. geometry and composition of the material it is made of) is, obviously, one of the prerequisite pieces of input information. Optimal design of a device necessitates the capability to predict the effect of modifying any of the various process steps involved in device fabrication. One principle barrier for predictive device simulation is the uncertainty of the results of process models due to still inadequate understanding of their underlying solid-state physics and chemistry. Particularly in the development of devices for integrated circuits and their technology, the need for process models is growing dramatically due to the tight coupling of two and three dimensional device effects with the doping profile [3.18]. Owing to these purposes, many computer programs capable of modeling quite generally the various processing steps of device fabrication have been developed, and they have proven to be extremely valuable tools, e. g. ICECREM [3.64], [3.67]; LADIS [3.76]; MEMBRE [3.54]; RECIPE [3.73]; SUPRA [3.15], [3.38], [3.39], [3.50] and the extraordinarily well established SUPREM program [3.4], [3.5], [3.50], [3.51], [3.59].
Apart from lithography which may be viewed as a fixed process that simply provides flexibility in layout [3.56], [3.57], the primary fabrication processes which determine the electrical characteristics of semiconductor devices, in particular silicon devices, are ion implantation, diffusion and thermal oxidation. Epitaxy, etching and deposition can certainly play an essential role in device fabrication as well. However, as the field of process modeling is extraordinarily wide and difficult, only the above cited process steps will be discussed here. Furthermore, it should be noted that only a review of the most important models can be presented here due to the complexity of the underlying phenomena. The aim here is just to give a flavor on what problems have to be dealt with in providing this all-important input for device simulation. We also shall restrict ourselves to silicon processing.

3.1 Ion Implantation

Ion implantation is the most applied doping technique in the fabrication of silicon devices, particularly integrated devices. A highly energetic beam of ions strikes and penetrates into a target of coated or uncoated semiconductor. The final distribution of particles within the semiconductor will be discussed in this section. The only exact procedure one can imagine for calculating the distribution of implanted ions would be a Monte Carlo simulation of the implantation process itself [3.48] by invoking the laws of statistical mechanics, or perhaps at a minimum solving a Boltzmann transport equation – similar to the one describing the movement of electrons – [3.16] for the penetrating ions. Both methods allow a treatment based on first principles of the undesirable parasitic effects of ion implantation like lattice disorder and defects [3.16], [3.77], [3.78], back scattering and target sputtering [3.74] and channeling [3.24]. However, for many practical applications it is sufficient to assume a distribution function and to calculate or, even better, to measure its parameters so that good agreement between experimentally determined and simulated profiles is established. Such a procedure does, obviously, not contribute to the understanding of the underlying physical phenomena but it can prove – it has already proved – to be adequate for many engineering applications, although in some novel processing techniques (e. g. multilayered mask structures) the disadvantages become quite evident and lead to frustration.

We first shall consider only one dimensional distributions. In order to describe the distribution of implanted ions by means of distribution functions we have to remember some of the mathematical properties of probability functions. $F(x)$ is termed a univariate cummulative distribution function if:

a) $F(x)$ is non-decreasing, i.e.,

$$F(x1) \leq F(x2), \quad \text{for } x1 < x2 \tag{3.1-1}$$

b) $F(x)$ is everywhere continuous from the right, i.e.,

$$F(x) = \lim_{h \to 0+} F(x+h) \tag{3.1-2}$$

c) $F(x)$ fulfills:

$$F(-\infty) = 0, \quad F(\infty) = 1 \tag{3.1-3}$$

Only distribution functions with these properties are suitable for the description of implantation profiles. We further allow only continuous distributions so that $F(x)$ can be written as:

$$F(x) = \int_{-\infty}^{x} f(t) \cdot dt \tag{3.1-4}$$

$f(x)$ is termed the "probability density function" or "frequency function". With these definitions the real ion distribution is given as:

$$C(x) = N_d \cdot f(x) \tag{3.1-5}$$

N_d is the total implant dose per unit area. The properties of the distribution function $F(x)$ guarantee that the profile is consistent in a physical sense, e. g. the total dose must be incorporated in the target. The frequency function $f(x)$ is, in practice, assumed in its structure to be one of the well established functions from statistical mathematics, and its associated parameters are calculated with knowledge about some characteristic quantities of the distribution. These are the mean value or projected range R_p:

$$R_p = \int_{-\infty}^{\infty} x \cdot f(x) \cdot dx \tag{3.1-6}$$

the standard deviation σ_p:

$$\sigma_p = \sqrt{\int_{-\infty}^{\infty} (x - R_p)^2 \cdot f(x) \cdot dx} \tag{3.1-7}$$

the skewness γ_1:

$$\gamma_1 = \frac{\int_{-\infty}^{\infty} (x - R_p)^3 \cdot f(x) \cdot dx}{\sigma_p^{\ 3}} \tag{3.1-8}$$

and the excess or kurtosis β_2:

$$\beta_2 = \frac{\int_{-\infty}^{\infty} (x - R_p)^4 \cdot f(x) \cdot dx}{\sigma_p^{\ 4}} \tag{3.1-9}$$

These characteristic quantities can either be calculated or measured in order to fit an assumed frequency function to experimentally determined doping profiles [3.64]. The oldest theory is termed *LSS* theory after the authors Lindhard, Scharff and Schiøtt [3.40]. The detailed physics of the ion range theory are covered rather extensively in the literature. Hence, I shall only give some important references and some qualitative discussion here. Numerical values for the projected range R_p, the standard deviation σ_p and the skewness γ_1 after the *LSS* theory are tabulated in the book of Gibbons et al. [3.25]. Values for the kurtosis β_2 have additionally been calculated by, e.g., Winterbon [3.83]. The LSS theory has been thoroughly discussed in, e.g., [3.82]; it has been refined for multilayered targets in [3.16]. In Fig. 3.1-1 the projected range R_p is shown for the most frequently used dopants in silicon. Correspondingly, Fig. 3.1-2 and Fig. 3.1-3 show the standard deviation σ_p and the skewness γ_1. As results in form of tables or graphs are tedious to handle, some sort of functional fit might be very attractive. The easiest approach for that task is a simple polynomial fit [3.71].

$$R_p = \sum_{i=1}^{n} a_i \cdot E^i \tag{3.1-10}$$

$$\sigma_p = \sum_{i=1}^{n} b_i \cdot E^i \tag{3.1-11}$$

E denotes the implantation energy. Coefficients for such polynomials are given in Table 3.1-1, Table 3.1-2 for silicon as target, in Table 3.1-3, Table 3.1-4 for silicon dioxide and in Table 3.1-5, Table 3.1-6 for silicon nitride (Si_3N_4) as target.

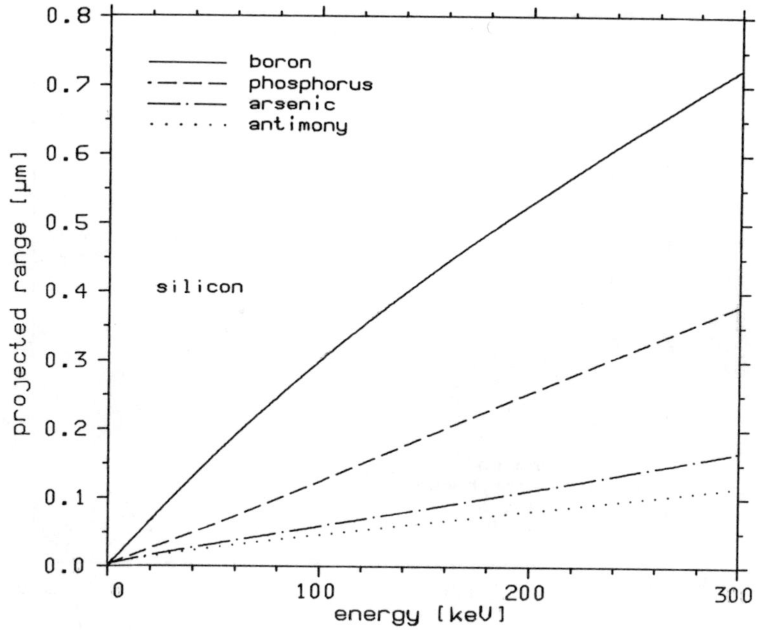

Fig. 3.1-1. Projected range of boron, phosphorus, arsenic and antimony in silicon

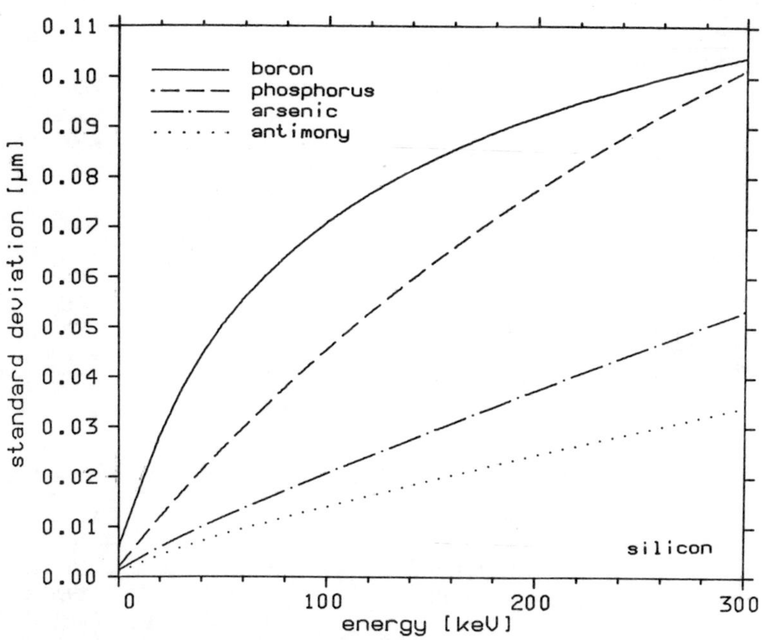

Fig. 3.1-2. Standard deviation of boron, phosphorus, arsenic and antimony in silicon

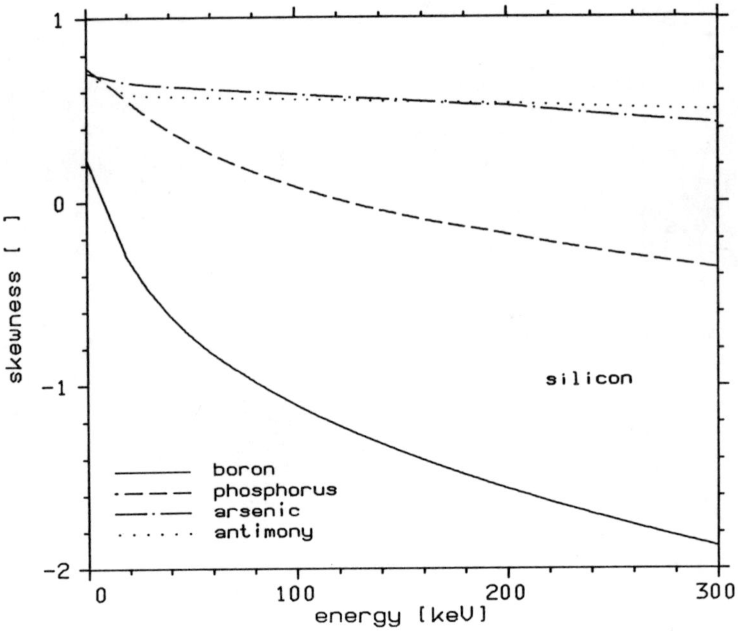

Fig. 3.1-3. Skewness of boron, phosphorus, arsenic and antimony in silicon

Table 3.1-1. *Coefficients for R_p in silicon*

Element	B	P	Sb	As
a_1	$3.338 \cdot 10^{-3}$	$1.259 \cdot 10^{-3}$	$8.887 \cdot 10^{-4}$	$9.818 \cdot 10^{-4}$
a_2	$-3.308 \cdot 10^{-6}$	$-2.743 \cdot 10^{-7}$	$-1.013 \cdot 10^{-5}$	$-1.022 \cdot 10^{-5}$
a_3		$1.290 \cdot 10^{-9}$	$8.372 \cdot 10^{-8}$	$9.067 \cdot 10^{-8}$
a_4			$-3.056 \cdot 10^{-10}$	$-3.442 \cdot 10^{-10}$
a_5			$4.028 \cdot 10^{-13}$	$4.608 \cdot 10^{-13}$

Table 3.1-2. *Coefficients for σ_p in silicon*

Element	B	P	Sb	As
b_1	$1.781 \cdot 10^{-3}$	$6.542 \cdot 10^{-4}$	$2.674 \cdot 10^{-4}$	$3.652 \cdot 10^{-4}$
b_2	$-2.086 \cdot 10^{-5}$	$-3.161 \cdot 10^{-6}$	$-2.885 \cdot 10^{-6}$	$-3.820 \cdot 10^{-6}$
b_3	$1.403 \cdot 10^{-7}$	$1.371 \cdot 10^{-8}$	$2.311 \cdot 10^{-8}$	$3.235 \cdot 10^{-8}$
b_4	$-4.545 \cdot 10^{-10}$	$-2.252 \cdot 10^{-11}$	$-8.310 \cdot 10^{-11}$	$-1.202 \cdot 10^{-10}$
b_5	$5.525 \cdot 10^{-13}$		$1.084 \cdot 10^{-13}$	$1.601 \cdot 10^{-13}$

Table 3.1-3. *Coefficients for R_p in silicon dioxide*

Element	B	P	Sb	As
a_1	$3.258 \cdot 10^{-3}$	$9.842 \cdot 10^{-4}$	$7.200 \cdot 10^{-4}$	$7.806 \cdot 10^{-4}$
a_2	$-2.113 \cdot 10^{-6}$	$2.240 \cdot 10^{-7}$	$-8.054 \cdot 10^{-6}$	$-7.899 \cdot 10^{-6}$
a_3			$6.641 \cdot 10^{-8}$	$7.029 \cdot 10^{-8}$
a_4			$-2.422 \cdot 10^{-10}$	$-2.653 \cdot 10^{-10}$
a_5			$3.191 \cdot 10^{-13}$	$3.573 \cdot 10^{-13}$

Table 3.1-4. *Coefficients for* σ_p *in silicon dioxide*

Element	B	P	Sb	As
b_1	$1.433 \cdot 10^{-3}$	$4.591 \cdot 10^{-4}$	$2.018 \cdot 10^{-4}$	$2.637 \cdot 10^{-4}$
b_2	$-1.077 \cdot 10^{-5}$	$-1.983 \cdot 10^{-6}$	$-2.328 \cdot 10^{-6}$	$-2.762 \cdot 10^{-6}$
b_3	$4.190 \cdot 10^{-8}$	$8.383 \cdot 10^{-9}$	$1.917 \cdot 10^{-8}$	$2.373 \cdot 10^{-8}$
b_4	$-6.000 \cdot 10^{-11}$	$-1.382 \cdot 10^{-11}$	$-6.997 \cdot 10^{-11}$	$-8.899 \cdot 10^{-11}$
b_5			$9.211 \cdot 10^{-14}$	$1.193 \cdot 10^{-13}$

Table 3.1-5. *Coefficients for* R_p *in silicon nitride*

Element	B	P	Sb	As
a_1	$2.514 \cdot 10^{-3}$	$7.617 \cdot 10^{-4}$	$5.660 \cdot 10^{-4}$	$6.094 \cdot 10^{-4}$
a_2	$-1.618 \cdot 10^{-6}$	$1.681 \cdot 10^{-7}$	$-6.440 \cdot 10^{-6}$	$-6.213 \cdot 10^{-6}$
a_3			$5.323 \cdot 10^{-8}$	$5.516 \cdot 10^{-8}$
a_4			$-1.944 \cdot 10^{-10}$	$-2.080 \cdot 10^{-10}$
a_5			$2.563 \cdot 10^{-13}$	$2.799 \cdot 10^{-13}$

Table 3.1-6. *Coefficients for* σ_p *in silicon nitride*

Element	B	P	Sb	As
b_1	$1.115 \cdot 10^{-3}$	$3.542 \cdot 10^{-4}$	$1.516 \cdot 10^{-4}$	$2.035 \cdot 10^{-4}$
b_2	$-8.328 \cdot 10^{-6}$	$-1.488 \cdot 10^{-6}$	$-1.655 \cdot 10^{-6}$	$-2.092 \cdot 10^{-6}$
b_3	$3.228 \cdot 10^{-8}$	$6.204 \cdot 10^{-9}$	$1.345 \cdot 10^{-8}$	$1.787 \cdot 10^{-8}$
b_4	$-4.612 \cdot 10^{-11}$	$-1.019 \cdot 10^{-11}$	$-4.878 \cdot 10^{-11}$	$-6.678 \cdot 10^{-11}$
b_5			$6.401 \cdot 10^{-14}$	$8.932 \cdot 10^{-14}$

The dimensions of the coefficients a_i, b_i are micrometer per i-th power of the units used for the implantation energy, usually keV. The maximum error of R_p and σ_p calculated with these coefficients and formulae (3.1-10), (3.1-11) in the range [5,300] keV is only a few percent compared to the tabulated data in [3.25]. The skewness γ_1 has not been approximated in this way, although there is in principal no problem, but for the construction of distributions for which an accurate value of the skewness is required, one possibly has to modify the values obtained by the *LSS* theory (cf. [3.64)]. A slightly different approach to the *LSS* concept about ion range theory has been proposed by Biersack [3.10], [3.11]. Comparisons of measured and calculated quantities according to the Biersack theory look quite convincing [3.66]. Tabulated values are given in the book by Ryssel and Ruge [3.63]. Measured values for the parameters (3.1-6) to (3.1-9) for boron implanted into silicon have been given in [3.65]. A thorough discussion and comparison of various other concepts on ion range theory can be found in, e.g., [3.41].
The distributions which are most frequently used for describing doping profiles are the simple Gaussian or normal distribution, the joined half Gaussian distribution, Edgeworth asymptotic expansions of the Gaussian distribution and the Pearson type IV distribution.

The Gaussian distribution makes use only of the projected range R_p and the standard deviation σ_p. The frequency function for a Gaussian distribution reads:

$$f(x)=\frac{1}{\sqrt{2\cdot\pi}\cdot\sigma_p}\cdot\exp\left(-\frac{(x-R_p)^2}{2\cdot\sigma_p^2}\right) \tag{3.1-12}$$

A Gaussian distribution has a skewness $\gamma_1=0$ and a kurtosis $\beta_2=3$. The approximation of a true profile with a Gaussian distribution is only accurate to first order. However, the simplicity of the calculation justifies to some degree its use when the primary concern is the average location and average extent of a distribution.

If one wants to fit more accurately the asymetrical distributions usually found in practice, it is necessary to at least account for the skewness. Such an approach was suggested in [3.26] and it has been used quite extensively. The frequency function is defined to consist of two half Gaussian distributions that join at a modal projected range R_m. For distances $x<R_m$ the distribution has standard deviation σ_1, while for $x>R_m$ the distribution has standard deviation σ_2.

$$f(x)=\begin{cases} x<R_m & \dfrac{2}{\sqrt{2\cdot\pi}\cdot(\sigma_1+\sigma_2)}\cdot\exp\left(-\dfrac{(x-R_m)^2}{2\cdot\sigma_1^2}\right) \\ x\geq R_m & \dfrac{2}{\sqrt{2\cdot\pi}\cdot(\sigma_1+\sigma_2)}\cdot\exp\left(-\dfrac{(x-R_m)^2}{2\cdot\sigma_2^2}\right) \end{cases} \tag{3.1-13}$$

The modal projected range R_m and the "one-sided" standard deviations σ_1, σ_2 can be calculated by using the characteristic quantities (3.1-6), (3.1-7) and (3.1-8). With some amount of algebra these integrals evaluate to:

$$R_p=R_m+\sqrt{\frac{2}{\pi}}\cdot(\sigma_2-\sigma_1) \tag{3.1-14}$$

$$\sigma_p=\sqrt{(\sigma_1^2-\sigma_1\cdot\sigma_2+\sigma_2^2)-\frac{2}{\pi}\cdot(\sigma_2-\sigma_1)^2} \tag{3.1-15}$$

$$\gamma_1=\frac{\sqrt{\frac{2}{\pi}}\cdot(\sigma_2-\sigma_1)\cdot\left(\left(\frac{4}{\pi}-1\right)\cdot(\sigma_1^2+\sigma_2^2)+\left(3-\frac{8}{\pi}\right)\cdot\sigma_1\cdot\sigma_2\right)}{\sigma_p^3} \tag{3.1-16}$$

The kurtosis β_2 can not be used to match a profile, because one has only three parameters $(R_m, \sigma_1, \sigma_2)$ available which are already fully determined by R_p, σ_p and γ_1 through relations (3.1-14) to (3.1-16). The kurtosis β_2 of a joined half Gaussian distribution can be demonstrated to depend only on the square of the skewness. Fig. 3.1-4 shows the value of the kurtosis one automatically obtains for a given value of the skewness. It is satisfying that over the entire range of validity for the skewness γ_1, (cf. 3.1-17) the value for the kurtosis is reasonable.

It would be very attractive to have explicit formulae for R_m, σ_1 and σ_2. However, such formulae do not exist. With some algebra one will end up with a cubic equation for the difference $(\sigma_1-\sigma_2)$ and a quadratic equation for the sum $(\sigma_1+\sigma_2)$. These equations could, in principle, be solved analytically, however, this approach can not

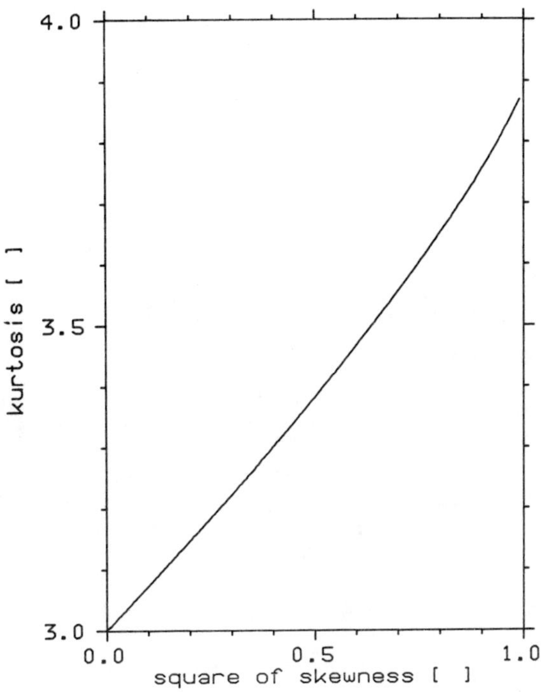

Fig. 3.1-4. Kurtosis of joined-half-Gaussian distribution versus square of skewness

be recommended for implementation in a computer program because of cancellation problems associated with the specific structure of the analytical solutions of cubic equations. An iterative method is much more feasible. It is to note that a joined half Gaussian distribution only exists for a restricted range of values for the skewness γ_1. In particular condition (3.1-17) must hold.

$$|\gamma_1| < \frac{4-\pi}{\pi-2} \cdot \sqrt{\frac{2}{\pi-2}} \cong 0{,}99527 \tag{3.1-17}$$

I feel obliged here to warn the reader about the various papers on joined half Gaussian distributions; in many of these the various constants are indisputably incorrect.

For distributions whose skewness has a magnitude larger than given in (3.1-17) (cf. Fig. 3.1-3) it is necessary to take into account higher order characteristic quantities of the distribution, minimally, the kurtosis β_2 (3.1-9). In [3.22] an Edgeworth asymptotic expansion has been suggested for that purpose. The first three terms of the Edgeworth expansion are:

$$f(y) = [g(y)] - \left[\frac{\gamma_1}{6} \cdot \frac{d^3 g}{d y^3}(y)\right] + \left[\frac{\beta_2 - 3}{24} \cdot \frac{d^4 g}{d y^4}(y) + \frac{\gamma_1{}^2}{72} \cdot \frac{d^6 g}{d y^6}(y)\right] + \dots \tag{3.1-18}$$

$f(y)$ is the frequency function to be expanded in terms of $g(y)$ and its derivatives. In our case $g(y)$ is the Gaussian frequency function.

$$g(y)=\frac{1}{\sqrt{2\cdot\pi}\cdot\sigma_p}\cdot\exp\left(-\frac{y^2}{2}\right) \tag{3.1-19}$$

$$y=\frac{x-R_p}{\sigma_p} \tag{3.1-20}$$

Terms in brackets in (3.1-18) are terms of the same order with respect to the expansion index. If the derivatives of $g(y)$ are evaluated and substituted into (3.1-18) we obtain:

$$\begin{aligned} f(y)=&\frac{1}{\sqrt{2\cdot\pi}\cdot\sigma_p}\cdot\exp\left(-\frac{y^2}{2}\right)\cdot\\ &\cdot\left(1+\frac{\beta_2-3}{8}-\frac{5\cdot\gamma_1{}^2}{24}-\frac{\gamma_1}{2}\cdot y-\left(\frac{\beta_2-3}{4}-\frac{5\cdot\gamma_1{}^2}{8}\right)\cdot y^2+\right.\\ &\left.+\frac{\gamma_1}{6}\cdot y^3+\left(\frac{\beta_2-3}{24}-\frac{5\cdot\gamma_1{}^2}{24}\right)\cdot y^4+\frac{\gamma_1{}^2}{72}\cdot y^6\right) \end{aligned} \tag{3.1-21}$$

If no values for the kurtosis β_2 are available, which is unfortunately very often the case, Gibbons et al. [3.25] suggest using relation (3.1-22) which guarantees that (3.1-21) is positive for $y=0$.

$$\beta_2\cong\frac{5}{3}\cdot\gamma_1{}^2+3 \tag{3.1-22}$$

The expansion (3.1-21) is only applicable for a limited range of values for γ_1, ρ_2 and also y. The condition which has to be fulfilled is that the multiplying polynomial in (3.1-21) is positive. The procedure of performing an Edgeworth asymptotic expansion into Gaussian frequency functions is, from a mathematical point of view, a very elegant way to introduce the influence of higher order characteristic quantities of the distribution function. However, due to the fact that only the very low order terms of the expansion can be accounted for, the frequency function exhibits an oscillatory behavior for distributions whose skewness γ_1 is of large magnitude.

An approach followed by Hofker [3.30] for fitting a frequency function to experimental data is to use a Pearson type IV distribution function. The whole family of Pearson distributions [3.26] is based on the differential equation

$$\frac{df}{dy}(y)=\frac{y-a}{b_0+b_1\cdot y+b_2\cdot y^2}\cdot f(y),\quad y=x-R_p \tag{3.1-23}$$

where $f(y)$ is, as defined before, the frequency function. The four constants a, b_0, b_1 and b_2 can be expressed in the four characteristic quantities projected range (3.1-6), standard deviation (3.1-7), skewness (3.1-8) and kurtosis (3.1-9) as we shall demonstrate in the following.

The *i*-th central moments of any distribution are defined as:

$$\mu_i = \int_{-\infty}^{\infty} y^i \cdot f(y) \cdot dy \tag{3.1-24}$$

These central moments are related to the characteristics which we prefer to use by:

$$R_p = \mu_1 \tag{3.1-25}$$

$$\sigma_p = \sqrt{\mu_2 - \mu_1^{\,2}} \tag{3.1-26}$$

$$\gamma_1 = \frac{\mu_3 - 3 \cdot \mu_2 \cdot \mu_1 + 2 \cdot \mu_1^{\,3}}{(\mu_2 - \mu_1^{\,2})^{3/2}} \tag{3.1-27}$$

$$\mu_2 = \frac{\mu_4 - 4 \cdot \mu_3 \cdot \mu_1 + 6 \cdot \mu_2 \cdot \mu_1^{\,3} - 3 \cdot \mu_1^{\,4}}{(\mu_2 - \mu_1^{\,2})^2} \tag{3.1-28}$$

We further remember that $\mu_0 = 1$ (cf. 3.1-3). Then we rearrange (3.1-23) after multiplying both sides by y^n.

$$y^n \cdot (b_0 + b_1 \cdot y + b_2 \cdot y^2) \cdot f'(y) = y^n \cdot (y - a) \cdot f(y) \tag{3.1-29}$$

By integrating both sides of (3.1-29) between $-\infty$ and ∞ and assuming that

$$\lim_{y \to \pm\infty} y^n \cdot f(y) = 0, \quad n < 6 \tag{3.1-30}$$

we obtain:

$$n \cdot b_0 \cdot \mu_{n-1} + ((n+1) \cdot b_1 - a) \cdot \mu_n + ((n+2) \cdot b_2 + 1) \cdot \mu_{n+1} = 0 \tag{3.1-31}$$

By putting $n = 0, 1, 2, 3$ in (3.1-31) we have four simultaneous linear equations for a, b_0, b_1 and b_2 with coefficients which are functions of the central moments. Note that we may introduce without loss of generality a coordinate transformation $y \to z$ prior to integrating (3.1-29), which eminently eases the calculus, such that:

$$\mu_i' = \begin{cases} i = 1 & 0 \\ i \neq 1 & \mu_i \end{cases} \tag{3.1-32}$$

The expressions for a, b_0, b_1 and b_2 are then evaluated to:

$$a = -\frac{\gamma_1 \cdot \sigma_p \cdot (\beta_2 + 3)}{10 \cdot \beta_2 - 12 \cdot \gamma_1^{\,2} - 18} \tag{3.1-33}$$

$$b_0 = -\frac{\sigma_p^{\,2} \cdot (4 \cdot \beta_2 - 3 \cdot \gamma_1^{\,2})}{10 \cdot \beta_2 - 12 \cdot \gamma_1^{\,2} - 18} \tag{3.1-34}$$

$$b_1 = -\frac{\sigma_p \cdot \gamma_1 \cdot (\beta_2 + 3)}{10 \cdot \beta_2 - 12 \cdot \gamma_1^{\,2} - 18} \tag{3.1-35}$$

$$b_2 = -\frac{2 \cdot \beta_2 - 3 \cdot \gamma_1^{\,2} - 6}{10 \cdot \beta_2 - 12 \cdot \gamma_1^{\,2} - 18} \tag{3.1-36}$$

The shape of $f(y)$ varies considerably with b_0, b_1 and b_2. Pearson has classified the different shapes into seven types. I shall give here a short resume of this classification because one can find some inconsistencies on that subject in some papers authored by engineers. I shall follow Pearson's numbering of the individual types of distributions, although it does not exhibit a clear systematic base [3.34]. The form of solution of (3.1-23) evidently depends on the nature of the roots of the equation:

$$b_0 + b_1 \cdot y + b_2 \cdot y^2 = 0 \qquad (3.1\text{-}37)$$

However, I first should like to note that if we have $b_1 = b_2 = 0$ which corresponds to:

$$\left.\begin{array}{l} \gamma_1 = 0 \\ \beta_2 = 3 \end{array}\right\} \text{Gaussian} \qquad (3.1\text{-}38)$$

the solution of (3.1-23) is the Gaussian frequency function which, in fact, is a limiting case for all types of Pearson distributions.
We have a Pearson type I distribution if the roots of (3.1-37) are real and of opposite sign. This is the case for:

$$\left.\begin{array}{l} \gamma_1 \neq 0 \\ \beta_2 < 3 + 1.5 \cdot \gamma_1{}^2 \end{array}\right\} \text{Type I} \qquad (3.1\text{-}39)$$

A degenerate case is the Pearson type II distribution.

$$\left.\begin{array}{l} \gamma_1 = 0 \\ \beta_2 < 3 \end{array}\right\} \text{Type II} \qquad (3.1\text{-}40)$$

The Pearson type III distribution corresponds to the case $b_2 = 0$ and $b_1 \neq 0$ which can be expressed as:

$$\left.\begin{array}{l} \gamma_1 \neq 0 \\ \beta_2 = 3 + 1.5 \cdot \gamma_1{}^2 \end{array}\right\} \text{Type III} \qquad (3.1\text{-}41)$$

The case when (3.1-37) does not have real roots corresponds to the Pearson type IV distribution. This case arises when:

$$\left\{\begin{array}{l} 0 < \gamma_1{}^2 < 32 \\ \beta_2 > \dfrac{39 \cdot \gamma_1{}^2 + 48 + 6 \cdot (\gamma_1{}^2 + 4)^{3/2}}{32 - \gamma_1{}^2} \end{array}\right\} \text{Type IV} \qquad (3.1\text{-}42)$$

If (3.1-37) is a perfect square we have the Pearson type V distribution.

$$\left\{\begin{array}{l} 0 < \gamma_1{}^2 < 32 \\ \beta_2 = \dfrac{39 \cdot \gamma_1{}^2 + 48 + 6 \cdot (\gamma_1{}^2 + 4)^{3/2}}{32 - \gamma_1{}^2} \end{array}\right\} \text{Type V} \qquad (3.1\text{-}43)$$

The Pearson type VI distribution corresponds to the case when the roots of (3.1-37) are real and of the same sign.

$$\left\{\begin{array}{l} \gamma_1 \neq 0 \\ 3 + 1.5 \cdot \gamma_1{}^2 < \beta_2 < \dfrac{39 \cdot \gamma_1{}^2 + 48 + 6 \cdot (\gamma_1{}^2 + 4)^{3/2}}{32 - \gamma_1{}^2} \end{array}\right\} \text{Type VI} \qquad (3.1\text{-}44)$$

Finally, the Pearson type VII distribution corresponds to the case when $b_0 > 0$, $b_1 = 0$ and $b_2 > 0$ which can also be expressed as:

$$\left\{\begin{matrix} \gamma_1 = 0 \\ \beta_2 > 3 \end{matrix}\right\} \quad \text{Type VII} \tag{3.1-45}$$

The conditions (3.1-38) to (3.1-45) are graphically summarized in Fig. 3.1-5. Only the types I, IV and VI correspond to areas in the (γ_1^2, β_2) plane. The remaining types correspond to lines and are sometimes called transition types. The Gaussian distribution which, as mentioned, is a limiting case for all distributions is denoted by an asterisk in Fig. 3.1-5. Note that for any distribution condition (3.1-46) holds as can easily be proved.

$$\beta_2 > 1 + \gamma_1^2 \tag{3.1-46}$$

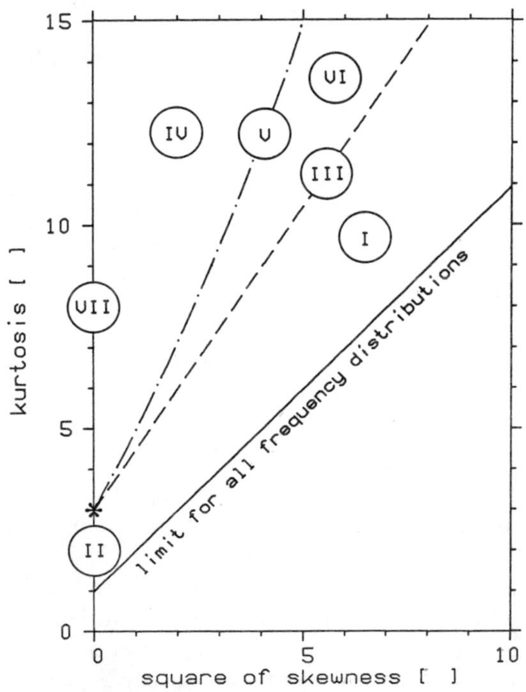

Fig. 3.1-5. Domains of validity for Pearson distributions

With regard to the conditions (3.1-38) to (3.1-45), respectively, it is just a matter of a simple calculus to solve the differential equation (3.1-23) for the frequency function of the specific type. However, for the description of implantation profiles only the Pearson type IV and type VII distribution can be generally applied. These frequency functions have a single maximum at $y = a$ and decay monotonously to zero on both sides. The type VII distribution is not skewed, which results in a very limited applicability, if any. I have seen some attempts to use other Pearson frequency functions, particularly type V and type VI, to fit in a piecewise manner implantation

profiles. Such approaches are inconsistent with the underlying mathematics and should be strictly avoided.
The general solution of the differential equation (3.1-23) is given in (3.1-47) when the restriction (3.1-42) which characterizes the Pearson type IV distribution is obeyed.

$$f(x)=K\cdot\left(-\left(b_0+b_1\cdot(x-R_p)+b_2\cdot(x-R_p)^2\right)\right)^{\left(\frac{1}{2\cdot b_2}\right)}\cdot \\ \cdot\exp\left(-\frac{b_1/b_2+2\cdot a}{\sqrt{4\cdot b_2\cdot b_0-b_1{}^2}}\cdot\operatorname{atan}\left(\frac{2\cdot b_2\cdot(x-R_p)+b_1}{\sqrt{4\cdot b_2\cdot b_0-b_1{}^2}}\right)\right) \quad (3.1\text{-}47)$$

The constant K is the normalization constant to fulfill (3.1-3). It can usually be determined only by numerical integration. Ryssel [3.64], [3.67] favours the Pearson type IV distribution very much. It is his experience that almost all practically arising profiles can be fitted rather accurately. A similar experience has been documented in [3.81].
Some authors, e.g. [3.4], [3.30], have added so-called exponential tails to the Pearson type IV frequency function.

$$\tilde{f}(x)=f(x)+A\cdot\exp(-\lambda\cdot(x-x_0{}^2)) \quad (3.1\text{-}48)$$

The constants A, λ and x_0 are determined by fitting the value, the first and the second derivative at a matching point. However, I feel such a procedure is not really to be recommended: on the one hand to use the elaborate mathematical approach with the Pearson type IV distribution function and on the other hand to modify the results with quite arbitrary extensions. A completely arbitrary distribution can be expected to work as well.
If no values for the kurtosis are available, a universal expression (3.1-49) is often used to overcome this problem (cf. [3.67]).

$$\beta_2\cong 2.8+2.4\cdot\gamma_1{}^2 \quad (3.1\text{-}49)$$

A similar approach has been suggested in, e.g., [3.64].
Certainly quite a few other frequency functions have also been suggested to describe doping distributions, e.g. the Gram-Charlier series [3.82] which is based on an expansion of the Gaussian frequency function into Hermite polynomials. However, the disadvantages of most of the more sophisticated frequency functions like the tendency to oscillate (which we have already mentioned for the Edgeworth expansions), or a very complicated calculus for the coefficient evaluation are usually too severe for practical application [3.29], [3.66].
In Fig. 3.1-6 a comparison of a Gaussian (dashed line), a joined half Gaussian (dot-dashed line), a Pearson type IV (solid line) distribution and an experimentally determined profile is given. The range parameters have been taken from results of the Biersack theory.
Fig. 3.1-7 shows the shape of the Pearson type IV distribution as it changes with the implantation energy as parameter.
Another problem one has to face is the implantation through a coating layer, typically an oxide mask. Let us assume that we have a semiconductor, material 2 in this context, covered by a layer of material 1 of thickness t_{mask}. All of the distribution functions which we have discussed above have to be modified in order to be able to

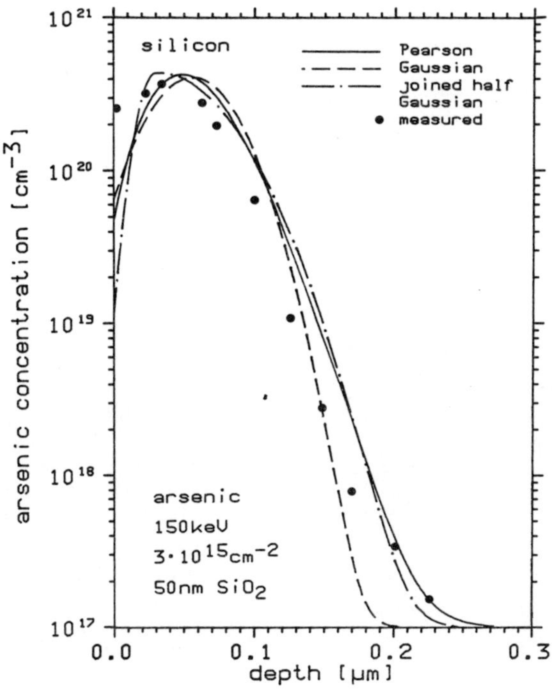

Fig. 3.1-6. Comparison of implantation models and experiment

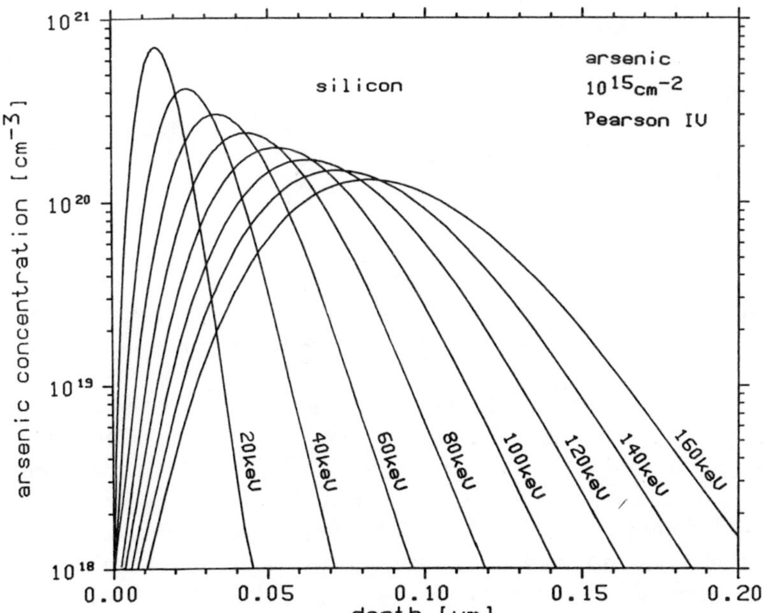

Fig. 3.1-7. Influence of the implantation energy on the shape of an arsenic profile

account for such a configuration. The easiest procedure one can carry out is to use transformed coordinates for the distribution in material 2.

$$C(x)=\begin{cases} x<t_{\text{mask}} & C_1(x) \\ x\geq t_{\text{mask}} & C_2(x-\delta) \end{cases} \tag{3.1-50}$$

$C_1(x)$ is the doping distribution in material 1; $C_2(x)$ in material 2. δ is a translation quantity which reflects a density transformation. It can be modeled after, e.g., [3.4] as:

$$\delta=t_{\text{mask}}\cdot\left(1-\frac{R_{p2}}{R_{p1}}\right) \tag{3.1-51}$$

or after, e.g., [3.67] as:

$$\delta=t_{\text{mask}}\cdot\left(1-\frac{\sigma_{p2}}{\sigma_{p1}}\right) \tag{3.1-52}$$

R_{p1}, R_{p2} and σ_{p1}, σ_{p2} are the projected range and the standard deviation in material 1, 2, respectively (cf. (3.1-6), (3.1-7)). Both approaches work comparatively well if the thickness of material 1 is sufficiently large. For thin coating layers none of these procedures is satisfactory. Ryssel [3.66] suggested, pragmatically, a procedure where just the distribution in the semiconductor (material 2) is needed.

$$C(x)=\begin{cases} x<t_{\text{mask}} & \dfrac{\sigma_{p2}}{\sigma_{p1}}\cdot C_2\left(\dfrac{\sigma_{p2}}{\sigma_{p1}}\cdot x\right) \\ x\geq t_{\text{mask}} & C_2\left(x-t_{\text{mask}}\cdot\left(1-\dfrac{\sigma_{p2}}{\sigma_{p1}}\right)\right) \end{cases} \tag{3.1-53}$$

This approach gives excellent results for thin coating layers if the distribution is concentrated in the substrate (cf. [3.67]).

The best procedure, from my personal point of view, has been suggested also by Ryssel [3.66], [3.67]. Here, the concentration $C_1(x)$ is calculated first for infinitely thick material 1. Then the total number of ions N_{d1} in the layer of thickness t_{mask} is calculated by integration. By assuming no coating layer, the concentration $C_2(x)$ is calculated in material 2, and the thickness t' which contains N_{d1} ions is determined. The final profile is composed of profile $C_1(x)$ in material 1 up to t_{mask} and profile $C_2(x)$ starting from t'. Thus, the resulting profile incorporates the total implantation dose N_d. This approach can be generalized to multi layer structures in a straightforward manner.

The profile obtained with any of the above given procedures is discontinuous at the interface $x=t_{\text{mask}}$. This seems to be in contrast to Monte Carlo simulations, however, the discontinuity is not very pronounced if the mass density of material 1 differs little from the mass density of material 2.

Several problems like recoil or knock-on implantation of ions arise with the implantation through masking layers. These problems and many more are dealt with in the specialized literature. The articles [3.41], [3.66], [3.74] and of a more general interest, the books [3.2], [3.25], [3.63], [3.80] can be recommended for a more in depth study.

The last problem regarding ion implantation which I would like to discuss here is the two dimensional distribution of implanted ions for a non-constant mask thickness t_{mask} across the lateral dimension (y-coordinate). The idea for the solution to this problem is based on the work of Furukawa et al. [3.50], who considered only Gaussian distributions near an infinitely steep mask edge. Runge [3.62] has refined that approach for arbitrarily shaped mask edges, but still considering only Gaussian distributions. However, the idea behind the approach is a simple convolution of a quasi-one dimensional profile $C(x; t_{\text{mask}}(y))$ with a Gaussian distribution in the y-direction. $t_{\text{mask}}(y)$ is to be considered as a parameter. A general formula can, therefore, be given.

$$C(x, y) = \frac{1}{\sqrt{2 \cdot \pi} \cdot \sigma_{py}} \cdot \int_{-\infty}^{\infty} C(x; t_{\text{mask}}(y')) \cdot \exp\left(-\frac{(y-y')^2}{2 \cdot \sigma_{py}^2}\right) \cdot dy' \quad (3.1\text{-}54)$$

σ_{py} is the lateral standard deviation. Tabulated values for σ_{py} are given in, e.g. [3.23], [3.25], [3.63]. Fig. 3.1-8 shows a graph of σ_{py} in silicon obtained from the *LSS* theory. This convolution (3.1-54) can usually be carried out only with numerical methods. In case of an infinitely high mask extending to the point $y = a$, the convolution can be performed analytically. We have:

$$C(x; t_{\text{mask}}(y)) = \begin{cases} y < a & 0 \\ y \geq a & C(x) \end{cases} \quad (3.1\text{-}55)$$

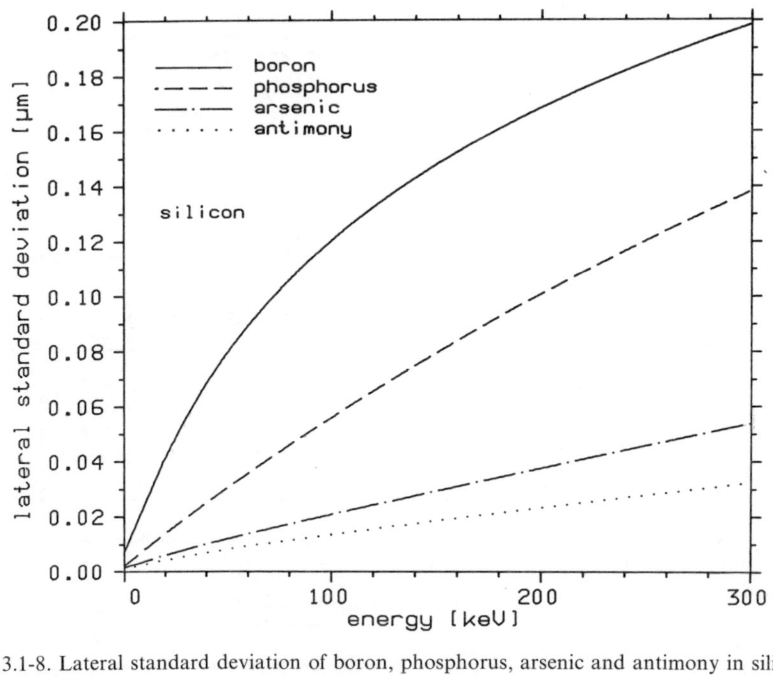

Fig. 3.1-8. Lateral standard deviation of boron, phosphorus, arsenic and antimony in silicon

The integral (3.1-54) evaluates to:

$$C(x, y) = C(x) \cdot \frac{\operatorname{erfc}\left(\dfrac{a-y}{\sqrt{2} \cdot \sigma_{py}}\right)}{2} \tag{3.1-56}$$

where erfc(x) denotes the complementary error function defined as:

$$\operatorname{erfc}(x) = \frac{2}{\sqrt{\pi}} \cdot \int_x^\infty e^{-t^2} \cdot dt \tag{3.1-57}$$

All two dimensional process modeling programs, to my knowledge, use the convolution integral (3.1-54) in a more or less simplified manner to calculate two dimensional distributions (cf. [3.50], [3.54], [3.66], [3.73], [3.75], [3.76]). One problem, however, associated with that formulation is the assumption that the lateral standard deviation σ_{py} is independent of the depth (x-coordinate). Considering two dimensional calculations of the damage distribution formed by ion implantation [3.47], which have been confirmed experimentally [3.36], one could speculate that this is a poor assumption. Examples of calculated two dimensional implantation profiles are shown in Fig. 3.1-9 and Fig. 3.1-10. A vertical mask edge has been assumed for Fig. 3.1-9, whereas a mask tapered at 45 degrees has been used for Fig. 3.1-10.

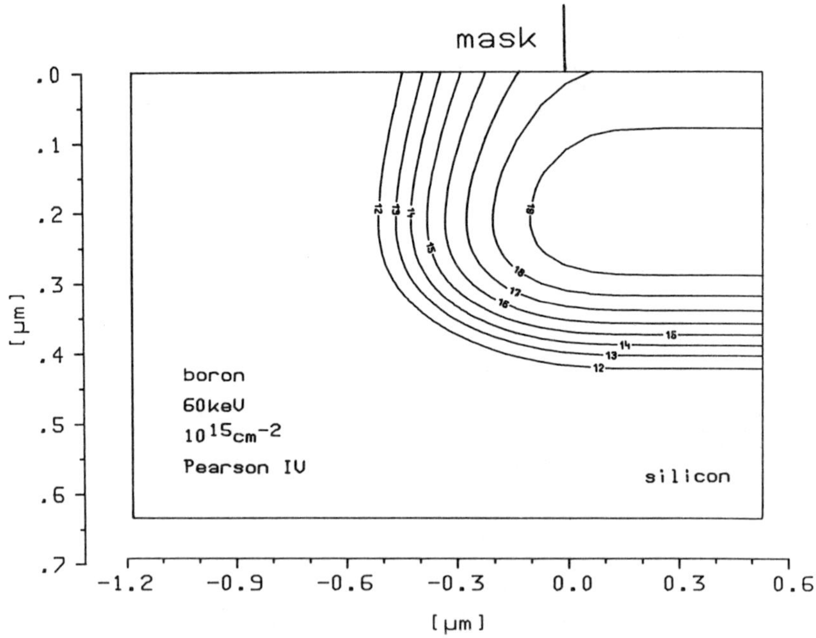

Fig. 3.1-9. Calculated two-dimensional implantation profile with vertical mask edge

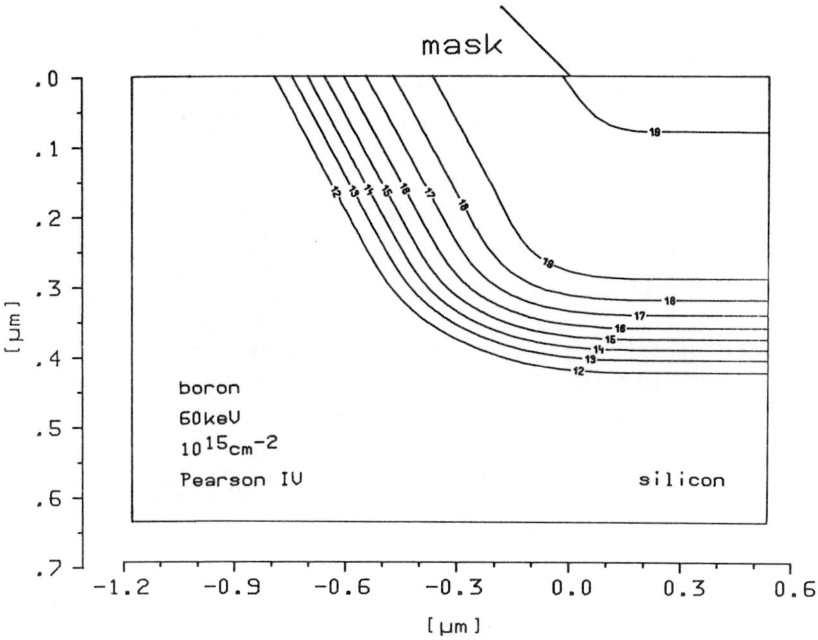

Fig. 3.1-10. Calculated two-dimensional implantation profile with mask tapered at 45 degrees

3.2 Diffusion

Diffusion is the physical mechanism which is responsible for the redistribution of impurity atoms in semiconductor processing. By means of diffusion processes one can obtain a desired shape of the distribution of dopants incorporated into the semiconductor by, e.g., ion implantation or which are deposited at the surface as a paste, fluid or gas of high concentration. The latter process is usually termed a predeposition process. The former process, diffusing a profile which has been produced by ion implantation, always has to be carried out in order to "recreate" the semiconductor lattice from the bombardment damage caused by ion implantation. In this context diffusion is usually termed "annealing".

The diffusion of dopants in semiconductors is described by the two laws of Fick, which read:

$$\vec{J}_i = -D_i \cdot \left(\operatorname{grad} Ct_i - Z_i \cdot \frac{q}{k \cdot T} \cdot (Ct_i - Cc_i) \cdot \operatorname{grad} \psi \right) \tag{3.2-1}$$

$$\frac{\partial Ct_i}{\partial t} + \operatorname{div} \vec{J}_i = 0 \tag{3.2-2}$$

Ct_i is the total concentration; Cc_i is the electrically inactive part of the concentration, i.e. the concentration of dopants which is not well incorporated in the silicon lattice and, thus, is not ionized (e.g. neutral clusters). $\vec{J}_i$ denotes the impurity flux; Z_i is the charge state of the impurity (+1 for singly ionized acceptors, −1 for

singly ionized donors). D_i represents the diffusion coefficient which depends, in general, on all sorts of quantities as we shall discuss later. ψ is the electrostatic potential. The index i of all above cited quantities denotes the i-th impurity type as there is usually more than one kind of impurity incorporated into the silicon when a diffusion process is performed.
By substituting (3.2-1) into (3.2-2) we obtain the classical form of the diffusion equation, a continuity equation, for the i-th dopant.

$$\frac{\partial Ct_i}{\partial t} = \operatorname{div}\left(D_i \cdot \left(\operatorname{grad} Ct_i - Z_i \cdot \frac{q}{k \cdot T} \cdot (Ct_i - Cc_i) \cdot \operatorname{grad} \psi\right)\right) \tag{3.2-3}$$

The electrostatic potential ψ is determined by the Poisson equation which we have discussed in Section 2.1.

$$\operatorname{div} \operatorname{grad} \psi = \frac{q}{\varepsilon} \cdot (n - p - C) \tag{3.2-4}$$

The quantity C represents the total net concentration of all ionized impurities. For a system with n impurities we have:

$$C = -\sum_{i=1}^{n} Z_i \cdot (Ct_i - Cc_i) \tag{3.2-5}$$

Note that in (3.2-5) the influence of, e.g., charged vacancies is neglected. In all process modeling programs I am familiar with the Poisson equation is not solved as an elliptic differential equation, but rather assuming vanishing space charge and Boltzmann statistics. The electrostatic potential can then be calculated explicitly (cf. Section 2.4).

$$\psi = \frac{k \cdot T}{q} \cdot \operatorname{arsinh}\left(\frac{C}{2 \cdot n_i}\right) \tag{3.2-6}$$

The electrostatic potential ψ is identical to the built-in potential which we have derived in Section 2.4, because the semiconductor structure is not electrically biased during a diffusion process. n_i denotes the intrinsic concentration at the process temperature. It may be modeled as depending on the concentration of dopants, thus, representing an effective intrinsic concentration (cf. Section 2.4). The assumption of vanishing space charge is very poor when considering the coupled diffusion in a structure with pn-junctions. Obviously, the exact location of the pn-junction, which is one of the most important results one should like to obtain from process modeling, will depend on the interaction of the space charge layer at the pn-junction with the electrostatic potential, particularly if one deals with steeply graded junctions. This problem is stressed also in [3.1].
In the literature one can often find that field enhanced diffusion is accounted for with a so-called field enhancement factor multiplying the diffusion coefficient. In the case of just one type of impurity – we can drop the index i – and under the assumption that the total concentration of impurities is electrically active we may take the electrostatic potential as:

$$\psi = \frac{k \cdot T}{q} \cdot \operatorname{arsinh}\left(-\frac{Z \cdot Ct}{2 \cdot n_i}\right) \tag{3.2-7}$$

The gradient of ψ reads:

$$\operatorname{grad}\psi = -\frac{k\cdot T}{q}\cdot\frac{1}{\sqrt{1+\left(\dfrac{Ct}{2\cdot n_i}\right)^2}}\cdot\frac{Z}{2\cdot n_i}\cdot\operatorname{grad} Ct \tag{3.2-8}$$

By substituting (3.2-8) into (3.2-3) and with a small amount of algebra we obtain:

$$\frac{\partial Ct}{\partial t} = \operatorname{div}(D_i\cdot df\cdot\operatorname{grad} Ct) \tag{3.2-9}$$

with the field enhancement factor df:

$$df = 1+\frac{Ct}{\sqrt{Ct^2+(2\cdot n_i)^2}} \tag{3.2-10}$$

It should be noted that the approach using a field enhancement factor is only correct if just one species of impurities is involved in the diffusion process. Although it has been used quite frequently, tracing the literature, I cannot see any plausible reason to introduce such a field enhancement factor.

In (3.2-7) we have assumed the validity of Boltzmann statistics for the description of the mobile carriers. This seems to be justified at typical process temperatures even for doping concentrations up to the solubility limit [3.64]. However, if one wants to avoid the assumption of Boltzmann statistics, it is just a matter of the complexity of the calculus to do so (cf. Section 2.4). This problem has been treated in, e.g. [3.33], [3.55].

In the following I should like to discuss models for the diffusion coefficient D_i. It is well established that the diffusion vehicles are the intrinsic point defects of the lattice, i.e. vacancies and interstitials [3.68]. In Section 3.3 evidence will be given showing both kinds of defects are important for the diffusion of dopants in silicon. However, at this time there is a lack of mathematical models describing the diffusion by interstitials. Therefore, the following considerations are based on the vacancy mechanism. Hence, the diffusion coefficient D_i is assumed to be the sum of several diffusivities [3.8], where each accounts for the impurity interactions with different charge states of lattice vacancies.

$$D_i = D_i^o + D_i^- \cdot V^- + D_i^= \cdot V^= + D_i^+ \cdot V^+ \tag{3.2-11}$$

D_i^o is the diffusion coefficient for the dopants of the i-th species diffusing with neutral vacancies, D_i^- for those diffusing with singly negative charged vacancies, $D_i^=$ for those diffusing with doubly negative charged vacancies, and D_i^+ for those diffusing with singly positive charged vacancies. Other types of configurations are certainly also imaginable; however, the cited ones are considered to be most relevant. V^-, $V^=$ and V^+ are the concentrations of singly negative, doubly negative and singly positive charged vacancies normalized by the concentration of neutral vacancies. These concentrations can be modeled under the validity of Boltzmann statistics as:

$$V^- = \frac{n}{n_i},\quad V^= = \left(\frac{n}{n_i}\right)^2,\quad V^+ = \frac{p}{n_i} \tag{3.2-12}$$

n and p denote the electron and hole concentration, respectively. The individual diffusion coefficients D_i^x are usually given as expressions in Arrhenius form:

$$D_i^x = D_{oi}^x \cdot \exp\left(-\frac{E_{ai}^x}{k \cdot T}\right) \qquad (3.2\text{-}13)$$

Numerical values for the prefactors D_{oi}^x and the so-called activation energies E_{ai}^x are summarized in Table 3.2-1. These data have been compiled from [3.19], [3.20]; they are also nicely summarized in [3.59]. More numerical values for different dopants are reviewed in [3.72].

Table 3.2-1. *Diffusion coefficients*

element	D_o^o [cm^2 s^{-1}]	E_a^o [eV]	D_o^- [cm^2 s^{-1}]	E_a^- [eV]	$D_o^=$ [cm^2 s^{-1}]	$E_a^=$ [eV]	D_o^+ [cm^2 s^{-1}]	E_a^+ [eV]
B	0.037	3.46					0.72	3.46
P	3.85	3.66	4.44	4.00	44.20	4.37		
Sb	0.214	3.65	15.0	4.08				
As	0.066	3.44	12.0	4.05				

Boron mainly diffuses with neutral and positively charged vacancies, the latter being the dominant effect. Fig. 3.2-1 shows the total diffusivity of boron versus doping concentration for various temperatures.

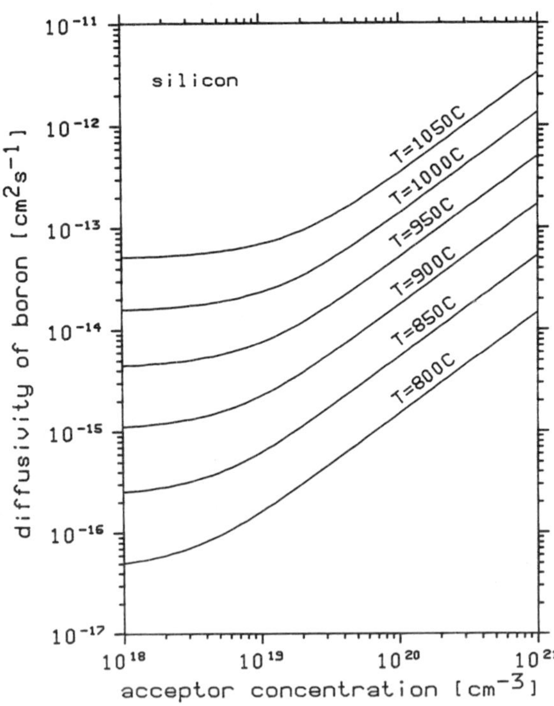

Fig. 3.2-1. Diffusivity of boron versus acceptor concentration in silicon

The examination of the diffusivity of phosphorus is most difficult. One tends to assume that phosphorus diffuses with neutral, singly negative and doubly negative charged vacancies. For low concentrations the diffusion with neutral vacancies is dominant, whereas for high concentrations the doubly negative charged vacancies are considered to dominate the overall diffusivity. Particularly for high concentrations the diffusion of phosphorus shows various unusual phenomena [3.21], like kink formation, enhanced tail diffusion, enhancement of the diffusion of other impurities (e.g. boron). No unified treatment of the diffusion of phosphorus has been found so far, and many fairly detailed modifications to the effective diffusion coefficient have to be undertaken to obtain acceptable simulation results [3.59]. Some of the features of phosphorus diffusion have been discussed in, e.g., the work of Matsumoto et al. [3.44], [3.45], [3.46]. A plot of the total diffusivity of phosphorus after (3.2-11) versus doping concentration is shown in Fig. 3.2-2.

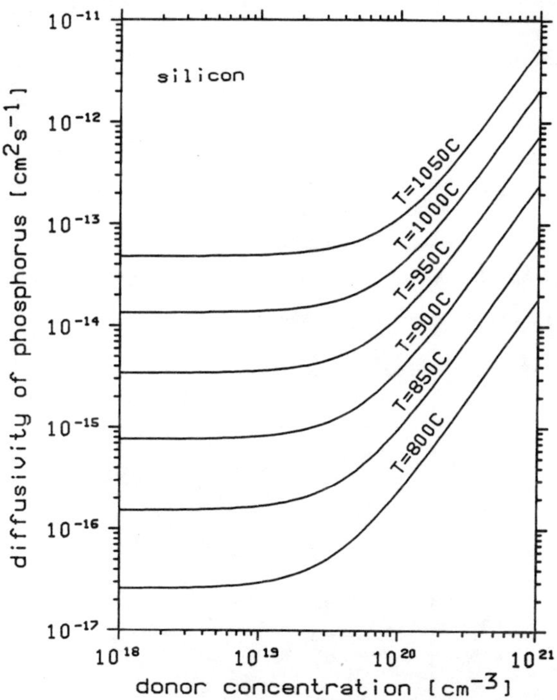

Fig. 3.2-2. Diffusivity of phosphorus versus donor concentration in silicon

Antimony diffuses with neutral and singly negative charged vacancies. At process temperatures the diffusion with the negatively charged vacancies is somewhat greater. Fig. 3.2-3 shows the total diffusivity of antimony versus doping concentration.

Arsenic as a donor diffuses primarily with neutral and singly negative charged vacancies. Both mechanisms are comparable throughout the process temperature range. A plot of the diffusivity of arsenic versus doping concentration is given in Fig. 3.2-4.

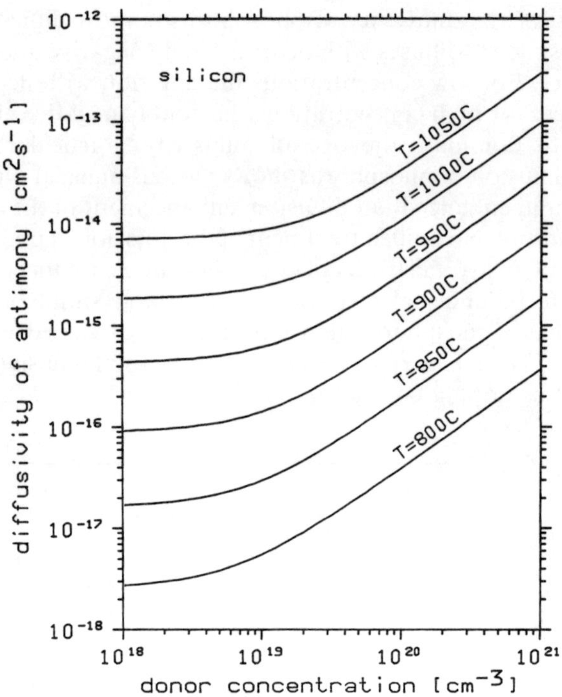

Fig. 3.2-3. Diffusivity of antimony versus donor concentration in silicon

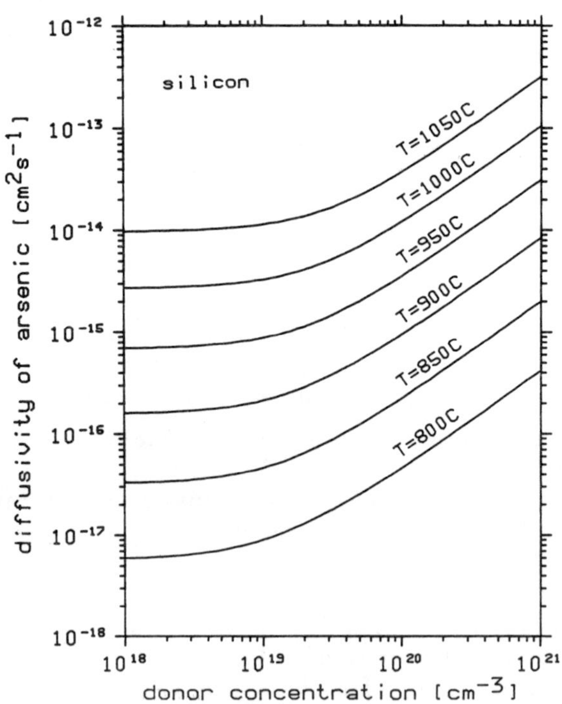

Fig. 3.2-4. Diffusivity of arsenic versus donor concentration in silicon

In some papers one can find a slightly different form for the diffusion coefficient (3.2-11).

$$D_i = D_i^{\mathrm{intr}} \cdot \frac{1 + \beta_i^- \cdot \frac{n}{n_i} + \beta_i^= \cdot \left(\frac{n}{n_i}\right)^2 + \beta_i^+ \cdot \frac{p}{n_i}}{1 + \beta_i^- + \beta_i^= + \beta_i^+} \tag{3.2-14}$$

D_i^{intr} denotes the diffusion coefficient for intrinsic conditions which exist at relatively low doping concentrations. The β_i^x can be understood as parameters describing the effectiveness of charged vacancies relative to neutral ones in impurity diffusion. The expressions (3.2-11) and (3.2-14) can, obviously, be made equivalent by a proper choice of the parameters. However, I feel that (3.2-11) is more reasonable.
Fig. 3.2-5 shows the intrinsic diffusion coefficient ($n = p = n_i$) for boron (fully drawn line), phosphorus (dashed line), arsenic (dot-dashed line) and antimony (dotted line) versus temperature.

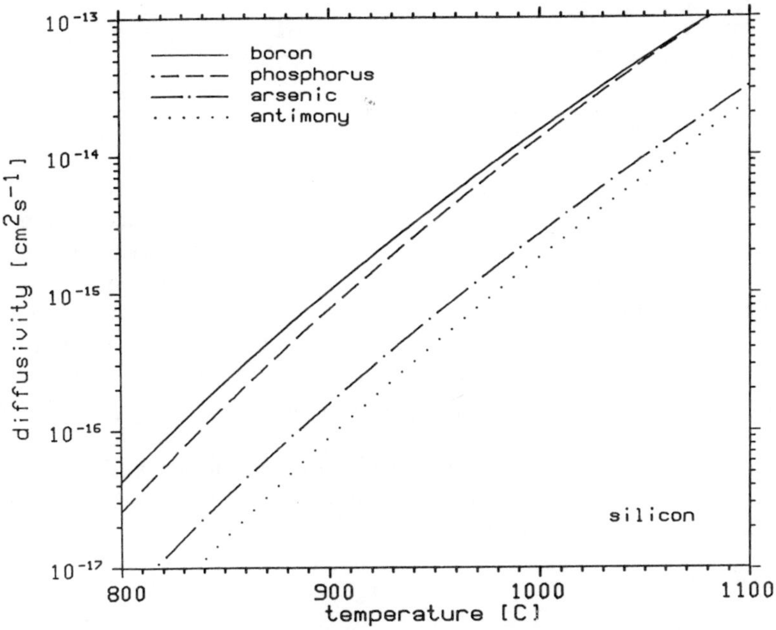

Fig. 3.2-5. Intrinsic diffusion coefficient of boron, phosphorus, arsenic and antimony in silicon versus temperature

Another formulation which has been used especially in older work is based on a different formulation of the diffusion equation:

$$\frac{\partial Ct_i}{\partial t} = \mathrm{div}\,\mathrm{grad}\,(D_i^? \cdot Ct_i) \tag{3.2-15}$$

Here $D_i^?$ is supposed to be an effective diffusion coefficient which accounts by a proper model for field enhancement as well as for interaction phenomena between mobile carriers and dopants and between dopants of different species. This formulation, however, is not at all physically reasonable, and it should therefore be avoided for careful simulations. In the case when one can use a constant diffusion coefficient this question is, obviously, irrelevant.

Although the model (3.2-11) for the diffusion coefficient is already quite sophisticated, it has to be applied very carefully. An additional modification, usually an enhancement, of the diffusivity takes place in oxidizing ambients as will be discussed in the next section. If the dopant concentration becomes so high that it approaches its solubility limit in silicon — this is the case in many practical applications — the impurities are considered to precipitate or to cluster, and they will, supposingly, not diffuse. However, quantitative statements are very difficult to make at the moment. The interested reader should carefully check the, hopefully, forthcoming literature on that and related problems. Currently, the most frequently used model which describes the relation between the total concentration Ct and the electrically inactive (e.g. cluster) concentration Cc is based on the following differential equation.

$$\frac{\partial Cc}{\partial t} = m \cdot kc \cdot (Ct - Cc)^m - kd \cdot Cc \tag{3.2-16}$$

kc and kd are the clustering and declustering rate, respectively. These are usually assumed to be temperature dependent. m is the cluster size, i.e., the number of impurity atoms which form an electrically inactive complex, the cluster. However, in [3.79] it is explained, particularly for arsenic, that the allowance for electrically charged clusters seems to improve the agreement with experimental results. Different types of charged and uncharged clusters are further considered in [3.28]. These effects become significant when the dopant concentration reaches the solubility limit (e.g. $3 \cdot 10^{20}\,cm^{-3}$ for arsenic at 1000° Celsius). I should like to speculate that in essence these statements are correct, but in order to derive models which are applicable for engineering purposes much more investigatory work has still to be carried out.

Very often it is assumed that the effect of dynamic clustering and declustering is negligible. Then we obtain an algebraic, equilibrium cluster relation between the total and the electrically active concentration.

$$Ct = (Ct - Cc) + \beta c \cdot (Ct - Cc)^m \tag{3.2-17}$$

$$\beta c = m \cdot \frac{kc}{kd} \tag{3.2-18}$$

Numerical values and the temperature dependence of the equilibrium cluster coefficient βc are presented in the report [3.59].

So far we have discussed diffusion models, the complexity of which grew in turn as each detail was considered in just slightly more depth. We have ended up with a model capable of quite accurately describing diffusion in silicon, but it requires the solution of a system of equations for the coupled diffusion of n dopants, which is composed of n parabolic partial differential equations (3.2-3), n ordinary differential equations (3.2-16) and the Poisson equation, an elliptic partial differential equation (3.2-4). This system represents a mixed initial boundary value problem, and it,

unfortunately, has incorporated an enormous amount of very uncertain parameters in the physical sense. Particularly the models for the boundary conditions (e.g. segregation), which we shall not discuss here but only sketch in Section 3.3, are fairly poor because of the lack of understanding in the underlying physics and chemistry. A thorough discussion of the great many problems associated with diffusion is beyond the scope of this text which is primarily devoted to device modeling. In the following we shall only consider simplifications to the diffusion models which have proved to be quite valuable in the context of engineering problems.
If the diffusion coefficient can be treated as a constant and if field enhancement of the diffusivity can be neglected, which is the case only for low doping concentrations, the diffusion equation simplifies to:

$$\frac{\partial C}{\partial t} = D \cdot \left(\frac{\partial^2 C}{\partial x^2} + \frac{\partial^2 C}{\partial y^2} \right) \tag{3.2-19}$$

For an inert diffusion we can assume at the surface of the semiconductor the boundary condition:

$$\left. \frac{\partial C}{\partial x} \right|_{x=0} = 0 \tag{3.2-20}$$

This boundary condition guarantees that no impurity atoms diffuse through the surface. It is just correct to first order. However, by assuming the applicability of (3.2-19) and (3.2-20), the solution of the diffusion problem can be carried out analytically for a unit impulse source $\delta(x', y')$ as the initial condition.

$$\hat{C}(x, y, 0) = \delta(x', y') \tag{3.2-21}$$

$\delta(x', y')$ denotes the Dirac delta function. The solution of (3.2-19) in the half-plane $x \in [0, \infty]$, $y \in [-\infty, \infty]$ with boundary condition (3.2-20) and initial condition (3.2-21) is a classical result.

$$\begin{aligned} \hat{C}(x, y, t) = {} & \frac{1}{2 \cdot \pi \cdot D \cdot t} \cdot \exp\left(-\frac{(y-y')^2}{4 \cdot D \cdot t} \right) \cdot \\ & \cdot \left(\exp\left(-\frac{(x-x')^2}{4 \cdot D \cdot t} \right) + \exp\left(-\frac{(x+x')^2}{4 \cdot D \cdot t} \right) \right) \end{aligned} \tag{3.2-22}$$

With (3.2-22) we can solve the diffusion problem for an arbitrary initial condition $C(x, y, 0)$ by convolution.

$$C(x, y, t) = \int_{-\infty}^{\infty} \int_{0}^{\infty} C(x', y', 0) \cdot \hat{C}(x, y, t) \cdot dx' \cdot dy' \tag{3.2-23}$$

In some cases (3.2-23) can be solved in closed form, for instance, for a Gaussian implantation profile.

$$C(x, y, 0) = \begin{cases} y < a & 0 \\ y \geq a & \dfrac{N_d}{\sqrt{2 \cdot \pi} \cdot \sigma_p} \cdot \exp\left(-\dfrac{(x-R_p)^2}{2 \cdot \sigma_p^2} \right) \end{cases} \tag{3.2-24}$$

For the sake of simplicity, we ignore the lateral spread of the implanted profile (3.2-24). After substitution of the initial condition (3.2-24) and the unit impulse source solution (3.2-22) into (3.2-23) the diffused profile reads:

$$C(x,y,t)=\frac{N_d}{\sqrt{2\cdot\pi}\cdot 4\cdot\pi\cdot\sigma_p\cdot D\cdot t}\cdot\int_a^{\infty}\exp\left(-\frac{(y-y')^2}{4\cdot D\cdot t}\right)\cdot dy'\cdot$$

$$\cdot\left(\int_0^{\infty}\exp\left(-\frac{(x'-R_p)^2}{2\cdot\sigma_p^2}-\frac{(x-x')^2}{4\cdot D\cdot t}\right)\cdot dx'+\right. \tag{3.2-25}$$

$$\left.+\int_0^{\infty}\exp\left(-\frac{(x'-R_p)^2}{2\cdot\sigma_p^2}-\frac{(x+x')^2}{4\cdot D\cdot t}\right)\cdot dx'\right)$$

The two integrals in (3.2-25) can be evaluated with some algebra so that the final solution becomes:

$$C(x,y,t)=\frac{N_d}{4\cdot\sqrt{2\cdot\pi\,(\sigma_p^2+2\cdot D\cdot t)}}\cdot\operatorname{erfc}\left(\frac{a-y}{2\cdot\sqrt{D\cdot t}}\right)\cdot(H(x,t)+H(-x,t)) \tag{3.2-26}$$

with:

$$H(z,t)=\exp\left(-\frac{(z-R_p)^2}{2\cdot\sigma_p^2+4\cdot D\cdot t}\right)\cdot \tag{3.2-27}$$

$$\cdot\operatorname{erfc}\left(-\frac{R_p}{\sigma_p}\cdot\sqrt{\frac{D\cdot t}{\sigma_p^2+2\cdot D\cdot t}}-\sqrt{\frac{\sigma_p^2}{4\cdot D\cdot t\cdot(\sigma_p^2+2\cdot D\cdot t)}}\cdot z\right)$$

This solution strategy was introduced by Lee et al. [3.37], [3.38]. It has been refined to account for more general initial conditions [3.14], [3.15], [3.39], [3.50] and to account, qualitatively, for a non-constant diffusivity [3.73]. There have also been published slightly different approaches for analytic solutions of (3.2-19) with different boundary conditions at the semiconductor surface, e.g. [3.12], [3.35], however, these are more complex.

3.3 Oxidation

The thermal oxidation of silicon is one of the most important processing steps for the fabrication of modern devices. All existing models for oxide growth are based on the work of Deal and Grove in 1965 [3.17]. Their basic idea was the assumption of a steady state situation between three fluxes.

$$F1=h\cdot(C^*-C^o) \tag{3.3-1}$$

$$F2=-D\cdot\frac{\partial C}{\partial x}\cong D\cdot\frac{C^o-C^i}{x_{ox}} \tag{3.3-2}$$

$$F3=k_s\cdot C^i \tag{3.3-3}$$

$F1$ is the flux of oxidant from the bulk of the gas to the gas-oxide interface. C^o is the concentration of the oxidant at the oxide surface; C^* is the concentration of the oxidant in the oxide, which will be in equilibrium with the partial pressure in the bulk of the gas; and h is the gas phase mass transfer coefficient.
$F2$ denotes the flux across the oxide, which is assumed to be purely diffusive. C^i is the oxidant concentration in the oxide at the oxide-silicon interface; x_{ox} represents the oxide thickness.
$F3$ is the flux corresponding to the oxidation reaction at the oxide-silicon interface. k_s represents the chemical surface reaction rate [3.27]. In the steady state condition these three fluxes are identical and can be expressed:

$$F = \frac{k_s}{1 + \frac{k_s}{h} + \frac{k_s \cdot x_{ox}}{D}} \cdot C^* \tag{3.3-4}$$

The flux of oxidant reaching the oxide-silicon interface is described by the differential equation:

$$N_1 \cdot \frac{dx_{ox}}{dt} = F \tag{3.3-5}$$

N_1 is the number of oxidant molecules incorporated into a unit volume of oxide. The solution of (3.3-5) is:

$$x_{ox}(t) = \sqrt{\left(\frac{A}{2} + x_{ox}(0)\right)^2 + B \cdot t} - \frac{A}{2} \tag{3.3-6}$$

with:

$$A = 2 \cdot D \cdot \left(\frac{1}{k_s} + \frac{1}{h}\right) \tag{3.3-7}$$

$$B = \frac{2 \cdot D \cdot C^*}{N_1} \tag{3.3-8}$$

(3.3-6) is frequently written in a slightly different form:

$$x_{ox}(t)^2 + A \cdot x_{ox}(t) = B \cdot (t + \tau) \tag{3.3-9}$$

with:

$$\tau = \frac{x_{ox}(0)^2 + A \cdot x_{ox}(0)}{B} \tag{3.3-10}$$

B is refered to as the parabolic growth rate coefficient because for large t (3.3-9) approaches:

$$x_{ox}(t)^2 = B \cdot t, \; t \gg \frac{A^2}{4 \cdot B} \tag{3.3-11}$$

For small time we observe that B/A describes a linear growth rate:

$$x_{ox}(t) = \frac{B}{A} \cdot (t+\tau), \quad t \ll \frac{A^2}{4 \cdot B} - \tau \tag{3.3-12}$$

By proper modeling of the growth rate coefficients, many ambient attributes can be accounted for (composition, pressure, ...). However, for very thin oxides the flux models (3.3-1) to (3.3-3) appear to be oversimplified and have to be modified [3.32], [3.59]. An empirical formula for thin oxide thicknesses corresponding to (3.3-5) reads:

$$\frac{dx_{ox}}{dt} = \frac{B + K_1 \cdot \exp\left(-\frac{t}{\tau_1}\right) + K_2 \cdot \exp\left(-\frac{t}{\tau_2}\right)}{2 \cdot x_{ox} + A} \tag{3.3-13}$$

A and B are defined in (3.3-7) and (3.3-8), respectively. The two supplementary terms compared to (3.3-5) involve functions decaying exponentially in time which dominate, as it has been confirmed by observation of an extensive collection of experimental data [3.59], oxide growth in the 2 nm and 20 nm regime, respectively. One can expect, however, that many more modifications of this type will be introduced in order to account heuristically for effects which can be seen experimentally but have not been understood completely on a theoretical basis. As excellent reviews on the many problems of oxidation [3.59], [3.60] can be recommended. A more fundamental treatment of the kinetics of oxide growth, which is based on the solution of the Navier-Stokes hydrodynamic equation, has fortunately begun [3.13]; however, to keep track with present and future device technology many advances in understanding the underlying kinetics still have to be made. A worthwhile approach, particularly for thin oxides, should be a microscopic simulation of oxide growth with Monte Carlo methods [3.53].

The most complete models for the growth rate coefficients B/A and B have been summarized in [3.52], [3.59]. In these models the oxidant pressure dependence, the substrate doping dependence, and the dependence on the orientation of the silicon surface are accounted for in dry and wet ambients with and without HCl.

Another effect which has to be considered in the context of oxidation is the impact on the diffusion coefficient. It has been observed by several authors, e.g. [3.3], [3.42], [3.43], that the diffusivity is enhanced. This enhancement is, most plausibly, based on diffusion mechanisms additional to the vacancy diffusion mechanisms which we have outlined in the last section. The additional mechanism is due to intrinsic interstitials emitted from the oxidized surface as suggested by Hu [3.31] and proved experimentally by, e.g. Antoniadis and Moskowitz [3.6], [3.7]. Some theoretical considerations on this subject can be found in [3.49]. It is not clear at present how the vacancy and the interstitial mechanism interact, or which one dominates, in the temperature range [800, 1000] C [3.21]. Therefore, we have restricted ourselves solely to the vacancy diffusion mechanism in the last section. However, by postulating an enhancement of the interstitial concentration and their self diffusion during oxidation [3.3], [3.8], [3.9] an increase in the diffusivity of dopants is sound. Taniguchi et al. [3.75] have suggested modifying the diffusivity of boron and phosphorus, particularly, with the following empirical expression:

$$D = D_v + \kappa \cdot \left(\frac{\partial x_{ox}}{\partial t}\right)^{0.3} \cdot \exp\left(-\frac{x}{25\,\mu\text{m}} - \frac{2.08\,\text{eV}}{k \cdot T}\right) \tag{3.3-14}$$

$$\kappa = \begin{cases} \langle 100 \rangle & 3.08 \cdot 10^{-3}\,\dfrac{\text{cm}^{1.7}}{\text{s}^{0.7}} \\ \langle 111 \rangle & 2.67 \cdot 10^{-3}\,\dfrac{\text{cm}^{1.7}}{\text{s}^{0.7}} \end{cases}$$

D_v is the effective diffusion coefficient for inert ambients; x denotes the distance to the oxide-silicon interface; κ is a constant derived from fits to experimental data. Watch the dimension of κ because in all publications I am aware of it is given in a very sloppy manner. The qualitative dependence of the oxidation enhanced diffusion coefficient upon the oxide growth rate, the distance to the interface and temperature is plausible. Therefore, pragmatically, such a model can prove to be very valuable, but it clearly demonstrates how poorly the diffusion kinetics are understood. It should be noted that oxidation retarded diffusion has also been observed experimentally, e.g. [3.59]. This effect can be made plausible by the allowance of vacancy consumption by interstitials during oxidation. Similar models to (3.3-14) can be found in [3.5], [3.59], [3.64], [3.67].

From a numerical analyst's point of view the simulation of diffusion in oxidizing ambients is quite a tough problem. As particular difficulty the moving oxide-silicon interface boundary is evident. For a one dimensional simulation that problem is usually solved by means of a reorganization of the simulation domain during time integration. For a two dimensional simulation, however, a reorganization of the domain is rather difficult [3.61] and costly in terms of computer resources. To overcome this problem Penumalli [3.58] has introduced a coordinate transformation which maps the moving boundary in the physical domain into a stationary boundary in the transformed domain. Consider the diffusion equation:

$$\frac{\partial C}{\partial t} = \frac{\partial}{\partial x} \cdot \left(D \cdot \frac{\partial C}{\partial x}\right) + \frac{\partial}{\partial y} \cdot \left(D \cdot \frac{\partial C}{\partial y}\right) \tag{3.3-15}$$

in the domain:

$$\left\{\begin{matrix} X(y,t) \leq x < \infty \\ -\infty < y < \infty \end{matrix}\right\} \tag{3.3-16}$$

which is a function of time. $x = X(y,t)$ denotes the oxide-silicon interface in parameterized form. A drift term has been neglected in (3.3-15) only for the sake of simplicity. The coordinate transformation:

$$\begin{pmatrix} \xi \\ \eta \\ \tau \end{pmatrix} = \begin{pmatrix} x - X(y,t) \\ y \\ t \end{pmatrix} \tag{3.3-17}$$

will change (3.3-15) into:

$$\frac{\partial C}{\partial \tau} = \left(1 + \left(\frac{\partial X}{\partial \eta}\right)^2\right) \cdot \frac{\partial}{\partial \xi} \cdot \left(D \cdot \frac{\partial C}{\partial \xi}\right) + \left(\frac{\partial X}{\partial \tau} - D \cdot \frac{\partial^2 X}{\partial \eta^2}\right) \cdot \frac{\partial C}{\partial \xi} +$$
$$+ \frac{\partial}{\partial \eta} \cdot \left(D \cdot \frac{\partial C}{\partial \eta}\right) -$$
$$- \frac{\partial X}{\partial \eta} \cdot \left(\frac{\partial}{\partial \xi} \cdot \left(D \cdot \frac{\partial C}{\partial \eta}\right) + \frac{\partial}{\partial \eta} \cdot \left(D \cdot \frac{\partial C}{\partial \xi}\right)\right) \tag{3.3-18}$$

(3.3-16) will be transformed into the time invariant domain (3.3-19).

$$\left\{ \begin{matrix} 0 \leq \xi < \infty \\ -\infty < \eta < \infty \end{matrix} \right\} \tag{3.3-19}$$

The proper treatment of the boundary condition for oxidation at the silicon-oxide interface can be found in, e.g. [3.50], [3.69]. For the treatment of the lateral oxidation under a mask edge which gives rise to the "Bird's Beak", no established theory is available at this time. The function $X(y, t)$ is usually modeled as (cf. [3.58]):

$$X(y, t) = b \cdot x_{ox}(t) \cdot \frac{1}{2} \cdot \operatorname{erfc}\left(\frac{\sqrt{2} \cdot (y-a)}{k_1 \cdot x_{ox}(t)}\right) \tag{3.3-20}$$

b is the amount of silicon consumed to produce one unit of oxide. $x_{ox}(t)$ is the oxide thickness as a function of time given by the one dimensional theory which we have sketched above. a determines the location of the mask edge ($y > a$ is the free, oxidizing surface); k_1 denotes the ratio of lateral to vertical oxidation and is considered to be a function of the mask thickness.

The cross derivatives in (3.3-18) have been introduced by the coordinate transformation (3.3-17) because of the lack of orthogonality between lines of constant ξ and lines of constant η. They do not cause in principal a complication of the solution of (3.3-18) but they have a non-negligible impact on the efficiency of numerical procedures. Murphy et al. [3.54] have estimated that the treatment of the cross derivatives nearly triples the required computer resources. However, the coordinate transformation of Penumalli [3.58] seems to be a very feasible approach to the two dimensional simulation of oxidation at the moment (cf. [3.54]); it has also been used successfully by several other authors, e.g. [3.54], [3.70], [3.76].

3.4 References

[3.1] Anthony, P. J.: Alteration of Diffusion Profiles in Semiconductors due to p-n Junctions. Solid-State Electron. *25,* No. 10, 1003–1009 (1982).

[3.2] Antognetti, P., Antoniadis, D. A., Dutton, R. W., Oldham, W. G.: Process and Device Simulation for MOS-VLSI Circuits. The Hague: Martinus Nijhoff 1983.

[3.3] Antoniadis, D. A., Gonzales, A. G., Dutton, R. W.: Boron in Near-Intrinsic (100) and (111) Silicon under Inert and Oxidizing Ambients. J. Electrochem. Soc. *125,* 813–819 (1978).

[3.4] Antoniadis, D. A., Hansen, S. E., Dutton, R. W.: SUPREM II – a Program for IC Process Modeling and Simulation. Report 5019-2, Stanford University, 1978.

[3.5] Antoniadis, D. A., Dutton, R. W.: Models for Computer Simulation of Complete IC Fabrication Process. IEEE J. Solid State Circuits *SC-14,* No. 2, 412–422 (1979).

[3.6] Antoniadis, D. A., Moskowitz, I.: Diffusion of Indium in Silicon Inert and Oxidizing Ambients. J. Appl. Phys. *53,* No. 12, 9214–9216 (1982).

[3.7] Antoniadis, D. A., Moskowitz, I.: Diffusion of Substitutional Impurities in Silicon at Short Oxidation Times: An Insight into Point Defect Kinetics. J. Appl. Phys. *53,* No. 10, 6788–6796 (1982).

[3.8] Antoniadis, D. A.: Diffusion in Silicon. In: Process and Device Simulation for MOS-VLSI Circuits, pp. 1–47. The Hague: Martinus Nijhoff 1983.

[3.9] Antoniadis, D. A.: One-Dimensional Simulation of IC Fabrication Processes. In: Process and Device Simulation for MOS-VLSI Circuits, pp. 226–263. The Hague: Martinus Nijhoff 1983.

[3.10] Biersack, J. P.: Calculation of Projected Ranges – Analytical Solutions and a Simple General Algorithm. Nuclear Instruments and Methods, No. 182/183, 199–206 (1981).

[3.11] Biersack, J. P.: New Projected Range Algorithm as Derived from Transport Equations. Z. Phys. *A305,* 95–101 (1982).

[3.12] Cherednichenko, D. I., Gruenberg, H., Sarkar, T., K.: Solution to a Diffusion Problem with Mixed Boundary Conditions. Solid-State Electron. *17,* 315–318 (1974).

[3.13] Chin, D., Oh, S. Y., Hu, S. M., Dutton, R. W., Moll, J. L.: Two-Dimensional Oxidation Modeling. IEEE Trans. Electron Devices *ED-30,* No. 7, 744–749 (1983).

[3.14] Chin, D. J., Kump, M. R., Lee, H. G., Dutton, R. W.: Process Design Using Coupled 2 D Process and Device Simulators. Proc. Int. Electron Devices Meeting, 1981.

[3.15] Chin, D., Kump, M. R., Lee, H. G., Dutton, R. W.: Process Design Using Coupled Two-Dimensional Process and Device Simulators. IEEE Trans. Electron Devices *ED-29,* No. 2, 336–340 (1982).

[3.16] Christel, L. A., Gibbons, J. F., Mylroie, S.: An Application of the Boltzmann Transport Equation to Ion Range and Damage Distribution in Multilayered Targets. J. Appl. Phys. *51,* No. 12, 6176–6182 (1981).

[3.17] Deal, B. E., Grove, A. S.: General Relationship for the Thermal Oxidation of Silicon. J. Appl. Phys. *36,* 3770–3778 (1965).

[3.18] Dutton, R. W., Hansen, S. E.: Process Modeling of Integrated Circuit Device Technology. Proc. IEEE *69,* 1305–1320 (1981).

[3.19] Fair, R. B., Tsai, J. C. C.: A Quantitative Model for the Diffusion of Phosphorus in Silicon and the Emitter Dip Effect. J. Electrochem. Soc. *124,* No. 7, 1107–1117 (1977).

[3.20] Fair, R. B.: Concentration Profiles of Diffused Dopants. In: Impurity Doping Processes in Silicon, pp. 315–442. Amsterdam: North-Holland 1981.

[3.21] Frank, W., Gösele, U., Mehrer, H., Seeger, A.: Diffusion in Silicon and Germanium. In: Diffusion in Solids II. New York: Academic Press 1983.

[3.22] Furukawa, S., Ishiwara, H.: Range Distribution Theory Based on Energy Distribution of Implanted Ions. J. Appl. Phys. *43,* No. 3, 1268–1273 (1972).

[3.23] Furukawa, S., Matsumura, H., Ishiwara, H.: Theoretical Considerations on Lateral Spread of Implanted Ions. Jap. J. Appl. Phys. *11,* No. 2, 134–142 (1972).

[3.24] Gemmel, D. S.: Channeling and Related Effects in the Motion of Charged Particles through Crystals. Rev. Mod. Phys. *46,* No. 1, 129–227 (1974).

[3.25] Gibbons, J. F., Johnson, W. S., Mylroie, S. W.: Projected Range Statistics. Strandsberg: Halstead Press 1975.

[3.26] Gibbons, J. F., Mylroie, S.: Estimation of Impurity Profiles in Ion-Implanted Amorphous Targets Using Joined Half-Gaussian Distributions. Appl. Phys. Lett. *22,* No. 11, 568–569 (1973).

[3.27] Grove, A. S.: Physics and Technology of Semiconductor Devices. New York: Wiley 1967.

[3.28] Guerrero, E., Pötzl, H., Tielert, R., Grasserbauer, M., Stingeder, G.: Generalized Model for the Clustering of AS Dopants in Si. J. Electrochem. Soc. *129,* No. 8, 1826–1831 (1982).

[3.29] Haberger, K., Ryssel, H., Prinke, G., Hoffmann, K.: Diffusionsmechanismen in Halbleitern. Bericht RY1/6, pp. 1–46, Institut für Festkörpertechnologie, München, 1979.

[3.30] Hofker, W. K.: Concentration Profiles of Boron Implantations in Amorphous and Polycrystalline Silicon. Philips Research Reports *8,* 41–57 (1975).

[3.31] Hu, S. M.: Formation of Stacking Faults and Enhanced Diffusion in the Oxidation of Silicon. J. Appl. Phys. *45,* No. 4, 1567–1573 (1974).

[3.32] Hu, S. M.: New Oxide Growth Law and the Thermal Oxidation of Silicon. Appl. Phys. Lett. *42,* No. 10, 872–874 (1983).

[3.33] Jain, R. K., VanOverstraeten, R. J.: Accurate Theoretical Arsenic Diffusion Profiles in Silicon from Processing Data. J. Electrochem. Soc. *122,* No. 4, 552 – 557 (1975).

[3.34] Johnson, N. L., Kotz, S.: Continuous Univariate Distributions – 1. Boston: Houghton Mifflin 1970.

[3.35] Kennedy, D. P., O'Brien, R. R.: Analysis of the Impurity Atom Distribution Near the Diffusion Mask for a Planar p-n Junction. IBM J. Res. Dev. *9,* 179 – 186 (1965).

[3.36] Krimmel, E. F., Oppolzer, H., Runge, H.: Transmission Electron Microscopical Imaging of Lateral Implantation Effects Near Mask Edges in B+ Implanted Si Wafers. Revue de Physique Appliquee *13,* 791 – 795 (1978).

[3.37] Lee, H. G., Sansbury, J. D., Dutton, R. W., Moll, J. L.: Modeling and Measurement of Surface Impurity Profiles of Laterally Diffused Regions. IEEE J. Solid-State Circuits *SC-13,* 455 – 461 (1978).

[3.38] Lee, H. G.: Two-Dimensional Impurity Diffusion Studies: Process Models and Test Structures for Low-Concentration Boron Diffusion. Report G-201-8, Stanford University, 1980.

[3.39] Lee, H. G., Dutton, R. W.: Two Dimensional Low Concentration Boron Profiles: Modeling and Measurement. IEEE Trans. Electron Devices *ED-28,* No. 10, 1136 – 1147 (1981).

[3.40] Lindhard, J., Scharff, M., Schiøtt, H. E.: Range Concepts and Heavy Ion Ranges. Mat. Fys. Medd. Dan. Vid. Selsk. *33,* No. 14, 1 – 42 (1963).

[3.41] Maes, H., Vandervorst, W., VanOverstraeten, R.: Impurity Profile of Implanted Ions in Silicon. In: Impurity Doping Processes in Silicon, pp. 443 – 638. Amsterdam: North-Holland 1981.

[3.42] Masetti, G., Negrini, P., Solmi, S., Soncini, G.: Boron Drive-In in Silicon in Oxidizing Atmospheres, Alta Frequenza *42,* No. 8, 346 – 353 (1973).

[3.43] Masetti, G., Solmi, S., Soncini, G.: On Phosphorus Diffusion in Silicon under Oxidizing Atmospheres. Solid-State Electron. *16,* 1419 – 1421 (1973).

[3.44] Matsumoto, S., Niimi, T.: Phosphorus Diffusion under the Condition of Controlled Surface Concentration. Jap. J. Appl. Phys. *15,* No. 11, 2077 – 2082 (1976).

[3.45] Matsumoto, S., Arai, E., Nakamura, H., Niimi, T.: The Distribution of the Excess Vacancies in the Bulk at the Diffusion of Phosphorus into Silicon. Jap. J. Appl. Phys. *16,* No. 7, 1177 – 1185 (1977).

[3.46] Matsumoto, S., Akao, Y., Kohiyama, K., Niimi, T.: Effect of Diffusion Induced Strain and Dislocation on Phosphorus Diffusion Into Silicon. J. Electrochem. Soc. *125,* No. 11, 1840 – 1845 (1978).

[3.47] Matsumura, H., Furukawa, S.: Lateral Spread of Damage Formed by Ion Implantation. J. Appl. Phys. *47,* No. 5, 1746 – 1751 (1976).

[3.48] Mazzone, A. M.: Formation and Growth of a Damaged Layer During Implantation: A Three Dimensional Monte Carlo Simulation. Proc. NASECODE III Conf., pp. 179 – 184. Dublin: Boole Press 1983.

[3.49] McLellan, R. B.: Thermodynamics and Diffusion Behavior of Interstitial Solute Atoms in Crystals. Acta Metallurgica *27,* 1655 – 1663 (1979).

[3.50] Mei, L., Dutton, R. W., Hansen, S. E., Kump, M., Greenfield, J. A., Price, C. H.: An Overview of Process Models and Two-Dimensional Analysis Tools. Report G-201-13, Stanford University, 1982.

[3.51] Mei, L., Dutton, R. W.: A Process Simulation Model for Multilayer Structures Involving Polycrystalline Silicon. IEEE Trans. Electron Devices *ED-29,* No. 11, 1726 – 1734 (1982).

[3.52] Meindl, J. D., Dutton, R. W., Gibbons, J. F., Helms, C. R., Plummer, J. D. Tiller, W. A., Ho, C. P., Saraswat, K. C., Deal, B. E., Kamins, T. I.: Computer-Aided Engineering of Semiconductor Integrated Circuits. Report TRDXG501, ICL17-78, Stanford University, 1980.

[3.53] Murch, G. E.: Monte Carlo Calculation as an Aid in Teaching Solid-State Diffusion. Am. J. Phys. *47,* No. 1, 78 – 80 (1979).

[3.54] Murphy, W. D., Hall, W. F., Maldonado, C. D., Louie, S. A.: MEMBRE: An Efficient Two-Dimensional Process Code for VLSI. COMPEL *1,* No. 4, 219 – 239 (1982).

[3.55] Nuyts, W., VanOverstraeten, R.: Computer Calculations of Impurity Profiles in Silicon. Phys. Stat. Sol. *15,* 329 – 341 (1973).

[3.56] Oldham, W. G., Nandgaonkar, S. N., Neureuther, A. R., O'Toole, M.: A General Simulator for VLSI Lithography and Etching Processes: Part I – Application to Projection Lithography. IEEE Trans. Electron Devices *ED-26,* No. 4, 717 – 722 (1979).

[3.57] Oldham, W. G., Neureuther, A. R., Sung, C., Reynolds, J. L., Nandgaonkar, S. N.: A General

Simulator for VLSI Lithography and Etching Processes: Part II – Application to Deposition and Etching. IEEE Trans. Electron Devices *ED-27,* No. 8, 1455–1459 (1980).

[3.58] Penumalli, B. R.: Lateral Oxidation and Redistribution of Dopants. Proc. NASECODE II Conf., pp. 264–269. Dublin: Boole Press 1981.

[3.59] Plummer, J. D., Dutton, R. W., Gibbons, J. F., Helms, C. R., Meindl, J. D., Tiller, W. A., Christel, L. A., Ho, C. P., Mei, L., Saraswat, K. C., Deal, B. E., Kamins, T. I.: Computer-Aided Design of Integrated Circuit Fabrication Processes for VLSI Devices. Report TRDXG 501-82, ICL17-79, Stanford University, 1982.

[3.60] Plummer, J. D., Deal, B. E.: Thermal Oxidation: Kinetics, Charges, Physical Models, and Interaction with Other Processes in VLSI Devices. In: Process and Device Simulation for MOS-VLSI Circuits, pp. 48–87. The Hague: Martinus Nijhoff 1983.

[3.61] Poncet, A.: Two-Dimensional Simulation of Local Oxidation in VLSI Processes. Proc. VLSI Process and Device Modeling, pp. 1–45, Katholieke Universiteit Leuven, 1983.

[3.62] Runge, H.: Distribution of Implanted Ions under Arbitrarily Shaped Mask Edges. Phys. Stat. Sol. *(a) 39,* 595–599 (1977).

[3.63] Ryssel, H., Ruge, I.: Ionenimplantation. Stuttgart: Teubner 1978.

[3.64] Ryssel, H., Haberger, K., Hoffmann, K., Prinke, G., Dümcke, R., Sachs, A.: Simulation of Doping Processes. IEEE Trans. Electron Devices *ED-27,* 1484–1492 (1980).

[3.65] Ryssel, H., Prinke, G., Haberger, K., Hoffmann, K., Müller, K., Henkelmann, R.: Range Parameters of Boron Implanted into Silicon. J. Appl. Phys. *24,* 39–43 (1981).

[3.66] Ryssel, H., Hoffmann, K.: Ion Implantation. In: Process and Device Simulation for MOS-VLSI Circuits, pp. 125–179. The Hague: Martinus Nijhoff 1983.

[3.67] Ryssel, H.: Implantation and Diffusion Models for Process Simulation. Proc. VLSI Process and Device Modeling, pp. 1–41, Katholieke Universiteit Leuven, 1983.

[3.68] Seeger, A., Chick, K. P.: Diffusion Mechanisms and Point Defects in Silicon and Germanium. Phys. Stat. Sol. *29,* 455–542 (1968).

[3.69] Seidl, A.: Mathematical Implementation of Segregation Model for Two-Dimensional Process Simulation. IEEE Trans. Electron Devices *ED-30,* No. 6, 722–723 (1983).

[3.70] Seidl, A.: A Multi Grid Method for Solution of the Diffusion Equation in VLSI Process Modelling. IEEE Trans. Electron Devices *ED-30,* No. 9, 999–1004 (1983).

[3.71] Selberherr, S., Guerrero, E.: Simple and Accurate Representation of Implantation Parameters by Low Order Polynomals. Solid-State Electron. *24,* 591–593 (1981).

[3.72] Shaw, D.: Self- and Impurity Diffusion in Ge and Si. Phys. Stat. Sol. *(b) 72,* 11–39 (1975).

[3.73] Smith, G. E., Steckl, A. J.: RECIPE – A Two-Dimensional VLSI Process Modeling Program. IEEE Trans. Electron Devices *ED-29,* No. 2, 216–221 (1982).

[3.74] Stone, J. L., Plunkett, J. C.: Ion Implantation Processes in Silicon. In: Impurity Doping Processes in Silicon, pp. 56–146. Amsterdam: North-Holland 1981.

[3.75] Taniguchi, K., Kashiwagi, M., Iwai, H.: Two Dimensional Computer Simulation Models for MOSLSI Fabrication Processes. IEEE Trans. Electron Devices *ED-28,* No. 5, 574–580 (1981).

[3.76] Tielert, R.: Two-Dimensional Numerical Simulation of Impurity Redistribution in VLSI Processes. IEEE Trans. Electron Devices *ED-27,* No. 8, 1479–1483 (1980).

[3.77] Titov, A. I., Christodoulides, C. E., Carter, G., Nobes, M. J.: The Depth Distribution of Disorder Produced by Room Temperature 40 keV N^+ Ion Irradiation of Silicon. Radiation Effects *41,* 107–111 (1979).

[3.78] Troxell, J. R.: Ion-Implantation Associated Defect Production in Silicon. Solid-State Electron. *26,* No. 6, 539–548 (1983).

[3.79] Tsai, M. Y., Morehead, F. F., Baglin, J. E. E., Michel, A. E.: Shallow Junctions by High-Dose AS Implants in Si: Experiments and Modeling. J. Appl. Phys. *51,* No. 6, 3230–3235 (1980).

[3.80] Wang, F. F. Y.: Impurity Doping Processes in Silicon. Amsterdam: North-Holland 1981.

[3.81] Wilson, R. G., The Pearson IV Distribution and Its Application to Ion Implanted Depth Profiles. Radiation Effects *46,* 141–148 (1980).

[3.82] Winterbon, K. B., Sigmund, P., Sanders, J. B.: Spatial Distribution of Energy Deposited by Atomic Particles in Elastic Collisions. Mat. Fys. Medd. Dan. Vid. Selsk. *37,* No. 14, 1–73 (1970).

[3.83] Winterbon, K. B.: Ion Implantation Range and Energy Deposition Distributions, Vol. 2: Low Incident Ion Energies. New York: IFI/Plenum 1975.

4 The Physical Parameters

The basic semiconductor equations, the derivation of which we have thoroughly discussed in Chapter 2, just determine the structure of the set of equations which we shall have to solve in order to simulate the internal behavior of a device. Process modeling, as sketched in Chapter 3, delivers information about the geometry of a device and the distribution of dopants, which can also be considered to be a physical parameter. As we have already noticed, a couple of additional physical parameters are inherently associated with the basic semiconductor equations. Any quantitative, or even qualitative, simulation of a device relies heavily on applicable models for these parameters. In addition, a mathematical characterization of the problem of solving the basic semiconductor equations is only feasible with at least qualitative knowledge of the associated parameters (e. g. sign, smoothness, order of magnitude). Therefore, we shall discuss in this chapter the most important models for the physical parameters. A review has also been presented in [4.41].

In Section 4.1 models for the mobility of electrons and holes are summarized. Quantitatively accurate mobility values are required for the purpose of predictive simulation because of the multiplicative dependence of the current upon mobility, which is, obviously, one of the results most desired.

In Section 4.2 carrier generation/recombination phenomena are dealt with. These determine many essential effects associated with parasitic currents and device breakdown.

Section 4.3 and Section 4.4 are devoted to models for the thermal conductivity and heat generation, respectively. Their influence becomes apparent when accounting for electrothermal interaction phenomena by solving consistently the heat flow equation as part of the basic semiconductor equations.

4.1 Carrier Mobility Modeling

In Section 2.3 we have introduced relaxation times τ_n, τ_p in order to be able to derive the current relations. These relaxation times determine the rate at which electrons and holes are caused to change their momentum vector; therefore, they describe the average time between the scattering events electrons and holes undergo. We, additionally, have made use of the definitions of carrier mobilities because these are intuitively much easier to imagine than relaxation times. In this section we shall discuss the various mechanisms which determine the carrier mobilities and, obviously, also the relaxation times. In particular electrons and holes can be scattered by thermal lattice vibrations, ionized impurities, neutral impurities, vacancies, interstitials, dislocations, surfaces and electrons and holes themselves. A further mobility reduction is due to the saturation of the drift velocity of warm and hot carriers which is caused by lattice vibrations. Unfortunately, many of these mechanisms, especially their interactions, are extremely complicated and hence difficult to model exactly. Therefore, we shall discuss several approaches which have been published to model with phenomenological expressions the various experimentally observed mobility phenomena. A review on that subject can also be found in [4.58].

The most fundamental process by which carriers in a pure crystal are scattered is their interaction with the thermally generated vibrations of the atoms of the crystal. These lattice vibrations are a function of temperature. The theoretical result for the mobility caused by so-called "acoustic deformation potential lattice scattering" (μ_n^L, μ_p^L for electrons and holes) reads:

$$\mu_n^L = \frac{2 \cdot \sqrt{2 \cdot \pi}}{3} \cdot \frac{q \cdot \hbar^4 \cdot C_1}{(m_n^*)^{5/2} \cdot E_{ac}^2 \cdot (k \cdot T)^{3/2}} \tag{4.1-1}$$

$$\mu_p^L = \frac{2 \cdot \sqrt{2 \cdot \pi}}{3} \cdot \frac{q \cdot \hbar^4 \cdot C_1}{(m_p^*)^{5/2} \cdot E_{av}^2 \cdot (k \cdot T)^{3/2}} \tag{4.1-2}$$

C_1 is the average longitudinal elastic constant of the semiconductor; its numerical value lies in the order of 10^5 VAScm^{-3}. E_{ac} and E_{av} are the deformation potential constants of the conduction band and the valence band, respectively. They have a numerical value of a few eV. A concise derivation of (4.1-1), (4.1-2) can be found in, e.g., [4.108] and a rigorous treatment of lattice scattering has been published in, e.g. [4.55], [4.126]. Since silicon and germanium have a multivalley band structure and since high energetic phonons take part in the lattice scattering process (especially in galliumarsenide where this effect is even dominant), the behavior of the mobility cannot be correctly described by (4.1-1), (4.1-2). Band structure and optical phonons give rise to additional scattering mechanisms [4.15], [4.70], [4.108], [4.117]. A detailed discussion of these effects is beyond the scope of this text. For the purpose of simulation one usually takes a simple power law whose coefficients are obtained by fitting experimental mobility values.

$$\mu_n^L = \mu_n^o \cdot \left(\frac{T}{300\,\text{K}}\right)^{-\alpha_n} \tag{4.1-3}$$

$$\mu_p^L = \mu_p^o \cdot \left(\frac{T}{300\,\text{K}}\right)^{-\alpha_p} \tag{4.1-4}$$

The published numerical values for the constants in (4.1-3), (4.1-4) show some scatter. A menu of these coefficients compiled from various sources is summarized in Table 4.1-1. A judgement of the data and a recommendation is rather difficult.

Table 4.1-1. *Lattice mobility constants*

material	μ_n^o [$cm^2 V^{-1} s^{-1}$]	α_n [ ]	μ_p^o [$cm^2 V^{-1} s^{-1}$]	α_p [ ]	reference
Si	1388.157		467.729		[4.8]
	1448	2.33	473	2.23	[4.9]
	1438	2.42	465	2.2	[4.19]
	1330		495		[4.21]
	1407.3		467.73		[4.33]
	1430	2.2	495	2.2	[4.38]
	1360		520		[4.41]
	1350	2.5	480	2.5	[4.50]
	1450	2.6	500	2.3	[4.57]
	1448	2.42	479	2.2	[4.58]
	1400		500		[4.90]
	1354				[4.96]
	1400		480		[4.106]
	1240	2.5	438	2.7	[4.117]
	1500	2.5	600	2.7	[4.123]
	1600		600		[4.130], [4.138]
	1500	2.6	500	2.3	[4.135]
GaAs	8500				[4.6]
	8600		250		[4.50]
	7500				[4.80]
	8500	1	400	2.1	[4.123]
	9000	1	500	2.1	[4.135]
Ge	3900		1900		[4.50]
	3800	1.66	1800	2.33	[4.57]
	3900		1800		[4.90]
	2694	1.6	1818	1.3	[4.117]
	3900	1,66	1900	2.33	[4.123]
	3800	1.66	1820	2.33	[4.135]

Sah et al. have published a different model [4.102] which is claimed to predict reliably mobility values in silicon in the temperature range [4.2, 600] K.

$$\mu_n^L = \frac{1}{\dfrac{1}{4195\,\dfrac{cm^2}{Vs}\cdot\left(\dfrac{T}{300\,K}\right)^{-1.5}} + \dfrac{1}{2153\,\dfrac{cm^2}{Vs}\cdot\left(\dfrac{T}{300\,K}\right)^{-3.13}}} \tag{4.1-5}$$

$$\mu_p^L = \frac{1}{\dfrac{1}{2502\,\dfrac{cm^2}{Vs}\cdot\left(\dfrac{T}{300\,K}\right)^{-1.5}} + \dfrac{1}{591\,\dfrac{cm^2}{Vs}\cdot\left(\dfrac{T}{300\,K}\right)^{-3.25}}} \tag{4.1-6}$$

This model combines the theoretical lattice scattering mobility caused by acoustical phonons with a mobility component caused by optical and intervalley phonons by the simple Mathiessen's rule. It is stated in [4.102] that more elaborate formulae based on complicated theoretical models do not justify the additional effort for the purpose of simulation. I fully agree with that statement from my personal experience.

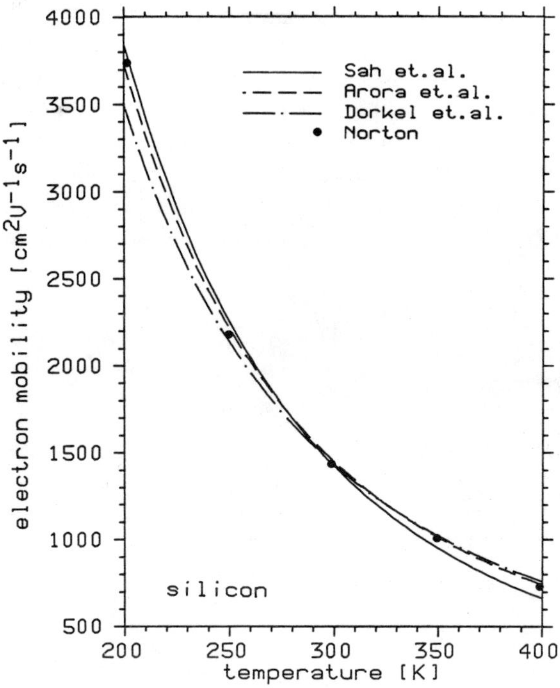

Fig. 4.1-1. Mobility of electrons due to lattice scattering in silicon versus temperature

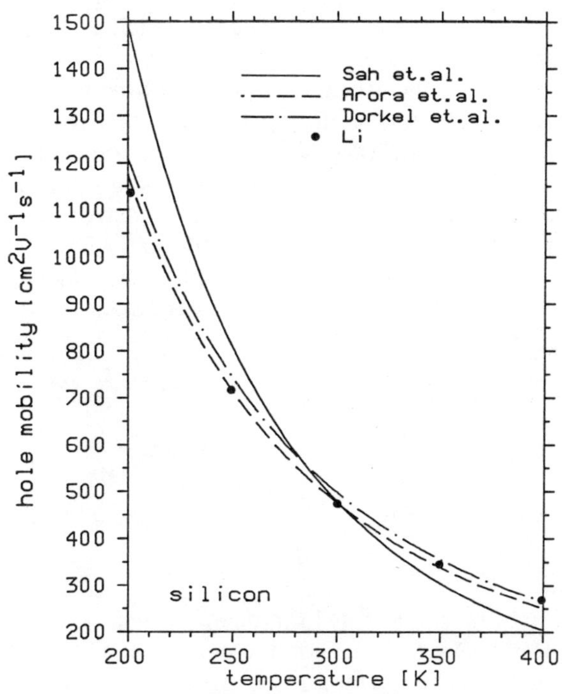

Fig. 4.1-2. Mobility of holes due to lattice scattering in silicon versus temperature

Fig. 4.1-1 and Fig. 4.1-2 show the lattice mobility for electrons and holes in silicon versus temperature after the model of Sah et al. (4.1-5), (4.1-6) (solid line), the model (4.1-3), (4.1-4) with parameters of Arora et al. [4.9] (dashed line) and with parameters of Dorkel et al. [4.38] (dot dashed line) together with experimental values of Norton et al. [4.81] and Li [4.70], respectively.

The next scattering mechanism we shall consider for mobility modeling is ionized impurity scattering. The first useful model which was derived by theoretical investigations has been published by Conwell and Weiskopf [4.28]. Their formula reads:

$$\mu_{n,p}^{I}=\frac{64\cdot\varepsilon^2\cdot(2\cdot k\cdot T)^{3/2}}{q^3\cdot CI}\cdot\sqrt{\frac{\pi}{m_{n,p}^{*}}}\cdot g_{CW}\left(\frac{12\cdot\pi\cdot\varepsilon\cdot k\cdot T}{q^2\cdot CI^{1/3}}\right) \qquad (4.1\text{-}7)$$

with:

$$g_{CW}(x)=\frac{1}{\ln(1+x^2)} \qquad (4.1\text{-}8)$$

CI is the sum of all ionized impurity species times the magnitude of their charge state.

$$CI=\sum_{i=1}^{n}|Z_i|\cdot C_i \qquad (4.1\text{-}9)$$

The dopants usually taken in semiconductor processing have a charge state with magnitude one. However, for some applications such as solar cells, zinc can be used [4.102] which has two acceptor levels, one of which is doubly ionized, thus corresponding to $Z=2$.

Expression (4.1-7) reflects a reciprocal dependence of the mobility upon the total concentration of ionized impurities for moderately high concentrations, which becomes weaker ($\sim CI^{-1/3}$) for very large concentrations. The function $g_{CW}(x)$ models the influence of "neighboring" ionized impurities which screen each other due to their Coulomb potential and, therefore, are inactive as scattering centers. However, the model for this effect is relatively crude (cf. [4.28]). A refinement of the influence of charge screening has been introduced by Brooks [4.17]. He has also accounted for the fact that free electrons and holes screen the impurities.

$$\mu_{n,p}^{I}=\frac{64\cdot\varepsilon^2\cdot(2\cdot k\cdot T)^{3/2}}{q^3\cdot CI}\cdot\sqrt{\frac{\pi}{m_{n,p}^{*}}}\cdot g_{B}\left(\frac{24\cdot m_{n,p}^{*}\cdot\varepsilon\cdot(k\cdot T)^2}{q^2\cdot\hbar^2\cdot(n+p)}\right) \qquad (4.1\text{-}10)$$

with:

$$g_B(x)=\frac{1}{\ln(1+x)-\dfrac{x}{1+x}} \qquad (4.1\text{-}11)$$

(4.1-10) is claimed to be more accurate for moderately heavy doped semiconductors ($CI<10^{19}\,\mathrm{cm}^{-3}$). For degenerate semiconductors, however, no applicable theoretical models have been published so far and heuristic models have to be used instead. (4.1-7) and (4.1-10) give nearly the same results as long as the free carrier concentration about equals the ionized impurity concentration. When the free carrier concentration is appreciably smaller, which is the case for compensated material [4.16] and space charge regions, (4.1-10) gives a lower mobility due to the fact that less screening will take place, and, therefore, the ionized impurities will scatter more efficiently (cf. [4.35]). Some more considerations on this subject have been given in, e. g. [4.94], [4.95], [4.103].

The mobility components due to lattice scattering and due to ionized impurity scattering have to be combined in some way to obtain an effective mobility. The Mathiessen rule is inappropriate for that purpose, because lattice scattering and ionized impurity scattering can not be considered to be fully independent mechanisms, which is a definite requirement for the applicability of the Mathiessen rule [4.108]. Debye and Conwell [4.35] have derived from theoretical reasoning the following expression for the combined mobility $\mu_{n,p}^{LI}$.

$$\mu_{n,p}^{LI} = \mu_{n,p}^{L} \cdot \left(1 + x^2 \cdot \left(Ci(x) \cdot \cos(x) + si(x) \cdot \sin(x)\right)\right) \tag{4.1-12}$$

with

$$x = \sqrt{\frac{6 \cdot \mu_{n,p}^{L}}{\mu_{n,p}^{I}}} \tag{4.1-13}$$

$$Ci(x) = -\int_{x}^{\infty} \frac{\cos(t)}{t} \cdot dt \tag{4.1-14}$$

$$si(x) = -\int_{x}^{\infty} \frac{\sin(t)}{t} \cdot dt \tag{4.1-15}$$

A discussion of that model can be found in [4.35], [4.108]. (4.1-12) is quite tedious to handle. In [4.38] an approximation to (4.1-12) has been presented, which is quite simple:

$$\mu_{n,p}^{LI} = \mu_{n,p}^{L} \cdot \left(\frac{1.025}{1 + \left(2.126 \cdot \frac{\mu_{n,p}^{L}}{\mu_{n,p}^{I}}\right)^{0.715}} - 0.025 \right) \tag{4.1-16}$$

This approximation is claimed to be accurate to within a 2% maximum error if:

$$\mu_{n,p}^{L} < 13.5 \cdot \mu_{n,p}^{I} \tag{4.1-17}$$

(4.1-17) is not really a serious restriction because the expressions for the ionized impurity mobility component (4.1-7) and (4.1-10) are already invalid if they produce results which would violate (4.1-17).

There is, unfortunately, some uncertainty which numerical values one should pick for the effective masses m_n^*, m_p^* in (4.1-10). Dorkel et al. [4.38] use in the case of silicon $0.953 \cdot m_o$ and $1.0048 \cdot m_o$ for m_n^* in the leading term and in the argument of g_b in (4.1-10), respectively; and, equivalently, $20.25 \cdot m_o$, which is rather large, and $0.413 \cdot m_o$ for m_p^*. Evaluating all the dependent constants with these recommendations, (4.1-10) will read:

$$\mu_n^I = \frac{2.4 \cdot 10^{21} \dfrac{1}{\text{cm Vs}} \cdot \left(\dfrac{T}{300\,\text{K}}\right)^{3/2}}{CI} \cdot g_B \left(\frac{1.37 \cdot 10^{20}\,\text{cm}^{-3} \cdot \left(\dfrac{T}{300\,\text{K}}\right)^2}{n+p}\right) \tag{4.1-18}$$

$$\mu_p^I = \frac{5.2 \cdot 10^{20} \dfrac{1}{\text{cm Vs}} \cdot \left(\dfrac{T}{300\,\text{K}}\right)^{3/2}}{CI} \cdot g_B \left(\frac{5.63 \cdot 10^{19}\,\text{cm}^{-3} \cdot \left(\dfrac{T}{300\,\text{K}}\right)^2}{n+p}\right) \tag{4.1-19}$$

A more pragmatic approach for modeling the combined lattice and ionized impurity mobility has been introduced by Caughey and Thomas [4.21]. They have used a Fermi like function, or hyperbolic tangent, to fit experimental data:

$$\mu_{n,p}^{LI} = \mu_{n,p}^{\min} + \frac{\mu_{n,p}^{L} - \mu_{n,p}^{\min}}{1 + \left(\dfrac{CI}{C_{n,p}^{\text{ref}}}\right)^{\alpha_{n,p}}} \tag{4.1-20}$$

(4.1-20) incorporates a saturation effect of the mobility reduction for high impurity concentrations, which has been observed by experimental investigations. Numerical values for the parameters for silicon at 300 K temperature which are involved in (4.1-20) are summarized in Table 4.1-2 for electrons and in Table 4.1-3 for holes.

Table 4.1-2. *Coefficients for ionized impurity scattering of electrons*

$\mu_n^{\min}$ [$\text{cm}^2\,\text{V}^{-1}\,\text{s}^{-1}$]	α_n [ ]	C_n^{ref} [cm^{-3}]	reference
55.24	0.733	$1.072 \cdot 10^{17}$	[4.8]
92	0.91	$1.3 \cdot 10^{17}$	[4.10], [4.41], [4.58]
65	0.72	$8.5 \cdot 10^{16}$	[4.21], [4.54], [4.72]
71.12	0.7291	$1.059 \cdot 10^{17}$	[4.33]
52.2	0.680	$9.68 \cdot 10^{16}$	[4.73] (arsenic)
68.5	0.711	$9.20 \cdot 10^{16}$	[4.73] (phosphorus)

Table 4.1-3. *Coefficients for ionized impurity scattering of holes*

$\mu_p^{\min}$ [$\text{cm}^2\,\text{V}^{-1}\,\text{s}^{-1}$]	α_p [ ]	C_p^{ref} [cm^{-3}]	reference
49.705	0.7	$1.606 \cdot 10^{17}$	[4.8], [4.33]
47.7	0.76	$6.3 \cdot 10^{16}$	[4.21], [4.54], [4.58], [4.72]
65	0.61	$2.4 \cdot 10^{17}$	[4.41]
44.9	0.719	$2.23 \cdot 10^{17}$	[4.73]
47.7	0.76	$1.9 \cdot 10^{17}$	[4.91]

The various publications again have some scatter in these data; however, in all of the references a perfect fit to experimental results has been claimed!
The saturation of mobility reduction for high impurity concentration has been treated more sophisticatedly than (4.1-17) by some authors. In [4.91], [4.119] the mobility of electrons is modeled with:

$$\mu_n^{LI} = \frac{65\,\frac{\mathrm{cm}^2}{\mathrm{Vs}}}{1+\left(\frac{CI}{5\cdot 10^{20}\,\mathrm{cm}^{-3}}\right)^{1.5}} + \frac{\mu_n^L - 65\,\frac{\mathrm{cm}^2}{\mathrm{Vs}}}{1+\left(\frac{CI}{8.5\cdot 10^{16}\,\mathrm{cm}^{-3}}\right)^{0.72}} \tag{4.1-21}$$

Roulston et al. [4.96] have used (4.1-22) for the electron mobility with good success:

$$\mu_n^{LI} = 86.5\,\frac{\mathrm{cm}^2}{\mathrm{Vs}}\cdot\left(1-\frac{CI}{6.18\cdot 10^{20}\,\mathrm{cm}^{-3}}\right) + \frac{1268\,\frac{\mathrm{cm}^2}{\mathrm{Vs}}}{\left(1+\frac{CI}{1.3\cdot 10^{17}\,\mathrm{cm}^{-3}}\right)^{0.91}} \tag{4.1-22}$$

A similar treatment can be found in, e. g. [4.73].
Another formula for modeling ionized impurity scattering in silicon at 300 K temperature with just two parameters has been proposed by Scharfetter and Gummel [4.106].

$$\mu_{n,p}^{LI} = \frac{\mu_{n,p}^L}{\sqrt{1+\frac{CI}{C_{n,p}^{\mathrm{ref}}+\frac{CI}{S_{n,p}}}}} \tag{4.1-23}$$

$$C_n^{\mathrm{ref}} = 3\cdot 10^{16}\,\mathrm{cm}^{-3},\quad S_n = 350$$

$$C_p^{\mathrm{ref}} = 4\cdot 10^{16}\,\mathrm{cm}^{-3},\quad S_p = 81$$

(4.1-23) has been widely used by, e. g. [4.47], [4.79], [4.130], [4.138], [4.139] and, again, excellent agreement between experimental and calculated results has been claimed. A discussion on a theoretical basis of this equation has been carried out by Thornber [4.128].
Arora et al. [4.9] have published formulae with a very similar structure to the Caughey and Thomas expression (4.1-20) with coefficients for silicon which depend on temperature.

$$\mu_n^{LI} = 88\,\frac{\mathrm{cm}^2}{\mathrm{Vs}}\cdot\left(\frac{T}{300\,\mathrm{K}}\right)^{-0.57} + \frac{1252\,\frac{\mathrm{cm}^2}{\mathrm{Vs}}\cdot\left(\frac{T}{300\,\mathrm{K}}\right)^{-2.33}}{1+\frac{CI}{1.432\cdot 10^{17}\,\mathrm{cm}^{-3}\cdot\left(\frac{T}{300\,\mathrm{K}}\right)^{2.546}}} \tag{4.1-24}$$

$$\mu_p^{LI} = 54.3 \frac{\text{cm}^2}{\text{Vs}} \cdot \left(\frac{T}{300\,\text{K}}\right)^{-0.57} + \frac{407 \frac{\text{cm}^2}{\text{Vs}} \cdot \left(\frac{T}{300\,\text{K}}\right)^{-2.23}}{1 + \frac{CI}{2.67 \cdot 10^{17}\,\text{cm}^{-3} \cdot \left(\frac{T}{300\,\text{K}}\right)^{2.546}}} \tag{4.1-25}$$

These formulae are supposed to be accurate to within a maximum error of 13% in a temperature range [250, 500] K and a total ionized impurity concentration range $[10^{13}, 10^{20}]\,\text{cm}^{-3}$.

The last approach for combined lattice and ionized impurity scattering in silicon which I should like to present here has been proposed by Sah et al. [4.102].

$$\mu_n^I = 90 \frac{\text{cm}^2}{\text{Vs}} \cdot \left(1 + \frac{2 \cdot 10^{18}\,\text{cm}^{-3}}{CI} \cdot \left(\frac{T}{300\,\text{K}}\right)\right) \tag{4.1-26}$$

$$\mu_p^I = 45 \frac{\text{cm}^2}{\text{Vs}} \cdot \left(1 + \frac{1.2 \cdot 10^{18}\,\text{cm}^{-3}}{CI} \cdot \left(\frac{T}{300\,\text{K}}\right)\right) \tag{4.1-27}$$

$$\mu_{n,p}^{LI} = \frac{1}{\frac{1}{\mu_{n,p}^L} + \frac{1}{\mu_{n,p}^I}} \tag{4.1-28}$$

(4.1-28) is the simple Mathiessen rule which is not applicable based on theoretical reasoning as described previously. For the lattice mobilities $\mu_{n,p}^L$ expressions (4.1-5), (4.1-6) are used. These formulae are claimed to be accurate for ionized impurity concentrations in the range $[10^{11}, 10^{20}]\,\text{cm}^{-3}$.

More theoretical considerations can be found in the various books on semiconductor physics, e.g. [4.15], [4.29], [4.61], [4.108]. It is to note that usually no difference is made between ionized impurity scattering of minority or majority carriers. For high impurity concentrations the thereby introduced error can be quite significant [4.14].

Fig. 4.1-3 and Fig. 4.1-4 show, respectively, the mobility of electrons and holes caused by combined lattice and ionized impurity scattering versus concentration of ionized impurities at 300 K temperature in silicon. The solid line corresponds to the model of Dorkel et al. (4.1-16), (4.1-18), (4.1-19); the dashed line denotes the model of Scharfetter and Gummel (4.1-23); the dot-dashed line corresponds to the Caughey and Thomas like model of Arora et al. (4.1-24), (4.1-25); and the dotted line denotes the model of Sah et al. (4.1-26), (4.1-27), (4.1-28). Experimental data have not been included in these figures because their scatter is even larger than the differences between the models. That can lead to the pragmatical conclusion that it does not make much difference which model to choose. More experimental and theoretical investigations have to be carried out in order to get rid of these quite significant uncertainties. Comparisons between experimental and model mobility values are given in, e.g. [4.58], [4.91], [4.119].

I am aware of no specific models for ionized impurity scattering in GaAs. Supposedly, there has not been an urgent need for such models. For germanium the

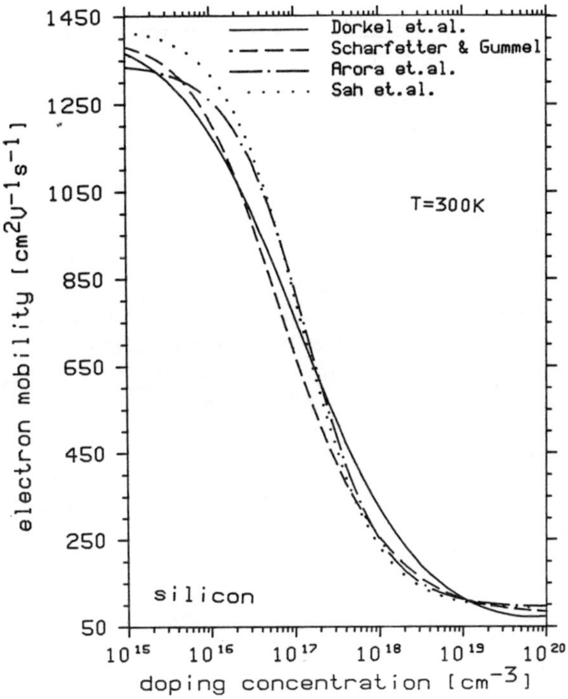

Fig. 4.1-3. Mobility of electrons due to ionized impurity scattering in silicon at 300 K temperature

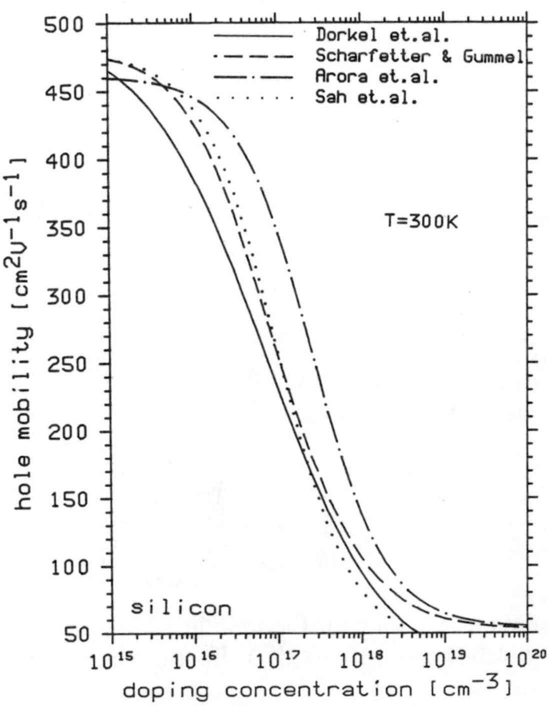

Fig. 4.1-4. Mobility of holes due to ionized impurity scattering in silicon at 300 K temperature

models for silicon should work quite well with minor changes in some constants. However, as there is presently little activity in the development of germanium devices except some very special power devices, the question of modeling in germanium is more or less irrelevant.

Another scattering mechanism which we have to consider for device modeling is carrier-carrier scattering. Particularly in power devices operating in the on-state this effect becomes pronounced because the free carrier concentrations may increase far above the doping concentration.

A very simple approach to account for carrier-carrier scattering has been suggested by Engl and Dirks [4.41]. They use the Caughey and Thomas expression (4.1-20) for ionized impurity scattering with an effective concentration of scattering centers depending not only on the ionized impurity concentration CI but also on the free carrier concentration $n+p$.

$$CI_{\text{eff}} = 0.34 \cdot CI + 0.66 \cdot (n+p) \tag{4.1-29}$$

In one paper Adler [4.1] has suggested a similar approach. He has simply added an extra term in the denominator of the Caughey and Thomas formula.

$$\mu_{n,p}^{LIC} = \mu_{n,p}^{\min} + \frac{\mu_{n,p}^{L} - \mu_{n,p}^{\min}}{1 + \left(\frac{CI}{C_{n,p}^{\text{ref}}}\right)^{\alpha_{n,p}} + \left(\frac{\sqrt{n \cdot p}}{14 \cdot C_{n,p}^{\text{ref}}}\right)^{\alpha_{n,p}}} \tag{4.1-30}$$

In another paper [4.3] Adler has suggested a different treatment as being more accurate. Here, the mobility component due to carrier-carrier scattering is modeled as:

$$\mu^{C} = \frac{1.428 \cdot 10^{20} \dfrac{1}{\text{cm Vs}}}{\sqrt{n \cdot p} \cdot \ln\left(1 + 4.54 \cdot 10^{11}\,\text{cm}^{-2} \cdot (n \cdot p)^{-1/3}\right)} \tag{4.1-31}$$

This component is combined with the original Caughey and Thomas model (4.1-20) for lattice and ionized impurity scattering with the simple Mathiessen rule.

$$\mu_{n,p}^{LIC} = \frac{1}{\dfrac{1}{\mu_{n,p}^{LI}} + \dfrac{1}{\mu^{C}}} \tag{4.1-32}$$

A structurally equivalent expression to (4.1-31) for μ^C has been proposed in [4.38] with temperature dependent coefficients.

$$\mu^{C} = \frac{1.04 \cdot 10^{21} \dfrac{1}{\text{cm Vs}} \cdot \left(\dfrac{T}{300\,\text{K}}\right)^{3/2}}{\sqrt{n \cdot p} \cdot \ln\left(1 + 7.45 \cdot 10^{13}\,\text{cm}^{-2} \cdot \left(\dfrac{T}{300\,\text{K}}\right)^{2} \cdot (n \cdot p)^{-1/3}\right)} \tag{4.1-33}$$

These authors combine (4.1-33) with the mobility component $\mu_{n,p}^{I}$ due to ionized impurity scattering (4.1-18), (4.1-19) to $\mu_{n,p}^{IC}$ using the Mathiessen rule. This result is then used in formula (4.1-16) instead of $\mu_{n,p}^{I}$ to calculate the global mobility $\mu_{n,p}^{LIC}$. This approach seems to be physically very sound.

Li and Thurber [4.69] have investigated carrier-carrier scattering at low injection levels for uncompensated silicon, where the free carrier concentration equals the ionized impurity concentration. In this approach both the mobility component due to lattice scattering and due to impurity scattering are multiplied with a correction factor.

$$\mu_{n,p}^{LC} = \mu_{n,p}^{L} \cdot \min\left(\max\left(0.88, 1.0133 - \frac{CI}{1.5 \cdot 10^{18}\,\mathrm{cm}^{-3}}\right), 1\right) \tag{4.1-34}$$

$$\mu_{n,p}^{IC} = \mu_{n,p}^{I} \cdot \min\left(\max\left(0.632, 1.0409 - \frac{CI}{4.89 \cdot 10^{17}\,\mathrm{cm}^{-3}}\right), 1\right) \tag{4.1-35}$$

These components are then combined with the Debye and Conwell formula (4.1-12) to obtain an effective mobility. However, as mentioned above, this approach is restricted to low injection conditions, where the influence of carrier-carrier scattering is relatively small and unimportant. Therefore, (4.1-34), (4.1-35) have to be seen only as an improvement to the theoretical model of ionized impurity scattering.

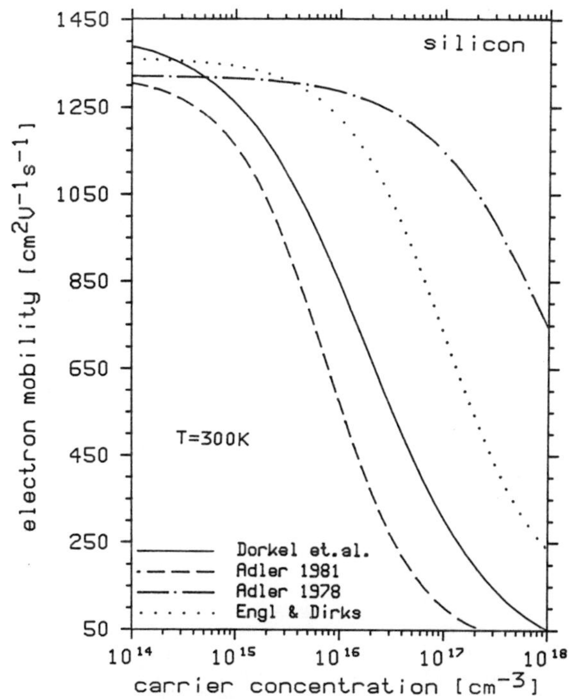

Fig. 4.1-5. Mobility of electrons due to carrier-carrier scattering in silicon at 300 K temperature

Fig. 4.1-5 and Fig. 4.1-6 show the mobility due to lattice, ionized impurity and carrier-carrier scattering versus free carrier concentration in silicon at 300 K temperature. The solid line corresponds to the model of Dorkel et al. (4.1-33); the dashed line denotes (4.1-31), (4.1-32); the dot-dashed line corresponds to (4.1-30); and the dotted line to (4.1-29). Electron and hole concentrations have been assumed

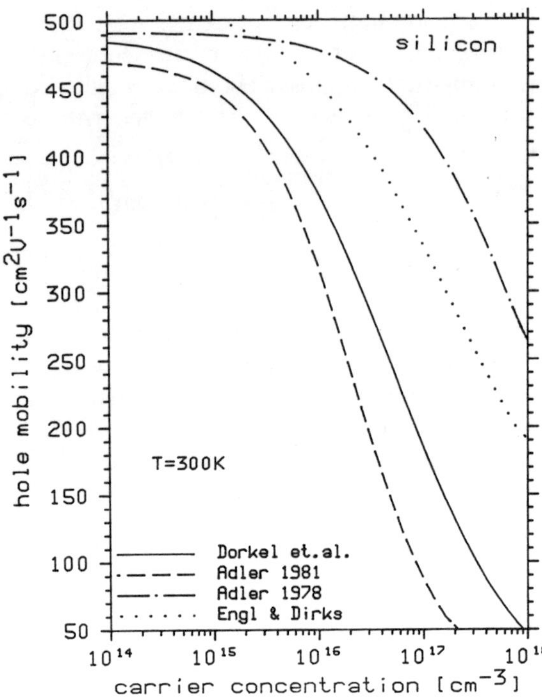

Fig. 4.1-6. Mobility of holes due to carrier-carrier scattering in silicon at 300 K temperature

to be equal which is usually the case when the free carrier concentration strongly exceeds the doping concentration. The ionized impurity concentration CI has been assumed to be $10^{14}\,\mathrm{cm}^{-3}$.

Another scattering mechanism which possibly has to be taken into account is neutral impurity scattering. This effect is not very pronounced at room temperature and can usually be ignored. However, it can become relevant for simulations at low temperatures ($T<77\,\mathrm{K}$). Early theoretical results predict that the mobility component due to neutral impurity scattering is temperature independent (cf. [4.16], [4.108]).

$$\mu_{n,p}^{N} = \frac{q \cdot m_{n,p}^{*}}{20 \cdot a_B \cdot \hbar \cdot m_o \cdot \varepsilon \cdot CN} \tag{4.1-36}$$

a_B is the Bohr radius ($5.2917706 \cdot 10^{-11}\,\mathrm{m}$) and CN denotes the concentration of neutral impurities. More recent investigations (cf. [4.69]) predict a weak temperature dependence for temperatures below 50 K.

$$\mu_{n,p}^{N} = \frac{0.041 \cdot q \cdot m_{n,p}^{*}}{a_B \cdot \hbar \cdot m_o \cdot \varepsilon \cdot CN} \cdot \left(\frac{2}{3} \cdot \sqrt{\frac{k \cdot T}{EN_{n,p}}} + \frac{1}{3} \cdot \sqrt{\frac{EN_{n,p}}{k \cdot T}}\right) \tag{4.1-37}$$

with:

$$EN_{n,p} = 0.71\,\mathrm{eV} \cdot \frac{m_{n,p}^{*}}{m_o} \cdot \left(\frac{\varepsilon_o}{\varepsilon}\right)^2 \tag{4.1-38}$$

The mobility component due to neutral impurity scattering can be combined with the mobility due to lattice, ionized impurity and carrier-carrier scattering with the Mathiessen rule. A more elaborate treatment is not known at the moment.

$$\mu_{n,p}^{LICN} = \frac{1}{\dfrac{1}{\mu_{n,p}^{LIC}} + \dfrac{1}{\mu_{n,p}^{N}}} \tag{4.1-39}$$

The next effect we shall consider for mobility modeling is the saturation of the drift velocity for high electric field. This effect has to be accounted for by a reduction of the effective mobility since the magnitude of the drift velocity is the product of the mobility and the electric field component in the direction of current flow.

$$|\vec{v}_n| = -\mu_n^* \cdot \frac{\vec{E} \cdot \vec{J}_n}{|\vec{J}_n|} \tag{4.1-40}$$

$$|\vec{v}_p| = \mu_p^* \cdot \frac{\vec{E} \cdot \vec{J}_p}{|\vec{J}_p|} \tag{4.1-41}$$

However, the above given statement holds only if the diffusion current component is negligibly small. Instead of the electric field component in the direction of current flow it is more appropriate to use the magnitude of the gradient of the respective quasi-Fermi potential, which to first order is the driving force for the carriers (cf. Section 2.3).

$$|\vec{v}_n| = \mu_n^* \cdot |\operatorname{grad} \varphi_n| \tag{4.1-42}$$

$$|\vec{v}_p| = \mu_p^* \cdot |\operatorname{grad} \varphi_p| \tag{4.1-43}$$

The gradient of the quasi-Fermi potential always points in the direction of the flow of the corresponding current density. Note also, that there arises a non-trivial problem when the inner product of electric field and current density is positive in (4.1-40), or negative in (4.1-41). Such a situation can locally arise in a device when the diffusion current component dominates the drift current component. The use of (4.1-42) and (4.1-43) has to be strongly recommended as, probably, first pointed out in the various publications of Engl et al., e.g. [4.41]. However, in most publications the electric field is considered to be relevant for the saturation of the drift velocity. The differences which arise in simulation results between taking the electric field or the gradient of the quasi-Fermi potentials as responsible for carrier heating are, luckily, not very pronounced in many applications because the electric field and the gradient of the quasi-Fermi potentials are almost parallel in critical device areas. Therefore, the question of which quantity to take is not really as important as it looks at first glance. In the following we shall use the symbols E_n and E_p for the magnitude of the driving forces for electrons and holes, respectively.

In 1951 Shockley [4.114] derived theoretically probably the first useful equation for the influence of carrier heating on the drift velocity and, thus, the mobility.

$$\mu_n^{LICNE} = \frac{\mu_n^{LICN}}{\sqrt{\dfrac{1}{2} + \dfrac{1}{2} \cdot \sqrt{1 + \dfrac{3 \cdot \pi}{8} \cdot \left(\dfrac{\mu_n^{LICN} \cdot E_n}{Cs}\right)^2}}} \tag{4.1-44}$$

Cs denotes the speed of longitudinal acoustic phonons ($\sim 1.66 \cdot 10^6\,\mathrm{cms}^{-1}$). Thornber [4.128] has stated more recently that this equation is very reasonable for not too large driving forces, which was Shockley's intention. The drift velocity associated with (4.1-44) does not saturate.

A widely used expression for mobility due to carrier heating reads:

$$\mu_{n,p}^{LICNE} = \frac{\mu_{n,p}^{LICN}}{\left(1 + \left(\frac{E_{n,p}}{E_{n,p}^{\mathrm{crit}}}\right)^{\beta_{n,p}}\right)^{1/\beta_{n,p}}} \tag{4.1-45}$$

Some numerical values which have been found in the literature for the critical fields $E_{n,p}^{\mathrm{crit}}$ and the exponent $\beta_{n,p}$ for silicon at 300 K temperature are summarized in Table 4.1-4.

We shall not differentiate between carrier heating taking place in the bulk of a semiconductor or in an inversion channel, which was done intuitively some years ago. Recent experiments, e.g. [4.101], indicate reliably that, if there is any difference at all, it is very small. The former opinion is the reason, I suspect, why some of the numerical constants published in the literature show an unexpectedly pronounced scatter.

Table 4.1-4. *Coefficients for velocity saturation of* (4.1-45)

E_n^{crit} [$\mathrm{V\,cm}^{-1}$]	β_n [ ]	E_p^{crit} [$\mathrm{V\,cm}^{-1}$]	β_p [ ]	reference
$8.57 \cdot 10^3$	1.11	$1.8 \cdot 10^4$	1.21	[4.19]
$8.7 \cdot 10^3$	2.9	$1.2 \cdot 10^4$	2.6	[4.25] for ⟨100⟩ material
$1.4 \cdot 10^4$	2.8			[4.25] for ⟨111⟩ material
$7.396 \cdot 10^3$	1	$2 \cdot 10^4$	1	[4.33]
$8.0 \cdot 10^3$	2	$1.95 \cdot 10^4$	1	[4.54], [4.72]
$2 \cdot 10^4$	1			[4.84], [4.85]
$7.22 \cdot 10^3$	2			[4.89]
$7.1 \cdot 10^3$	1.4			[4.134]

The temperature dependence of these coefficients for silicon has been investigated in [4.19].

$$\begin{gathered} E_n^{\mathrm{crit}} = 6.98 \cdot 10^3 \frac{\mathrm{V}}{\mathrm{cm}} \cdot \left(\frac{T}{300\,\mathrm{K}}\right)^{1.55}, \quad \beta_n = 1.11 \cdot \left(\frac{T}{300\,\mathrm{K}}\right)^{0.66} \\ E_p^{\mathrm{crit}} = 1.80 \cdot 10^4 \frac{\mathrm{V}}{\mathrm{cm}} \cdot \left(\frac{T}{300\,\mathrm{K}}\right)^{1.68}, \quad \beta_n = 1.21 \cdot \left(\frac{T}{300\,\mathrm{K}}\right)^{0.17} \end{gathered} \tag{4.1-46}$$

Thornber [4.128] has pointed out that (4.1-45) is incorrect to use, although the formula is exceedingly simple and might have proven to be useful. The value of the saturation velocity associated with (4.1-44) can be calculated as:

$$v_{n,p}^{\mathrm{sat}} = \mu_{n,p}^{LICN} \cdot E_{n,p}^{\mathrm{crit}} \tag{4.1-47}$$

This means that the saturation velocity depends on, e.g., impurity scattering which is totally implausible by physical reasoning. The experiments of Sabnis and Clemens

[4.101] have clearly demonstrated that the saturation velocity is almost independent of the doping concentration. However, this problem can be easily overcome [4.21] by eliminating in (4.1-45) the critical field with relation (4.1-47) and introducing instead the saturation velocity as new parameter.

$$\mu_{n,p}^{LICNE} = \frac{\mu_{n,p}^{LICN}}{\left(1 + \left(\dfrac{\mu_{n,p}^{LICN} \cdot E_{n,p}}{v_{n,p}^{\mathrm{sat}}}\right)^{\beta_{n,p}}\right)^{1/\beta_{n,p}}} \tag{4.1-48}$$

Numerical values for the coefficients $v_{n,p}^{\mathrm{sat}}$ and $\beta_{n,p}$ for silicon at 300 K temperature are given in Table 4.1-5.

Table 4.1-5. *Coefficients for velocity saturation of* (4.1-48)

v_n^{sat} [cm s^{-1}]	β_n [ ]	v_p^{sat} [cm s^{-1}]	β_p [ ]	reference
$1.1 \cdot 10^7$	2	$9.5 \cdot 10^6$	1	[4.21], [4.41]
$1.0 \cdot 10^7$	1			[4.34], [4.62]

Another formulation has been suggested by Scharfetter and Gummel [4.106] and has been successfully used by various authors, e.g. [4.47], [4.138], [4.130].

$$\mu^{LIE} = \frac{\mu^{LI}}{\sqrt{1 + (\mu^{LI})^2 \cdot \left(\dfrac{(E/(\mu^L \cdot A))^2}{\mu^L \cdot E/(\mu^L \cdot A) + F} + \left(\dfrac{E}{\mu^L \cdot B}\right)^2\right)}} \tag{4.1-49}$$

All quantities in (4.1-49) have to be imagined with index n or p for electrons or holes. For μ^{LI} expression (4.1-23) is used. For the constants A, F and B the following values have been recommended in [4.106] for silicon at 300 K temperature.

$$\begin{aligned} A_n &= 3.5 \cdot 10^3 \frac{\mathrm{V}}{\mathrm{cm}}, \quad F_n = 8.8, \quad B_n = 7.4 \cdot 10^3 \frac{\mathrm{V}}{\mathrm{cm}} \\ A_p &= 6.1 \cdot 10^3 \frac{\mathrm{V}}{\mathrm{cm}}, \quad F_p = 1.6, \quad B_p = 2.5 \cdot 10^4 \frac{\mathrm{V}}{\mathrm{cm}} \end{aligned} \tag{4.1-50}$$

Formula (4.1-49) is usually written in a different manner which might be more familiar to some readers:

$$\mu^{LIE} = \frac{\mu^{LI}}{\sqrt{1 + \dfrac{CI}{C^{\mathrm{ref}} + CI/S} + \dfrac{(E/A)^2}{E/A + F} + \left(\dfrac{E}{B}\right)^2}} \tag{4.1-51}$$

Again all quantities, except CI, have to be imagined with index n or p. However, (4.1-49) allows a nice interpretation of its parameters. As Thornber [4.128] has already shown, this formula is very attractive for several reasons. The term $\mu^L \cdot B$ can be interpreted as a saturation velocity which is independent of impurity scattering. Using the original data of [4.106] these saturation velocities evaluate to:

$$v_n^{\text{sat}} = 1.04 \cdot 10^7 \frac{\text{cm}}{\text{s}}$$
$$v_p^{\text{sat}} = 1.20 \cdot 10^7 \frac{\text{cm}}{\text{s}} \tag{4.1-52}$$

These values are quite plausible. One probably would expect a smaller value for the saturation velocity of holes. The term $\mu^L \cdot A$ represents a velocity which can be identified as an acoustic phonon velocity. Such a quantity is responsible for the behavior of warm carriers which is nicely reflected by (4.1-49). It can also be expected that this velocity is to first order independent of impurity scattering. The only modification which has to be recommended for (4.1-49) is to replace the term $\mu^L \cdot E$ by $\mu^{LI} \cdot E$ in order to stay consistent with the elaborate scattering scaling results of Thornber [4.128]. By rewriting (4.1-49) with the above obtained interpretative results we obtain:

$$\mu^{LIE} = \frac{\mu^{LI}}{\sqrt{1 + \left(\frac{\mu^{LI} \cdot E}{v^{ac}}\right)^2 \cdot \left(\frac{v^{ac}}{\mu^{LI} \cdot E + F \cdot v^{ac}}\right) + \left(\frac{\mu^{LI} \cdot E}{v^{\text{sat}}}\right)^2}} \tag{4.1-53}$$

F in (4.1-53) is not straightforward to interprete by physical reasoning; it should be considered as a fitting parameter. If carrier-carrier scattering has to be taken into account, μ^{LI} should be replaced by μ^{LICN} calculated with any of the previously discussed models of preference.
The temperature dependence of v^{ac} and v^{sat} is expected to be fairly weak. In the case of v^{ac} I am not aware of any results. By interpreting the temperature dependent critical fields (4.1-46) of [4.19] as the ratio of the saturation velocity v^{sat} over the lattice mobility μ^L we obtain:

$$v_n^{\text{sat}} = 10^7 \frac{\text{cm}}{\text{s}} \cdot \left(\frac{T}{300\,\text{K}}\right)^{-0.87}$$
$$v_p^{\text{sat}} = 8.37 \cdot 10^6 \frac{\text{cm}}{\text{s}} \cdot \left(\frac{T}{300\,\text{K}}\right)^{-0.52} \tag{4.1-54}$$

These values are very sound since, by theoretical investigations, one predicts a $T^{-0.5}$ dependence [4.61] in v^{sat} which is quite close to (4.1-54).
A different expression for the saturation velocity depending on temperature has been suggested in [4.58].

$$v_n^{\text{sat}} = \frac{2.4 \cdot 10^7 \frac{\text{cm}}{\text{s}}}{1 + 0.8 \cdot \exp\left(\frac{T}{600\,\text{K}}\right)} \tag{4.1-55}$$

Several other formulae can be found in the literature for modeling the influence of velocity saturation upon mobility. In [4.6], [4.59], [4.60] expression (4.1-56) has been suggested for electron mobility in silicon at 300 K temperature.

$$\mu_n^{LIE} = \frac{\mu_n^{LI}}{\dfrac{1}{2} + \dfrac{1}{2} \cdot \sqrt{1 + \left(\dfrac{2 \cdot \mu_n^{LI} \cdot E_n}{v_n^{sat}}\right)^2}} \tag{4.1-56}$$

$$v_n^{sat} = \begin{cases} 1.05 \cdot 10^7 \dfrac{\mathrm{V}}{\mathrm{cm}} & [4.6] \\ 1.18 \cdot 10^7 \dfrac{\mathrm{V}}{\mathrm{cm}} & [4.59], [4.60] \end{cases} \tag{4.1-57}$$

The associated saturation velocity (4.1-57) is properly attained by (4.1-56). In [4.93], quite a similar expression has been used:

$$\mu_n^{LIE} = \frac{\mu_n^{LI}}{\dfrac{1}{2} + \dfrac{1}{2} \cdot \sqrt{1 + \dfrac{2 \cdot \mu_n^{LI} \cdot E_n}{v_n^{sat}}}} \tag{4.1-58}$$

However, this equation, although it makes use of a parameter v_n^{sat}, does not attain a saturation velocity; from my point of view this equation does not make any sense at all.

In [4.97] an equation has been proposed which should take care of warm and hot carriers properly:

$$\mu_n^{LIE} = \frac{\mu_n^{LI}}{\dfrac{1}{2} + \dfrac{1}{2} \cdot \sqrt{1 + \dfrac{\mu_n^{LI} \cdot E_n}{2 \cdot v_o} + 2 \cdot \left(\dfrac{\mu_n^{LI} \cdot E_n}{v_o}\right)^2}} \tag{4.1-59}$$

It is interesting to note that, although unfortunately not stated in [4.97], the associated saturation velocity $v_n^{sat} = v_o \cdot \sqrt{2}$.

Fig. 4.1-7 and Fig. 4.1-8 show the mobility versus driving force E_n and E_p in silicon at 300 K temperature for electrons and holes, respectively. A lattice mobility of $1430\,\mathrm{cm^2\,V^{-1}\,s^{-1}}$ for electrons and $480\,\mathrm{cm^2\,V^{-1}\,s^{-1}}$ for holes has been assumed. Impurity scattering and carrier-carrier scattering has been assumed to be negligibly small. The solid line denotes the model of Scharfetter and Gummel (4.1-51); the dashed line corresponds to the model (4.1-48) with parameters of Caughey and Thomas [4.21]; and the dot-dashed line corresponds to the model (4.1-45) with parameters of Canali et al. [4.19].

For the GaAs the influence of velocity saturation is most frequently modeled with the following expression:

$$\mu_n^{LIE} = \frac{\mu_n^{LI} + v_n^{sat} \cdot \dfrac{(E_n)^3}{(E_n^{crit})^4}}{1 + \left(\dfrac{E_n}{E_n^{crit}}\right)^4} \tag{4.1-60}$$

Numerical values for the parameters of (4.1-60) at 300 K temperature have been compiled from literature data in Table 4.1-6.

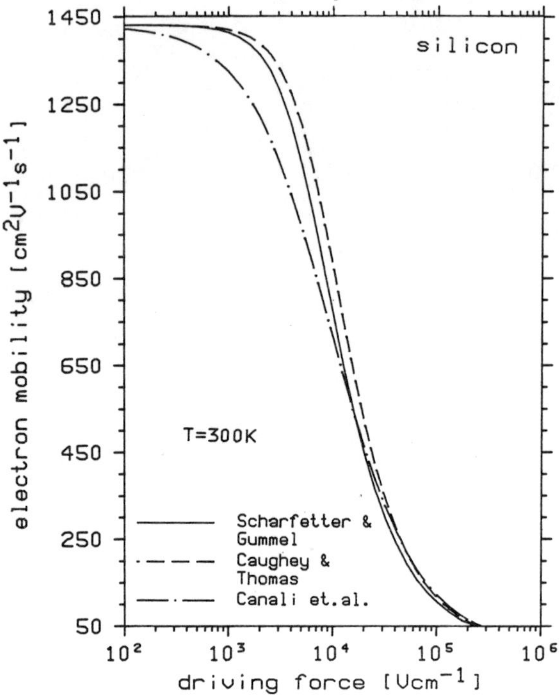

Fig. 4.1-7. Mobility of electrons due to velocity saturation in silicon at 300 K temperature

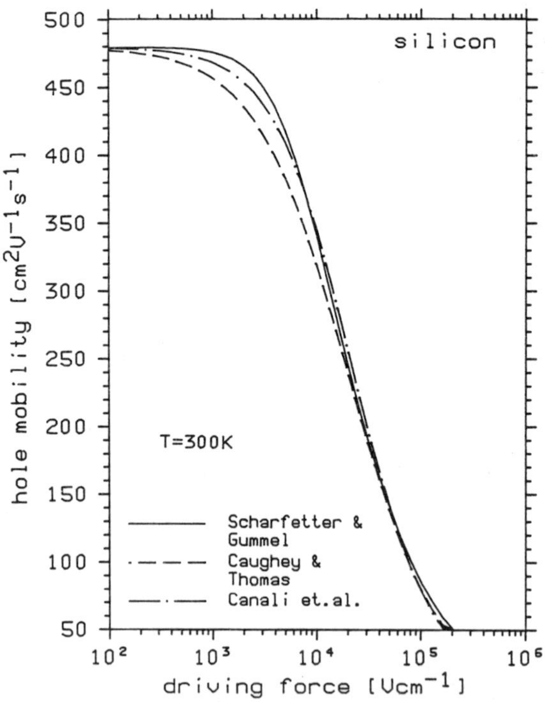

Fig. 4.1-8. Mobility of holes due to velocity saturation in silicon at 300 K temperature

Table 4.1-6. *Coefficients for velocity saturation of* (4.1-60)

v_n^{sat} [$cm\,s^{-1}$]	E_n^{crit} [$V\,cm^{-1}$]	reference
$8,5 \cdot 10^6$	$4.0 \cdot 10^3$	[4.13]
$1.0 \cdot 10^7$	$4.0 \cdot 10^3$	[4.32]
$1.0 \cdot 10^7$	$2.691 \cdot 10^3$	[4.66]
$1.12 \cdot 10^7$	$5.7 \cdot 10^3$	[4.67]
$7.5 \cdot 10^6$	$4.0 \cdot 10^3$	[4.80]

Alley [4.6] has used a slightly different relation.

$$\mu_n^{LIE} = \begin{cases} E_n < 11027\,\dfrac{V}{cm} & \mu_n^{LI} \cdot \dfrac{8 + 0.1 \cdot \left(\dfrac{E_n}{3600\,\dfrac{V}{cm}} \right)^5}{8 + \left(\dfrac{E_n}{3600\,\dfrac{V}{cm}} \right)^5} \\ E_n \geq 11027\,\dfrac{V}{cm} & \mu_n^{LI} \cdot \dfrac{1388\,\dfrac{V}{cm}}{E_n} \end{cases} \tag{4.1-61}$$

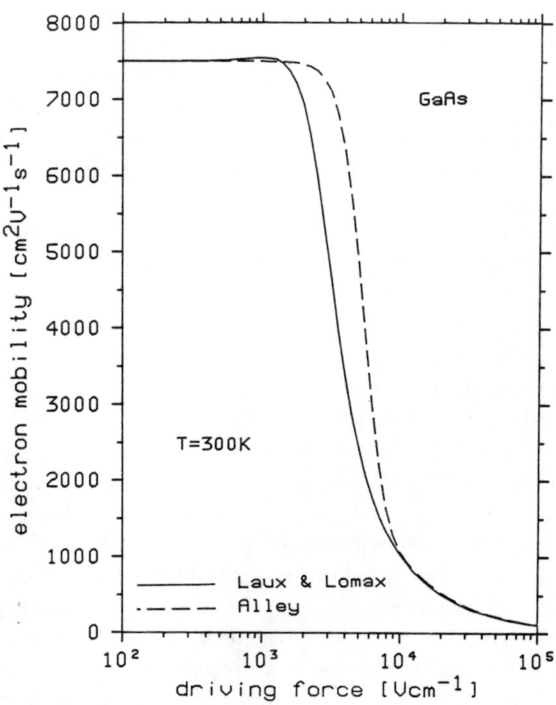

Fig. 4.1-9. Mobility of electrons due to velocity saturation in gallium-arsenide at 300 K temperature

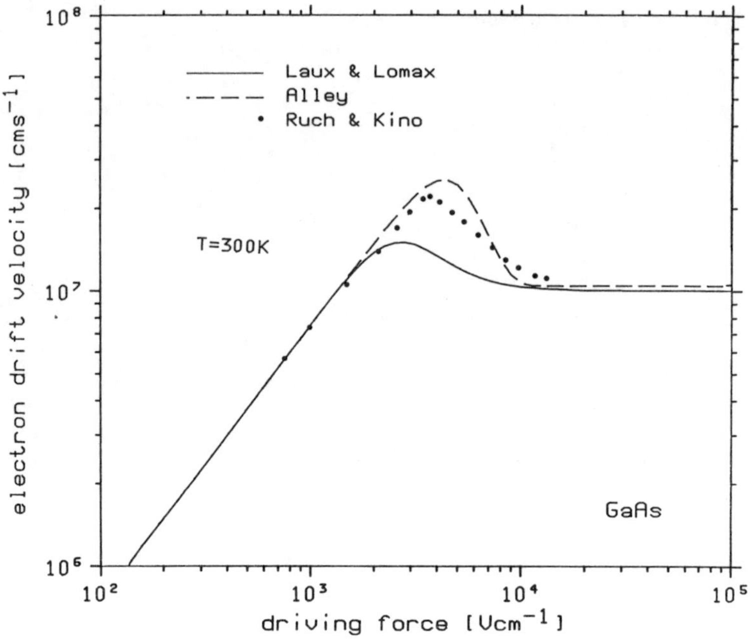

Fig. 4.1-10. Drift velocity in gallium-arsenide at 300 K temperature

Fig. 4.1-9 compares the model (4.1-60) with parameters of Laux and Lomax [4.66] (solid line) with the model (4.1-61) (dashed line). A zero field mobility of $7500\,cm^2\,V^{-1}\,s^{-1}$ has been assumed. Fig. 4.1-10 shows the drift velocity for the same data together with experimental values by Ruch and Kino [4.98], [4.99]. These experimental results have been verified with Monte Carlo calculations in [4.100].

Particularly for compound semiconductors like GaAs such heuristic models can only qualitatively reflect the underlying physics. It should not be expected that these models enable a satisfying prediction of device performance for many cases. As pointed out in Section 2.3 the models for the current relations will have to be refined in order to more accurately simulate miniaturized GaAs devices.

The last scattering mechanism we shall discuss here is termed surface scattering. This effect is of obvious fundamental importance in all devices where current flow takes place primarily along a surface of the devices. The most prominent devices of that category are the MOS transistors. Theoretically, surface scattering is comprised of a good many different mechanisms like surface roughness scattering, scattering by interface charges, scattering by surface phonons and various quantum effects. Although the application of MOS structures has received a great deal of attention in recent years, the problems associated with conduction at surfaces have not been investigated as deeply as one would expect. Many physics oriented investigations are carried out at low temperatures because the results can be interpreted much more easily. Therefore, all models which are presently used have been constructed on a fully empirical basis with a scope to reflect the main experimental findings as well as possible.

One of the earliest models has been suggested by Yamaguchi [4.138]. He has used the formula of Scharfetter and Gummel (4.1-51) for impurity scattering and mobility reduction due to velocity saturation and an additional factor to reduce the mobility in the presence of an electric field component perpendicular to the current flow direction.

$$\mu_{n,p}^{LIES} = \mu_{n,p}^{LIE} \cdot \frac{1}{\sqrt{1 + \frac{|\vec{E} \times \vec{J}_{n,p}|}{E_{n,p}^{crit} \cdot |\vec{J}_{n,p}|}}} \tag{4.1-62}$$

$$E_n^{crit} = 6.49 \cdot 10^4 \frac{\mathrm{V}}{\mathrm{cm}} \qquad E_p^{crit} = 1.87 \cdot 10^4 \frac{\mathrm{V}}{\mathrm{cm}} \tag{4.1-63}$$

This model has been used and recommended by many others, too, e.g. [4.33], [4.40], [4.41], [4.83], because it has been claimed that excellent agreement with experimental results is obtained. However, Thornber [4.128] has strongly criticized Yamaguchi's treatment using theoretical arguments. The saturation velocity $v_{n,p}^{sat}$ associated with $\mu_{n,p}^{LIE}$ is scaled with the same factor, obviously, as the mobility. Sabnis and Clemens [4.101] have experimentally proved that surface scattering is almost independent of the doping concentration. Cooper and Nelson [4.30] have shown with elaborate measurements that the influence of surface fields on the saturation velocity is relatively small, which is in contrast to former opinions (cf. [4.58]) but which is quite believable considering their experiments. Very careful measurements on that subject have been published in [4.120], too. Thornber [4.128] has made a suggestion, which is well accepted today, to use a relation of the form (4.1-64) for the total effective mobility μ^*.

$$\mu^* = \mu^* (\mu^{LICNS} (\mu^{LICN}, E_\perp), E_{||}, V^{sat}(E_\perp)) \tag{4.1-64}$$

The function μ^{LICNS} combines the mobility due to lattice, ionized impurity, carrier-carrier and neutral impurity scattering with the influence of surface scattering ($E_\perp$ denotes the field component responsible for surface scattering) to a cold carrier mobility, which is then combined with the driving force (here denoted by $E_{||}$) and the saturation velocity which might be a function of $E_\perp$ to the total effective mobility μ^*. μ^{LICNS} may have the functional form of (4.1-62) as long as good agreement between simulated and measured results is obtained. μ^* should be a function of the type (4.1-48) or (4.1-53). Yamaguchi, as a matter of fact, has revised his model in a later paper [4.139] by exactly following the suggestions of Thornber.

I have suggested the following expression for the influence of surface scattering [4.109], [4.110].

$$\mu_{n,p}^{LICNS} = \mu_{n,p}^{LICN} \cdot \frac{x + x_{n,p}^{ref}}{x + b_{n,p} \cdot x_{n,p}^{ref}} \tag{4.1-65}$$

x denotes the distance perpendicular to the interface. Directly at the interface ($x=0$) the mobility is reduced by a factor $1/b$; at a distance $x = x^{ref}$ it is reduced by the factor

$2/(1+b)$; and at greater distance from the surface it naturally follows that the reduction factor approaches unity. x^{ref} represents a characteristic length which describes the range of influence of the surface.

$$x_{n,p}^{ref} = \frac{x_{n,p}^{o}}{1 + \frac{E_{n,p}}{E_{n,p}^{crit}}} \tag{4.1-66}$$

$$x_n^o = 5 \cdot 10^{-7}\,\text{cm}, \quad E_n^{crit} = 10^4 \frac{\text{V}}{\text{cm}} \tag{4.1-67}$$

$$x_p^o = 4 \cdot 10^{-7}\,\text{cm}, \quad E_p^{crit} = 8 \cdot 10^3 \frac{\text{V}}{\text{cm}}$$

This range is modeled as a function of the carrier driving force (field component parallel to current flow or magnitude of the gradient of the corresponding quasi-Fermi potential). The formulation of x^{ref} produces a reduction in the range of influence of surface scattering for greater driving forces, thereby velocity saturation appears. Carriers already traveling with the saturation velocity can be considered not to experience the influence of the surface as much as cold carriers [4.71]. The parameter b in (4.1-63) describes the strength of the influence of surface scattering.

$$b_{n,p} = 2 + \frac{E_\perp}{E_{\perp n,p}^{crit}} \tag{4.1-68}$$

$$E_{\perp n}^{crit} = 1.8 \cdot 10^5 \frac{\text{V}}{\text{cm}}, \quad E_{\perp p}^{crit} = 3.8 \cdot 10^5 \frac{\text{V}}{\text{cm}} \tag{4.1-69}$$

It is modeled as a function of $E_\perp$ which can be the electric field component perpendicular to current flow, or the electric field component perpendicular to the interface, or, what I suggest, the projection of the electric field component perpendicular to the current flow direction onto the direction perpendicular to the surface. The formulation of b rests upon the consideration that the charge carriers are pressed against the surface by an electric field, which results in a greater scattering, in such a way that a greater mobility reduction occurs. Without any electric field we also observe a mobility reduction due to surface roughness scattering ($b=2$). However, I am absolutely aware that (4.1-63) is a fully phenomenological expression neither of which the structure nor the associated parameters may be claimed to be correct in a theoretical sense. It simply represents my experience which has been confirmed over several years by many users of our simulation tools that an expression with such a structure nicely reflects the experimental observations.

There are plenty more suggestions on how to treat surface scattering phenomenologically. The interested reader could have a look at, e.g. [4.4], [4.11], [4.25], [4.42], [4.120], [4.133].

4.2 Carrier Generation-Recombination Modeling

In Section 2.2 we have introduced in a very formal manner a quantity R which has been interpreted as a function describing the balance of generation and recombination of electrons and holes. This section is now devoted to the discussion of the physical phenomena which have to be considered to derive models for R. The various physical mechanisms responsible for generation/recombination will be phenomenologically described, and their expected contribution to R will be indicated.

Let us assume a homogeneously doped semiconductor which is in thermal equilibrium with its ambient. Due to the thermal energy the concentration of electrons and holes will continously fluctuate because of generative and recombinative processes. However, as the semiconductor is in equilibrium, there will be a dynamic balance between the generation and recombination rates, which leads to an equilibrium concentration n_0 of electrons and p_o of holes. These concentrations are related by:

$$n_o \cdot p_o = n_i^2 \tag{4.2-1}$$

n_i denotes the intrinsic concentration which we have already discussed in Section 2.4. (4.2-1) is only appropriate for non-degenerate semiconductors; for moderate degeneracy, however, n_i can be replaced by n_{ie}, an effective intrinsic concentration (cf. Section 2.4). When the semiconductor is excited by some external stimulus, the balance between generation and recombination is disturbed as the electron and hole concentrations depart from their equilibrium values n_o and p_o. If excess carriers have been generated, recombination will prevail, whereas, if carriers have been removed, generation will dominate, so that a steady state situation between generation/recombination and the external stimulus is established.

Generation/recombination phenomena can be seen from two different points of view: either from the energy levels between which the various mechanisms take place or, directly, from the underlying physical effect. Viewed in terms of energy levels, generation/recombination may take place either in one step, which is termed direct generation/recombination or in two or more steps which is called indirect generation/recombination. In consideration of the physical mechanisms we divide generation/recombination into primarily phonon transitions, photon transitions, Auger or three particle transitions and impact ionization. In principle several more mechanisms like transitions caused by plasma oscillations, excitons and spin waves do exist, but these are usually not so important as the ones cited above [4.77].

Phonon transitions take place primarily in two steps by way of defects (traps). A theory of this effect has been established by Shockley and Read [4.115] and Hall [4.51]. Therefore, the mechanism is most frequently termed Shockley-Read-Hall generation/recombination. In detail four partial processes are involved.

SRH.a) electron capture: an electron from the conduction band is trapped by an unoccupied defect which becomes occupied.

SRH.b) hole capture: an electron from an occupied trap moves to the valence band and neutralizes a hole. The trap becomes unoccupied.

SRH.c) hole emission: an electron from the valence band is trapped by a defect, thus leaving a hole in the valence band and an occupied trap.

SRH.d) electron emission: an electron from an occupied trap moves to the conduction band. The trap becomes unoccupied.

By assuming that process *SRH*.a) and *SRH*.b) take place sequentially, an electron-hole pair will only recombine with the help of a trap which is occupied during the processes. Analogously, by assuming that process *SRH*.c) and *SRH*.d) take place sequentially, an electron-hole pair is only generated with the help of a trap which, again, is occupied during the processes. We nicely see that the overall generation/recombination process takes place in two steps with the aid of traps.
The traps can be characterized as follows. They are defects with an energy level E_t, a concentration N_t and capture cross sections κ_n and κ_p for electrons and holes, respectively. A trap is most effective for generation/recombination if its energy level E_t is in about the middle of the gap between valence band and conduction band. This is easy to imagine by remembering the involved partial processes. To generate an electron-hole pair, thermal energy is first consumed to move an electron from the valence band to the trap (process *SRH*.c)) and energy is again required to move the electron from the trap to the conduction band (process *SRH*.d)). The total energy required for such a process is obviously independent of the energetic location of the trap, which is the width of the band. However, the maximum energy of the partial processes is obviously a minimum if the trap level is exactly in the middle of the band. We can, therefore, deduce that impurities for the purpose of doping are very ineffective as generation/recombination centers, because these are located energetically close to one of the band edges in order to be effective as doping centers. These impurities are frequently called "shallow" impurities whereas impurities which are put into a semiconductor to increase the recombination rate (like gold in power devices) are termed "deep" impurities. Some deep impurities lead to multiple energy levels but one of these levels dominates the carrier generation/recombination in most cases [4.92].
With the above given phenomenological description of the generation/recombination mechanism it is rather straightforward to derive for single level traps an expression for the purpose of simulation. Following the ideas of Shockley et al. we assume the rates C_{cn}^{SRH}, C_{en}^{SRH}, C_{cp}^{SRH} and C_{ep}^{SRH}. C_{cn}^{SRH}, C_{cp}^{SRH} are the capture rates for electrons and holes per electron and hole, respectively, when all traps are unoccupied. C_{en}^{SRH}, C_{ep}^{SRH} are the emission rates for electrons and holes per electron and hole, respectively. Let further f_t denote the fraction of traps which is occupied. Then we may write the capture rates per unit volume for electrons and holes as:

$$R_n^{SRH} = C_{cn}^{SRH} \cdot n \cdot (1 - f_t) \tag{4.2-2}$$

$$R_p^{SRH} = C_{cp}^{SRH} \cdot p \cdot f_t \tag{4.2-3}$$

Analogously we may write emission rates per unit volume for electrons and holes as:

$$G_n^{SRH} = C_{en}^{SRH} \cdot f_t \tag{4.2-4}$$

$$G_p^{SRH} = C_{ep}^{SRH} \cdot (1 - f_t) \tag{4.2-5}$$

The total generation/recombination rate evaluates to (4.2-6) since electrons and holes always act in pairs, and, thus, their net generation/recombination rate must be identical. This statement will not hold for transient situations where the carrier concentrations change rapidly [4.26]. However, as there are no results available on that subject which could be used for device modeling, we have to leave that problem open.

$$R^{SRH} = R_n^{SRH} - G_n^{SRH} = R_p^{SRH} - G_p^{SRH} \tag{4.2-6}$$

In thermal equilibrium where we have no net generation/recombination it follows that the capture rates of electrons and holes must be equal to the corresponding emission rates, which enables us to calculate, for instance, the emission rates:

$$C_{en}^{SRH} = C_{cn}^{SRH} \cdot n_o \cdot \frac{1-f_{to}}{f_{to}} \tag{4.2-7}$$

$$C_{ep}^{SRH} = C_{cp}^{SRH} \cdot p_o \cdot \frac{f_{to}}{1-f_{to}} \tag{4.2-8}$$

Index o indicates equilibrium quantities. It is very convenient to define concentrations n_1 and p_1:

$$n_1 = n_o \cdot \frac{1-f_{to}}{f_{to}} \tag{4.2-9}$$

$$p_1 = p_o \cdot \frac{f_{to}}{1-f_{to}} \tag{4.2-10}$$

With these definitions the net generation/recombination rates become:

$$R_n^{SRH} - G_n^{SRH} = C_{cn}^{SRH} \cdot (n \cdot (1-f_t) - n_1 \cdot f_t) \tag{4.2-11}$$

$$R_p^{SRH} - G_p^{SRH} = C_{cp}^{SRH} \cdot (p \cdot f_t - p_1 \cdot (1-f_t)) \tag{4.2-12}$$

These two rates are equal so that we can calculate the fraction of occupied traps f_t.

$$f_t = \frac{n \cdot C_{cn}^{SRH} + p_1 \cdot C_{cp}^{SRH}}{C_{cn}^{SRH} \cdot (n+n_1) + C_{cp}^{SRH} \cdot (p+p_1)} \tag{4.2-13}$$

The net generation/recombination rate R^{SRH} is therefore given by the following equation:

$$R^{SRH} = \frac{n \cdot p - n_i^2}{\tau_p \cdot (n+n_1) + \tau_n \cdot (p+p_1)} \tag{4.2-14}$$

with:

$$\tau_p = \frac{1}{C_{cp}^{SRH}} \tag{4.2-15}$$

$$\tau_n = \frac{1}{C_{cn}^{SRH}} \tag{4.2-16}$$

It is very common to define carrier lifetimes τ_n and τ_p as reciprocals of the corresponding capture rates per single carrier. The capture rates can be assumed quite generally to be expressed as:

$$C_{cn}^{SRH} = \kappa_n \cdot v_{th} \cdot N_t \tag{4.2-17}$$

$$C_{cp}^{SRH} = \kappa_p \cdot v_{th} \cdot N_t \tag{4.2-18}$$

κ_n and κ_p denote, as already mentioned, the capture cross sections for electrons and holes; v_{th} is the thermal velocity; and N_t is the concentration of traps. As long as the trap concentration N_t remains independent of doping, the lifetimes τ_n, τ_p do not vary with doping. However, at high doping concentrations additional generation/recombination centers can be created. In the literature one can find empirical expressions of the following type for the doping dependence of the lifetimes, which are claimed to fit experimental findings.

$$\tau_n = \frac{\tau_{no}}{1 + \dfrac{N_D + N_A}{N_n^{\text{ref}}}} \tag{4.2-19}$$

$$\tau_p = \frac{\tau_{po}}{1 + \dfrac{N_D + N_A}{N_p^{\text{ref}}}} \tag{4.2-20}$$

Numerical values for the parameters τ_{no}, τ_{po} and N_n^{ref}, N_p^{ref} are compiled from literature data in Table 4.2-1.

Table 4.2-1. *Coefficients for* (4.2-19), (4.2-20)

τ_{no} [s]	N_n^{ref} [cm^{-3}]	τ_{po} [s]	N_p^{ref} [cm^{-3}]	reference
$5.0 \cdot 10^{-5}$	$5.0 \cdot 10^{16}$	$5.0 \cdot 10^{-5}$	$5.0 \cdot 10^{16}$	[4.33]
$3.94 \cdot 10^{-4}$	$7.1 \cdot 10^{15}$	$3.94 \cdot 10^{-5}$	$7.1 \cdot 10^{15}$	[4.36]
$3.95 \cdot 10^{-4}$	$7.1 \cdot 10^{15}$	$3.52 \cdot 10^{-5}$	$7.1 \cdot 10^{15}$	[4.43], [4.136]
$4.0 \cdot 10^{-4}$	$7.1 \cdot 10^{15}$			[4.44]
$1.0 \cdot 10^{-5}$	$3.0 \cdot 10^{17}$	$1.0 \cdot 10^{-5}$	$3.0 \cdot 10^{17}$	[4.78]

Similar expressions for the doping dependence of the lifetimes like:

$$\tau_{n,p} = \tau_{n,p}^{o} \cdot \left(\frac{N_{n,p}^{\text{ref}}}{N_D + N_A}\right)^{\alpha_{n,p}}, \quad 0.3 < \alpha_{n,p} < 0.6 \tag{4.2-21}$$

can also be found in the literature, e.g. [4.1], [4.41].
In the derivation of (4.2-14) it has been assumed that the number of available traps is much larger than the number of carriers involved in a generation/recombination process. Furthermore, the time of readjustment of an electron in a trap once it is trapped has been assumed to be negligibly small. Dhariwal, Kothari and Jain [4.37] have investigated these problems and they have obtained a very nice result.

$$R^{DKJ} = \frac{n \cdot p - n_i^2}{\tau_p^* \cdot (a \cdot n + b \cdot n_1) + \tau_n^* \cdot (b \cdot p + a \cdot p_1) + c \cdot (n \cdot p - n_i^2)} \tag{4.2-22}$$

with:

$$a = 1 + \frac{\delta t_n}{\delta t'_n}$$

$$b = 1 + \frac{\delta t_p}{\delta t'_p} \quad (4.2\text{-}23)$$

$$c = \frac{\delta t_n + \delta t_p}{N_t}$$

I shall not review in detail the derivation of (4.2-22), but we shall perform a qualitative discussion. δt_n, δt_p denote the times of transition for an excited electron in a state close to the conduction, valence band to the trap level. $\delta t'_n$, $\delta t'_p$ are the times for transition in the reverse direction. If the transitions are infinitely fast, we obtain $a = b = 1$ and $c = 0$ (cf. [4.37]) and (4.2-22) is identical to (4.2-14), the original Shockley-Read-Hall formula. τ_p^*, τ_n^* are then also equivalent to (4.2-15), (4.2-16). The general formulae for the lifetimes τ_p^*, τ_n^* are more complex; however, as one preferably uses experimental data for calibration due to the uncertainty in the values of various "theoretical" quantities involved in the formulae, a discussion will be skipped. The most attractive feature of (4.2-22) compared to (4.2-14) is the saturation of recombination for large carrier concentrations. Such a saturation will, obviously, happen; for instance, it can play a role in devices where breakdown is retarded by extraordinarily strong recombination [4.105]. However, I am not aware of any simulations which have corroborated this effect.

The next physical mechanisms we have to consider for generation/recombination are photon transitions. This mechanism takes place primarily in one step; it is thus a direct generation/recombination mechanism. There are two partial processes involved.

OPT.a) an electron loses energy on the order of the band gap, which is emitted as a photon, and moves from the conduction band to the valence band (radiative recombination).

OPT.b) an electron gains energy from incident photons and moves from the valence band to the conduction band (optical generation).

This effect is important for narrow gap semiconductors and semiconductors whose specific band structure allows direct transitions like GaAs [4.57]. In silicon and germanium band to band generation/recombination is insignificant for all imaginable conditions. An expression for modeling is easy to derive. By assuming a capture rate C_c^{OPT} and an emission rate C_e^{OPT}, the involved partial processes can be written:

$$R_{np}^{OPT} = C_c^{OPT} \cdot n \cdot p \quad (4.2\text{-}24)$$

$$G_{np}^{OPT} = C_e^{OPT} \quad (4.2\text{-}25)$$

These rates must be equal in thermal equilibrium so that:

$$C_e^{OPT} = C_c^{OPT} \cdot n_i^2 \quad (4.2\text{-}26)$$

The total band to band generation/recombination is the difference of the partial rates, which evaluates to:

$$R^{OPT} = C_c^{OPT} \cdot (n \cdot p - n_i^2) \quad (4.2\text{-}27)$$

Some guidelines on how to evaluate the capture rate C_c^{OPT} can be found in, e.g., [4.117].
The next physical mechanisms for generation/recombination we shall consider here are Auger or three particle transitions. The partial processes involved in Auger recombination are still a matter of investigation (cf. [4.45], [4.64], [4.131]) and only more or less qualitatively understood. However, qualitatively the partial processes will take place about as follows.

AU.a) electron capture: an electron from the conduction band moves to the valence band, transmitting the excess energy to another electron in the conduction band. In the valence band the electron recombines with a hole.

AU.b) hole capture: an electron from the conduction band moves to the valence band transmitting the excess energy to a hole in the valence band, which moves away from the valence band edge. The electron recombines with a hole.

AU.c) electron emission: an electron from the valence band moves to the conduction band by consuming the energy of a high energetic electron in the conduction band and leaving a hole in the valence band.

AU.d) hole emission: an electron from the valence band moves to the conduction band by consuming the energy of a high energetic hole in the valence band. A hole is left at the valence band edge.

In any of these partial processes three carriers are involved. In process AU.a) and AU.c) these are two electrons and one hole; in process AU.b) and AU.d) these are one electron and two holes. However, these processes describe only the direct band to band Auger generation/recombination. More recent investigations [4.107] indicate that trap assisted partial processes have a higher probability than direct band to band processes. This changes the partial processes AU.a) to AU.d) such that the carriers move from one of the bands to a trap instead to the other band. As additional partial processes the Shockley-Read-Hall transitions which we have formerly discussed will interfere. A fully consistent treatment of the interaction of trap assisted Auger generation/recombination and Shockley-Read-Hall generation/recombination has been carried out in [4.45]. The influence of a specific band structure of a semiconductor upon Auger generation/recombination has been reviewed in [4.27].
The partial processes AU.c) and AU.d) are referred to in many textbooks as impact ionization which is stated to be the antagonism of Auger recombination. However, this is wrong or at least enormously inexact. To clarify the situation we have to consider the rates per unit volume of the partial processes AU.a) to AU.d).

$$R_n^{AU} = C_{cn}^{AU} \cdot n^2 \cdot p \tag{4.2-28}$$

$$R_p^{AU} = C_{cp}^{AU} \cdot n \cdot p^2 \tag{4.2-29}$$

$$G_n^{AU} = C_{en}^{AU} \cdot n \tag{4.2-30}$$

$$G_p^{AU} = C_{ep}^{AU} \cdot p \tag{4.2-31}$$

Rate R_n^{AU} corresponding to the partial process AU.a) is proportional to the square of the electron concentration times the hole concentration with the Auger capture coefficient for electrons. Similarly R_p^{AU} corresponding to partial processes AU.b) is proportional to the electron concentration times the square of the hole concentration with the Auger capture coefficient for holes. Rate G_n corresponding to partial process AU.c) is proportional only to the electron concentration with the Auger emission coefficient for electrons. Here is a big difference to impact ionization, the corresponding generation rate of which would be proportional to the electron current density instead of the electron concentration. That means Auger generation may take place in regions with a high concentration of mobile carriers with negligible current flow, whereas impact ionization requires non-negligible current flow as a prerequisite. Some practical implications of that statement on the operation of MOS devices have been given in [4.71]. Finally, rate G_p^{AU} corresponding to partial process AU.d) is proportional only to the hole concentration with the Auger emission coefficient for holes. The same statement of comparison between Auger generation of holes and impact ionization of holes is valid as given above for electrons. However, I have to admit that viewed microscopically there is no difference between Auger generation and impact ionization. The enormous difference lies in the source of energy for the partial processes as I have outlined above.

In thermal equilibrium no generation/recombination exists so that we can calculate the Auger emission rate by equating (4.2-28) to (4.2-30) and (4.2-29) to (4.2-31).

$$C_{en}^{AU} = C_{cn}^{AU} \cdot n_i^2 \tag{4.2-32}$$

$$C_{ep}^{AU} = C_{cp}^{AU} \cdot n_i^2 \tag{4.2-33}$$

The total net Auger generation/recombination rate is the sum of the net rates for electrons and holes.

$$R^{AU} = R_n^{AU} - G_n^{AU} + R_p^{AU} - G_p^{AU} \tag{4.2-34}$$

If we substitute the partial rates (4.2-28) to (4.2-31) and the Auger emission coefficients (4.2-32), (4.2-33) into (4.2-34) we obtain:

$$R^{AU} = (C_{cn}^{AU} \cdot n + C_{cp}^{AU} \cdot p) \cdot (n \cdot p - n_i^2) \tag{4.2-35}$$

The numerical values of the Auger capture coefficients C_{cn}^{AU} and C_{cp}^{AU} show a remarkable scatter in the literature. Quite frequently used in modeling programs are the data of Dziewior and Schmid [4.39] which are summarized in Table 4.2-2.

Table 4.2-2. *Auger coefficients in silicon*

temperature [K]	C_{cn}^{AU} [$cm^6 s^{-1}$]	C_{cp}^{AU} [$cm^6 s^{-1}$]
77	$2.3 \cdot 10^{-31}$	$7.8 \cdot 10^{-32}$
300	$2.8 \cdot 10^{-31}$	$9.9 \cdot 10^{-32}$
400	$2.8 \cdot 10^{-31}$	$1.2 \cdot 10^{-31}$

The temperature dependence of the Auger coefficients is fairly weak as we can deduce from Table 4.2-2. An extensive collection of Auger coefficients from literature data can be found in [4.131]. The dependence on doping concentration is speculated also to be weak.

As excellent reviews about the present understanding of the generation/recombination mechanisms we have treated so far, [4.45], [4.52], [4.131] can be recommended.

In addition to generation/recombination in the bulk of a semiconductor, electrons and holes may also be generated/recombined at surfaces. The rate of surface generation/recombination can even be much greater under some conditions than the bulk generation/recombination rate. For the purpose of modeling one usually assumes a formula which is structurally equivalent to the Shockley-Read-Hall expression for bulk generation/recombination.

$$R^{SURF} = \frac{n \cdot p - n_i^2}{\frac{1}{s_p} \cdot (n + n_1) + \frac{1}{s_n} \cdot (p + p_1)} \cdot \delta(\vec{x}) \tag{4.2-36}$$

s_p and s_n denote the surface recombination velocities for electrons and holes, respectively. Their numerical values are on the order of $100\,\mathrm{cm}^2/\mathrm{s}$. It is more established to use velocities instead of lifetimes for surface generation/recombination. $\delta(\vec{x})$ is the Dirac delta function and $\vec{x} = 0$ denotes the surface. This means surface generation/recombination is only existing exactly at the surface. The transition between surface generation/recombination and bulk generation/recombination has not been implemented in any modeling program, as far as I know. However, theoretically it is not fully clear what one should expect; some considerations are given in [4.90], [4.117]. My suggestion, which a few others agree to, e.g. [4.65], is that one should not treat generation/recombination phenomena at surfaces differently than in bulk material, but one should care more thoroughly about the various generation/recombination parameters at surfaces. (4.2-36) obviously reflects an indirect generation/recombination mechanism with just a different interpretation of the capture rates (cf. (4.2-15) to (4.2-18)) in comparison with (4.2-14).

The last physical mechanism we shall discuss in the context of generation/recombination modeling is impact ionization. This effect is a pure generation process which, as already mentioned, is seen to be microscopically identical to the Auger generation process. Two partial processes have to be considered.

II.a) electron emission: an electron from the valence band moves to the conduction band by consuming the energy of a high energetic electron in the conduction band and leaving a hole in the valence band.

II.b) hole emission: an electron from the valence band moves to the conduction band consuming the energy of a high energetic hole in the valence band. A hole is left at the valence band edge.

These partial rates can be written:

$$G_n^{II} = \alpha_n \cdot \frac{|\vec{J}_n|}{q} \tag{4.2-37}$$

$$G_p^{II} = \alpha_p \cdot \frac{|\vec{J}_p|}{q} \tag{4.2-38}$$

The total rate is simply the sum.

$$R^{II} = -G_n^{II} - G_p^{II} \tag{4.2-39}$$

α_n and α_p are the ionization rates for electrons and holes defined as generated electron-hole pairs per unit length of travel and per electron and hole, respectively. For instance, an electron generates over a distance $1/\alpha_n$ one electron-hole pair on average. The energy which is consumed from the ionizing carrier is termed ionization energy or threshold energy for ionization E_i. This quantity has been a topic of many investigations in the past years with absolutely non-unique results (cf. summary in Table 4.2-5, Table 4.2-6). Similarly, the theoretical results for the ionization rates α_n, α_p are not unique. However, both theoretical and experimental investigations indicate a good approximation to be an exponential dependence of the ionization rates upon the electric field component E in direction of current flow.

$$\alpha_n = \alpha_n^\infty \cdot \exp\left(-\left(\frac{E_n^{\mathrm{crit}}}{E}\right)^{\beta_n}\right) \tag{4.2-40}$$

$$\alpha_p = \alpha_p^\infty \cdot \exp\left(-\left(\frac{E_p^{\mathrm{crit}}}{E}\right)^{\beta_p}\right) \tag{4.2-41}$$

The exponents β_n, β_p are found in the range [1, 2]. As a matter of fact fairly early theoretical considerations by Shockley [4.116] predict the exponents to be one, which has also been the very old experimental finding by Chynoweth [4.24]. A different treatment by Wolff [4.137] predicts the exponents to be two. Numerical values for the coefficients of (4.2-40) and (4.2-41) compiled from literature data are summarized in Table 4.2-3 for electrons and in Table 4.2-4 for holes.

Table 4.2-3. *Constants for impact ionization of electrons*

material	α_n^∞ [cm^{-1}]	E_n^{crit} [$\mathrm{V\,cm}^{-1}$]	β_n [ ]	reference
Si	$1.0 \cdot 10^6$	$1.66 \cdot 10^6$	1	[4.39] $1.1 \cdot 10^5$ V/cm $< E < 5 \cdot 10^5$ V/cm
	$6.2 \cdot 10^5$	$1.08 \cdot 10^6$	1	[4.49] $2.4 \cdot 10^5$ V/cm $< E < 5.3 \cdot 10^5$ V/cm
	$1{,}28 \cdot 10^6$	$2.54 \cdot 10^6$	1	[4.63] $E < 5 \cdot 10^5$ V/cm
	$1 \cdot 10^6$	$5.87 \cdot 10^6$	1	[4.63] $E > 5 \cdot 10^5$ V/cm
	$1.6 \cdot 10^6$	$1.65 \cdot 10^6$	1	[4.76] $2 \cdot 10^5$ V/cm $< E < 6.7 \cdot 10^5$ V/cm
	$3.8 \cdot 10^6$	$1.75 \cdot 10^6$	1	[4.122], [4.123]
	$2.2 \cdot 10^6$	$1.54 \cdot 10^6$	1	[4.124]
	$7 \cdot 10^5$	$1.4 \cdot 10^6$	1	[4.130]
	$7.03 \cdot 10^5$	$1.231 \cdot 10^6$	1	[4.132] $1.75 \cdot 10^5$ V/cm $< E < 6 \cdot 10^5$ V/cm
GaAs	$2.0 \cdot 10^6$	$2.0 \cdot 10^6$	1	[4.6]
	$2.994 \cdot 10^5$	$6.848 \cdot 10^5$	1.6	[4.18] $2.22 \cdot 10^5$ V/cm $< E < 6.25 \cdot 10^5$ V/cm
	$3.5 \cdot 10^5$	$6.85 \cdot 10^5$	2	[4.122]
	$1.34 \cdot 10^6$	$2.03 \cdot 10^6$	2	[4.123]
Ge	$1.55 \cdot 10^7$	$1.56 \cdot 10^5$	1	[4.118]
	$1.55 \cdot 10^7$	$1.56 \cdot 10^6$	1	[4.122], [4.123]

Table 4.2-4. *Constants for impact ionization of holes*

material	α_p^∞ [cm^{-1}]	E_p^{crit} [V cm^{-1}]	β_p [ ]	reference
Si	$2.0 \cdot 10^6$	$1.98 \cdot 10^6$	1	[4.49] $2.4 \cdot 10^5$ V/cm $< E < 5.3 \cdot 10^5$ V/cm
	$5.5 \cdot 10^5$	$1.65 \cdot 10^6$	1	[4.76] $5 \cdot 10^5$ V/cm $< E < 8 \cdot 10^5$ V/cm
	$1.0 \cdot 10^6$	$1.66 \cdot 10^6$	1	[4.82] $1.1 \cdot 10^5$ V/cm $< E < 5 \cdot 10^5$ V/cm
	$2.25 \cdot 10^7$	$3.26 \cdot 10^6$	1	[4.122], [4.123]
	$1.0 \cdot 10^6$	$2.22 \cdot 10^6$	1	[4.124]
	$1.3 \cdot 10^6$	$2.09 \cdot 10^6$	1	[4.130] $E \leq 6.07 \cdot 10^5$ V/cm
	$4.4 \cdot 10^5$	$1.4 \cdot 10^6$	1	[4.130] $E > 6.07 \cdot 10^5$ V/cm
	$1.582 \cdot 10^6$	$2.036 \cdot 10^6$	1	[4.132] $1.75 \cdot 10^5$ V/cm $< E \leq 4 \cdot 10^5$ V/cm
	$6.71 \cdot 10^5$	$1.693 \cdot 10^6$	1	[4.132] $4 \cdot 10^5$ V/cm $< E < 6 \cdot 10^5$ V/cm
GaAs	$2.215 \cdot 10^5$	$6.57 \cdot 10^5$	1.75	[4.18] $2.22 \cdot 10^5$ V/cm $< E < 6.25 \cdot 10^5$ V/cm
	$3.5 \cdot 10^5$	$6.85 \cdot 10^5$	2	[4.122]
	$1.34 \cdot 10^6$	$2.03 \cdot 10^6$	2	[4.123]
Ge	$1.0 \cdot 10^7$	$1.28 \cdot 10^6$	1	[4.118], [4.122]
	$1.0 \cdot 10^6$	$1.28 \cdot 10^6$	1	[4.123]

Investigations by Baraff [4.12] have predicted that these theories can be interpreted as the two limiting cases of a much more rigorous model. For low fields Shockley's model is more appropriate, whereas for high fields Wolff's model is asymptotically correct. Baraff's results can, unfortunately, not be given in closed form; they have been obtained by a numerical solution of the Boltzmann transport equation, however, restricted to the assumption of an unrealistic band structure. However, a universal plot for both electrons and holes has been presented, which shows

$$\alpha \cdot \lambda = f\left(\frac{E_r}{E_i}, \frac{E_i}{q \cdot \lambda \cdot E}\right) \tag{4.2-42}$$

λ is the mean free path between collisions with high energetic phonons; E_r is the average loss of energy defined per such collision; and E_i denotes the ionization energy, as already defined. Numerical values collected from miscellaneous publications are summarized in Table 4.2-5 for electrons and in Table 4.2-6 for holes. Baraff's universal curves have been approximated with compact formulae so that an application for the purpose of simulation is facilitated. Crowell and Sze [4.31] have proposed the following expression:

$$\alpha \cdot \lambda = \exp(C_o(r) + C_1(r) \cdot x + C_2(r) \cdot x^2) \tag{4.2-43}$$

with:

$$\begin{aligned} C_0(r) &= -1.92 + 75.5 \cdot r - 757 \cdot r^2 \\ C_1(r) &= 1.75 \cdot 10^{-2} - 11.9 \cdot r + 46 \cdot r^2 \\ C_2(r) &= 3.9 \cdot 10^{-4} - 1.17 \cdot r + 11.5 \cdot r^2 \end{aligned} \tag{4.2-44}$$

where:

$$r = \frac{E_r}{E_i}, \qquad x = \frac{E_i}{q \cdot \lambda \cdot E} \tag{4.2-45}$$

Table 4.2-5. *Constants for impact ionization of electrons*

material	λ [nm]	E_r [eV]	E_i [eV]	reference
Si	6	0.063	1.68	[4.5]
			1.1	[4.7]
	6.2	0.063		[4.31]
			1.4	[4.53]
	6.9	0.61	1.6	[4.68]
	7.0		1.5	[4.75]
			1.8	[4.76]
	4.8	0.051	1.1	[4.86], [4.87]
	4.8	0.053	1.1	[4.88]
	5.0	0.063	1.1	[4.116]
			1.8	[4.125]
GaAs	3.5	0.035	2.15	[4.5]
			1.7	[4.7]
	3.5	0.036		[4.31]
			1.5	[4.53]
	3.3	0.022	1.7	[4.86], [4.87], [4.88]
			2.0	[4.125]
Ge	6.5	0.037	1.01	[4.5]
			0.8	[4.7]
	6.5	0.036		[4.31]
			0.91	[4.53]
	3.6	0.019	0.8	[4.86]
	3.9	0.019	0.8	[4.87]
	3.9	0.022	0.8	[4.88]

Table 4.2-6. *Constants for impact ionization of holes*

material	λ [nm]	E_r [eV]	E_i [eV]	reference
Si			1.8	[4.7]
	3.8	0.063		[4.31]
			1.6	[4.53]
	4.4	0.61	1.6	[4.68]
	1.0		3.5	[4.75]
			2.4	[4.76]
	4.4	0.051	1.8	[4.86]
	4.7	0.051	1.8	[4.87]
	4.7	0.053	1.8	[4.88]
		0.063	1.1	[4.116]
GaAs			1.4	[4.7]
	3.5	0.036		[4.31]
			2.7	[4.53]
	3.3	0.022	1.7	[4.86], [4.87], [4.88]
Ge			0.9	[4.7]
	6.5	0.036		[4.31]
			1.3	[4.53]
	4.7	0.019	0.9	[4.86]
	5.1	0.019	0.9	[4.87]
	5.1	0.022	0.9	[4.88]

This approximation is claimed to be accurate within two percent maximum error over the range $r \in [0.01, 0.06]$ and $x \in [5, 16]$. A more accurate approximation has been given in [4.121]:

$$\alpha \cdot \lambda = \exp(C_0(r) + C_1(r) \cdot x + C_2(r) \cdot x^2 + C_3(r) \cdot x^3) \tag{4.2-46}$$

with:

$$\begin{aligned} C_0(r) &= -7.238 \cdot 10^{-2} - 51.5 \cdot r + 239.6 \cdot r^2 + 3357 \cdot r^3 \\ C_1(r) &= -0.4844 + 12.45 \cdot r + 363 \cdot r^2 - 5836 \cdot r^3 \\ C_2(r) &= 2.982 \cdot 10^{-2} - 7.571 \cdot 10^{-2} \cdot r - 148.1 \cdot r^2 + 1627 \cdot r^3 \\ C_3(r) &= -1.841 \cdot 10^{-5} - 0.1851 \cdot r + 10.41 \cdot r^2 - 95.65 \cdot r^3 \end{aligned} \tag{4.2-47}$$

This approximation fits Baraff's curves perfectly over the range $r \in [0.01, 0.07]$ and $x \in [3.14]$. The result is shown in Fig. 4.2-1. The question obviously, remains if these theoretically obtained results agree with experimental results. An answer to this question is, most unfortunately, not easy. The measured results of Lee et al. [4.94], for instance, agree nicely with Baraff's results when the numerical values for λ, E_r and E_i are taken from [4.31]. The experimental results of, e.g. Van Overstraeten et al. [4.132] and Grant [4.49], would require totally unrealistic values for λ, E_r and E_i to agree with Baraff's results. On the other hand, in, e.g. [4.104], [4.118], [4.127] the ionization rates of Van Overstraeten et al. have been used, whereby good agreement between experimental and simulation results on device breakdown

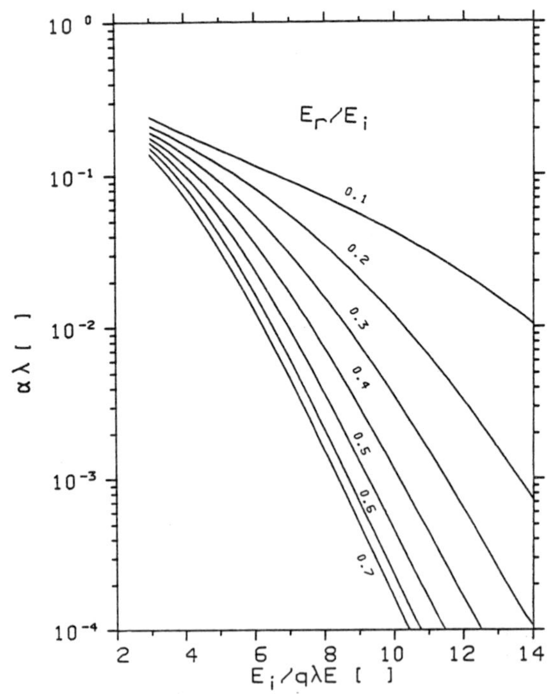

Fig. 4.2-1. Universal "Baraff" plot for ionization rates

phenomena has been obtained. The influence of the models for the ionization coefficients upon simulated device performance can be indeed very pronounced [4.2].

Chwang et al. [4.23] have rigorously obtained the same results as Baraff with a completely different approach for the calculation, however, with essentially the same assumptions.

Thornber [4.129] has suggested an empirical expression which has been proved to be consistent with an elaborate momentum and energy scaling theory.

$$\alpha = \frac{E}{E_i} \cdot \exp\left(-\frac{F_j}{\dfrac{k \cdot T \cdot F_j}{E_i} + E + \dfrac{E^2}{F_r}}\right) \tag{4.2-48}$$

F_j and F_r are interpreted as threshold fields describing the field at which the ionization energy is reached in one mean free path and the field at which the phonon energy is reached in one mean free path (cf. [4.129]). (4.2-48) includes the asymptotic behavior of the ionization rate, which has been predicted by Baraff. Thornber believes that (4.2-48) is the first simple, physical, analytical expression for the ionization coefficient valid for all fields. However, by fitting this expression to the experimental data of, e.g. [4.49], [4.132], Thornber himself has obtained unexpected large values for the ionization energies.

Okuto and Crowell [4.88] have proposed an empirical expression which is supposed to fit the theoretical results of Baraff as well as measurements.

$$\alpha = a_{300} \cdot \left(1 + c \cdot \left(\frac{T}{\mathrm{K}} - 300\right)\right) \cdot E \cdot \exp\left(-\left(\frac{b_{300} \cdot \left(1 + d \cdot \left(\dfrac{T}{\mathrm{K}} - 300\right)\right)}{E}\right)^2\right) \tag{4.2-49}$$

The coefficients of formula (4.2-49), which is temperature dependent, are summarized in Table 4.2-7.

Table 4.2-7. *Coefficients for* (4.2-49)

material	a_{300} [V^{-1}]	b_{300} [$V\,cm^{-1}$]	c [ ]	d [ ]	
Si	0.426	$4.81 \cdot 10^5$	$3.05 \cdot 10^{-4}$	$6.86 \cdot 10^{-4}$	electrons
	0.243	$6.53 \cdot 10^5$	$5.35 \cdot 10^{-4}$	$5.87 \cdot 10^{-5}$	holes
GaAs	0.294	$5.86 \cdot 10^5$	$8.5 \cdot 10^{-4}$	$7.17 \cdot 10^{-4}$	electrons and holes
Ge	0.569	$3.32 \cdot 10^5$	$6.33 \cdot 10^{-4}$	$9.34 \cdot 10^{-4}$	electrons
	0.559	$2.72 \cdot 10^5$	$7.87 \cdot 10^{-4}$	$8.82 \cdot 10^{-4}$	holes

Fig. 4.2-2 and Fig. 4.2-3 show the ionization rates for electrons and holes in silicon at 300 K temperature, respectively. The solid line corresponds to the data of Van Overstraten et al. [4.132]; the dashed line denotes the data of Grant [4.49]; the dot-dashed line corresponds to the model of Okuto and Crowell (4.2-29); and the dotted line has been calculated with the approximation (4.2-46) to Baraff's model with parameters from [4.31] and [4.53].

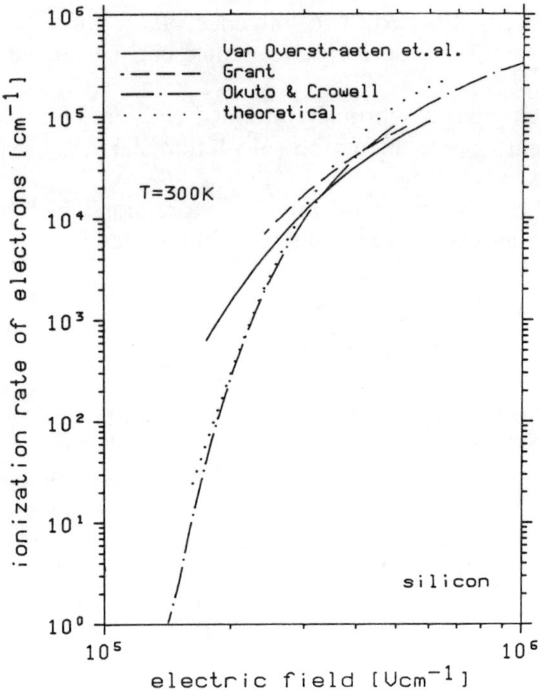

Fig. 4.2-2. Ionization rate for electrons in silicon at 300 K temperature

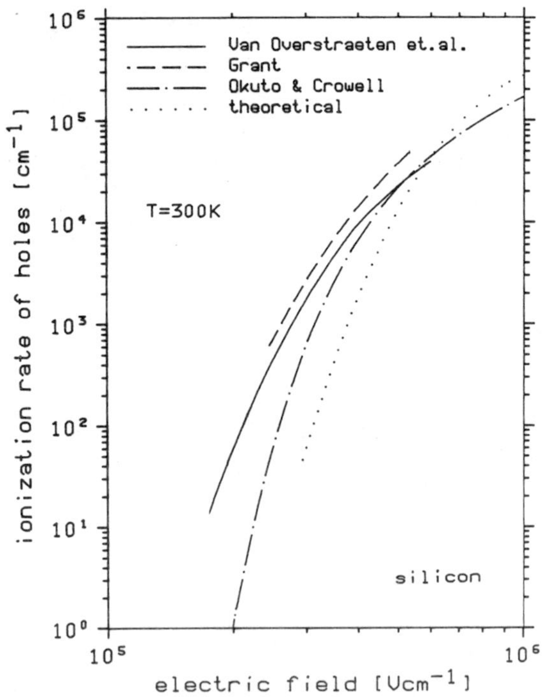

Fig. 4.2-3. Ionization rate for holes in silicon at 300 K temperature

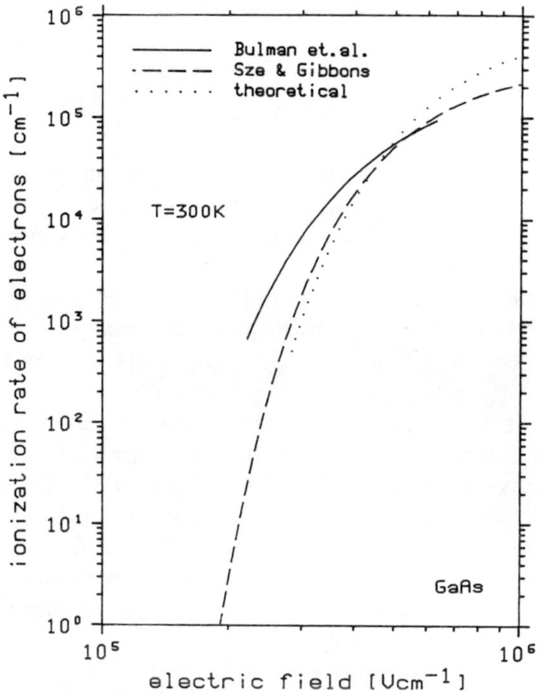

Fig. 4.2-4. Ionization rate for electrons in gallium-arsenide at 300 K temperature

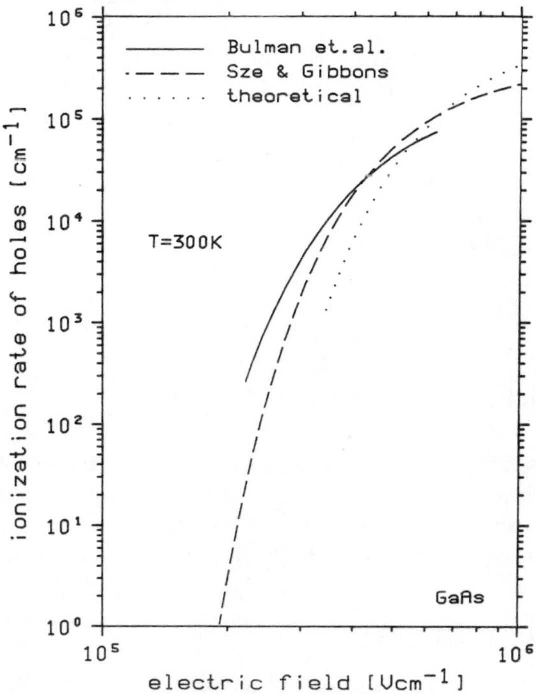

Fig. 4.2-5. Ionization rate for holes in gallium-arsenide at 300 K temperature

Fig. 4.2-4 and Fig. 4.2-5 show the ionization rates for electrons and holes in GaAs at 300 K temperature. The solid line corresponds to the data of Bulman et al. [4.18]; the dashed line denotes the data of Sze and Gibbons [4.122]; and the dotted line has been calculated, again, with the approximation (4.2-46) to Baraff's model with parameters from [4.7] and [4.31].
Recently attempts have been made to calculate the impact ionization coefficients by Monte Carlo methods taking into account a realistic band structure [4.112], [4.113], [4.125]. This interesting work, however, is subject to considerable controversy [4.20], [4.56].
One problem which arises in the context of impact ionization in very small devices and for low bias applications is the dark space phenomenon. In regions of a device with a large gradient of the electric field component parallel to current flow all models of impact ionization which we have discussed can be expected to overestimate the ionization rates. The carriers first have to gain an energy on their path through the device which is larger than the threshold energy for ionization before impact ionization can take place. The ionization rate at a specific place, thus, will depend, qualitatively spoken, not only on the local field, but also upon the field distribution in that vicinity. Therefore, the dark space phenomenon is frequently termed a non-local effect, e.g. [4.87]. A rigorous treatment of this and related phenomena [4.111] with models consisting of pure differential equations is impossible; it is also inconsistent with the usually assumed current relations because for their derivation moderate gradients for the electric field have been assumed (cf. Section 2.3). It has to be stated that much more experimental and theoretical investigations have to be carried out in order to understand the many detailed effects of impact ionization.
Finally, it should be noted that for the purpose of device modeling the individual generation/recombination rates are usually added up in the most simple manner.

$$R = R^{SRH} + R^{OPT} + R^{AU} + R^{SURF} + R^{II} \tag{4.2-50}$$

Interaction phenomena between the various mechanisms which obviously do exist are fully ignored in (4.2-50). A more concise treatment of interaction phenomena as demonstrated in [4.45] for Shockley-Read-Hall and trap assisted Auger generation/recombination can be expected to contribute to a great extent to answering many of the open questions.

4.3 Thermal Conductivity Modeling

Most currently available models for the thermal conductivity of silicon and germanium are based on the early measurements of Glassbrenner and Slack [4.48]. The same authors have also presented theoretical investigations which led to the following formula for the thermal conductivity in semiconductors.

$$k(T) = \frac{1}{a + b \cdot T + c \cdot T^2} \tag{4.3-1}$$

The constants a, b and c are summarized for silicon and germanium in Table 4.3-1. The agreement between measured thermal conductivity values and the results of

(4.3-1) is within five percent for silicon and germanium in the range $T \in [250, 1000]$ K and $T \in [50, 700]$ K, respectively.

Table 4.3-1. *Coefficients for* (4.3-1)

	Si	Ge	dimension
a	0.03	0.17	$V^{-1} A^{-1}$ cm K
b	$1.56 \cdot 10^{-3}$	$3.95 \cdot 10^{-3}$	$V^{-1} A^{-1}$ cm
c	$1.65 \cdot 10^{-6}$	$3.38 \cdot 10^{-6}$	$V^{-1} A^{-1}$ cm K^{-1}

Quite frequently one can find for the purpose of device simulation a simple power law for the thermal conductivity in silicon, e.g. [4.1], [4.22], [4.46].

$$k(T) = 1.5486 \frac{\mathrm{VA}}{\mathrm{cm\,K}} \cdot \left(\frac{T}{300\,\mathrm{K}}\right)^{-4/3} \tag{4.3-2}$$

(4.3-2) is a fairly good approximation to (4.3-1) as can be seen from Fig. 4.3-1. The solid line corresponds to (4.3-1); and the dashed line denotes (4.3-2).
The thermal conductivity of heavily doped semiconductors ($> 10^{19}$ cm^{-3}) might be expected to be higher than of pure material because of the large number of carriers available for transporting heat. The opposite is true (cf. [4.74]). Measurements indicate that the thermal conductivity is lowered by as much as thirty percent in the presence of a high doping concentration.

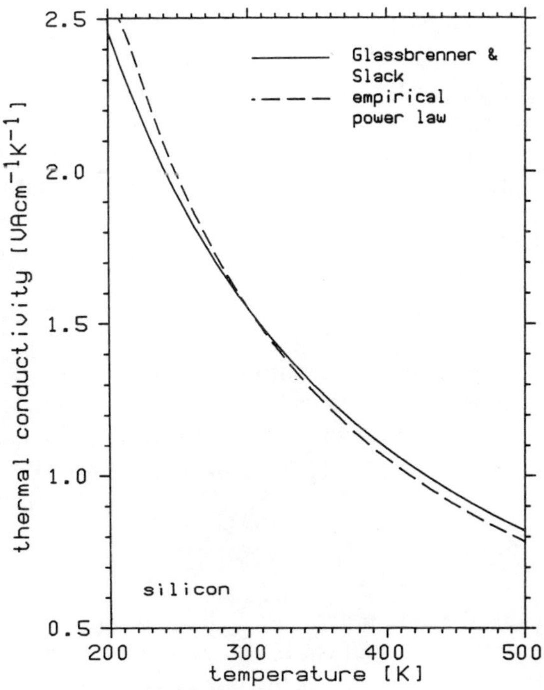

Fig. 4.3-1. Thermal conductivity in silicon versus temperature

A critical review of published data on the thermal conductivity of silicon, germanium and III-V compound semiconductors has been presented by Maycock [4.74].

4.4 Thermal Generation Modeling

Heat generation has been modeled in various ways in the literature. The simplest form has been used by, e.g., Gaur and Navon [4.46].

$$H=(\vec{J}_n+\vec{J}_p)\cdot\vec{E} \tag{4.4-1}$$

However, this expression is inappropriate for general application because it predicts heat sinks in device regions where the inner product of total current density and electric field is negative.

Adler [4.1] has suggested a formulation which is more sound.

$$H=\operatorname{div}\left(\frac{E_c}{q}\cdot\vec{J}_n+\frac{E_v}{q}\cdot\vec{J}_p\right) \tag{4.4-2}$$

E_c and E_v are the conduction band edge energy and the valence band edge energy, respectively. This formulation takes into account the energy loss/gain to the lattice through recombination/generation as one can see by expanding the "div" operator.

$$H=\frac{1}{q}\cdot\vec{J}_n\cdot\operatorname{grad}E_c+\frac{1}{q}\cdot\vec{J}_p\cdot\operatorname{grad}E_v+R\cdot(E_c-E_v) \tag{4.4-3}$$

R stands for the recombination/generation rate; and the difference E_c-E_v is the local band gap. Adler's formulation of the heat generation takes also into account the effects of band gap narrowing caused by heavy doping where the gradients of the band edges are not necessarily equal. For non-degenerate material (4.4-3) simplifies to:

$$H=(\vec{J}_n+\vec{J}_p)\cdot\vec{E}+R\cdot E_g \tag{4.4-4}$$

The heat generated/consumed by recombination/generation becomes nicely apparent in (4.4-4) compared to (4.4-1).

A different formulation for heat generation has been recommended by Chryssafis and Love [4.22]. Their considerations are based on the fact that the total power dissipation in a device is equivalent to the Joule heating. This quantity has been expressed in [4.22] as an integral over the entire surface of the device.

$$P=-\int_A(\varphi_n\cdot\vec{J}_n+\varphi_p\cdot\vec{J}_p)\cdot d\vec{A} \tag{4.4-5}$$

φ_n and φ_p are the quasi-Fermi potentials of electrons and holes, respectively. The major assumption in (4.4-5) is that electrons and holes contribute additively to the total power dissipation in a device. Using one of Green's theorems (4.4-5) can be transformed into an integral over the volume of the device.

$$P = -\int_V (\varphi_n \cdot \operatorname{div} \vec{J}_n + \operatorname{grad} \varphi_n \cdot \vec{J}_n + \varphi_p \cdot \operatorname{div} \vec{J}_p + \operatorname{grad} \varphi_p \cdot \vec{J}_p) \cdot dV \tag{4.4-6}$$

The integrand of (4.4-6) is obviously the heat generation per unit volume, thus, it denotes H. By rewriting the integrand we obtain:

$$H = -\vec{J}_n \cdot \operatorname{grad} \varphi_n - \vec{J}_p \cdot \operatorname{grad} \varphi_p + q \cdot R \cdot (\varphi_p - \varphi_n) \tag{4.4-7}$$

(4.4-7) does account for heavy doping effects as no specific assumptions have been made. Microscopically seen (4.4-2) is more attractive as the energy involved in a generation/recombination process is on the order of the band gap. The last term in (4.4-7) also predicts a heat consumption for dominant thermal recombination, which is absolutely implausible. (4.4-5), which is the basis for (4.4-7), has to be stated to be at least somewhat non-obvious. One should give preference to (4.4-2).

4.5 References

[4.1] Adler, M. S.: Accurate Calculations of the Forward Drop and Power Dissipation in Thyristors. IEEE Trans. Electron Devices *ED-25,* No. 1, 16 – 22 (1978).

[4.2] Adler, M. S., Temple, V. A. K., Rustay, R. C.: Theoretical Basis for Field Calculations on Multi-Dimensional Reverse Biased Semiconductor Devices. Solid-State Electron. *25,* No. 12, 1179 – 1186 (1982).

[4.3] Adler, M. S.: Accurate Numerical Models for Transistors and Thyristors. In: Introduction to the Numerical Analysis of Semiconductor Devices and Integrated Circuits, pp. 5 – 8. Dublin: Boole Press 1981.

[4.4] Akers, L. A., Portnoy, W. M.: Numerical Analysis of the Steady-State Behaviour of a MOS Field Effect Transistor. Int. J. Num. Meth. Eng. *15,* 1221 – 1238 (1980).

[4.5] Albrecht, H., Müller, R.: The Anomalous Breakdown. IEEE Trans. Electron Devices *ED-27, 970* – 977 (1980).

[4.6] Alley, G. D.: High-Voltage Two-Dimensional Simulations of Permeable Base Transistors. IEEE Trans. Electron Devices *ED-30,* No. 1, 52 – 60 (1983).

[4.7] Anderson, C. L., Crowell, C. R.: Threshold Energies for Electron-Hole Pair Generation by Impact Ionization in Semiconductors. Physical Review *B5,* 2267 – 2272 (1972).

[4.8] Antoniadis, D. A., Gonzales, A. G., Dutton, R. W.: Boron in Near-Intrinsic (100) and (111) Silicon under Inert and Oxidizing Ambients. J. Electrochem. Soc. *125,* 813 – 819 (1978).

[4.9] Arora, N. D., Hauser, J. R., Roulston, D. J.: Electron and Hole Mobilities in Silicon as a Function of Concentration and Temperature. IEEE Trans. Electron Devices *ED-29,* 292 – 295 (1982).

[4.10] Baccarani, G., Ostoja, P.: Electron Mobility Empirically Related to the Phosphorus Concentration in Silicon. Solid-State Electron. *18,* 579 – 580 (1975).

[4.11] Baccarani, G., Wordeman, M. R.: Transconductance Degradation in Thin-Oxide MOSFET's. Proc. Int. Electron Devices Meeting, pp. 278 – 281 (1982).

[4.12] Baraff, G. A.: Distribution Functions and Ionization Rates for Hot Electrons in Semiconductors. Physical Review *128,* 2507 – 2517 (1962).

[4.13] Barnes, J. J., Lomax, R. J., Haddad, G. I.: Finite-Element Simulation of GaAs MESFET's with Lateral Doping Profiles and Submicron Gates. IEEE Trans. Electron Devices *ED-23,* No. 9, 1042 – 1048 (1976).

[4.14] Bennett, H. S.: Improved Concepts for Predicting the Electrical Behavior of Bipolar Structures in Silicon. IEEE Trans. Electron Devices *ED-30,* No. 8, 920 – 927 (1983).

[4.15] Blatt, F. J.: Physics of Electronic Conduction in Solids. New York: McGraw-Hill 1968.

[4.16] Bourgoin, J., Lannoo, M.: Point Defects in Semiconductors II. Berlin-Heidelberg-New York: Springer 1983.

[4.17] Brooks, H.: Scattering by Ionized Impurities in Semiconductors. Physical Review *83,* 879 (1951).

[4.18] Bulman, G. E., Robbins, V. M., Brennan, K. F., Hess, K., Stillman, G. E.: Experimental Determination of Impact Ionization Coefficients in (100) GaAs. IEEE Electron Device Lett. *EDL-4,* No. 6, 181 – 185 (1983).

[4.19] Canali, C., Majni, G., Minder, R., Ottaviani, G.: Electron and Hole Drift Velocity Measurements in Silicon and Their Empirical Relation to Electric Field and Temperature. IEEE Trans. Electron Devices *ED-22,* 1045 – 1047 (1975).

[4.20] Capasso, F., Pearsall, T. P., Thornber, K. K.: The Effect of Collisional Broadening on Monte Carlo Simulations of High-Field Transport in Semiconductor Devices. IEEE Electron Device Lett. *EDL-2,* 295 (1981).

[4.21] Caughey, D. M., Thomas, R. E.: Carrier Mobilities in Silicon Empirically Related to Doping and Field. Proc. IEEE *52,* 2192 – 2193 (1967).

[4.22] Chryssafis, A., Love, W.: A Computer-Aided Analysis of One-Dimensional Thermal Transients in n-p-n Power Transistors. Solid-State Electron. *22,* 249 – 256 (1979).

[4.23] Chwang, R., Kao, C.-W., Crowell, C. R.: Normalized Theory of Impact Ionization and Velocity Saturation in Nonpolar Semiconductors via a Markov Chain Approach. Solid-State Electron. *22,* 599 – 620 (1979).

[4.24] Chynoweth, A. G.: Ionization Rates for Electrons and Holes in Silicon. Physical Review *109,* 1537 – 1540 (1958).

[4.25] Coen, R. W., Muller, R. S.: Velocity of Surface Carriers in Inversion Layers on Silicon. Solid-State Electron. *23,* 35 – 40 (1980).

[4.26] Collins, T. W., Churchill, J. N.: Exact Modeling of the Transient Response of an MOS Capacitor. IEEE Trans. Electron Devices *ED-22,* No. 3, 90 – 101 (1975).

[4.27] Conradt, R.: Auger-Rekombination in Halbleitern. In: Festkörperprobleme XII, pp. 449 – 464. Braunschweig: Vieweg 1972.

[4.28] Conwell, E., Weisskopf, V. F.: Theory of Impurity Scattering in Semiconductors. Physical Review *77,* No. 3, 388 – 390 (1950).

[4.29] Conwell, E. M.: High Field Transport in Semiconductors. New York: Academic Press 1967.

[4.30] Cooper, J. A., Nelson, D. F.: Measurement of the High-Field Drift Velocity of Electrons in Inversion Layers on Silicon. IEEE Electron Device Lett. *EDL-2,* No. 7, 171 – 173 (1981).

[4.31] Crowell, C. R., Sze, S. M.: Temperature Dependence of Avalanche Multiplication in Semiconductors. Appl. Phys. Lett. *9,* 242 – 244 (1966).

[4.32] Curtice, W. R.: Direct Comparison of the Electron Temperature Model with the Particle Mesh (Monte-Carlo) Model for the GaAs MESFET. IEEE Trans. Electron Devices *ED-29,* No. 12, 1942 – 1943 (1982).

[4.33] D'Avanzo, D. C., Vanzi, M., Dutton, R. W.: One-Dimensional Semiconductor Device Analysis (SEDAN). Report G-201-5, Stanford University, 1979.

[4.34] Dang, L. M., Konaka, M.: A Two-Dimensional Computer Analysis of Triode-Like Characteristics of Short-Channel MOSFET's. IEEE Trans. Electron Devices *ED-27,* 1533 – 1539 (1980).

[4.35] Debye, P. P., Conwell, E. M.: Electrical Properties of N-Type Germanium. Physical Review *93,* 693 – 706 (1954).

[4.36] Dhanasekaran, P. C., Gopalam, B. S. V.: The Physical Behaviour of an n^+p Silicon Solar Cell in Concentrated Sunlight. Solid-State Electron. *25,* No. 8, 719 – 722 (1982).

[4.37] Dhariwal, S. R., Kothari, L. S., Jain, S. C.: On the Recombination of Electrons and Holes at Traps with Finite Relaxation Time. Solid-State Electron. *24,* No. 8, 749 – 752 (1981).

[4.38] Dorkel, J. M., Leturcq, Ph.: Carrier Mobilities in Silicon Semi-Empirically Related to Temperature, Doping and Injection Level. Solid-State Electron. *24,* 821 – 825 (1981).

[4.39] Dziewior, J., Schmid, W.: Auger Coefficients for Highly Doped and Highly Excited Silicon. Appl. Phys. Lett. *31,* 346 – 348 (1977).

[4.40] Engl, W. L., Dirks, H. K., Meinerzhagen, B.: Device Modeling. Proc. IEEE *71,* No. 1, 10 – 33 (1983).

[4.41] Engl, W. L., Dirks, H.: Models of Physical Parameters. In: Introduction to the Numerical Analysis of Semiconductor Devices and Integrated Circuits, pp. 42 – 46. Dublin: Boole Press 1981.

[4.42] Ezawa, H.: Inversion Layer Mobility with Intersubband Scattering. Surface Science *58*, 25–32 (1976).

[4.43] Fossum, J. G.: Computer-Aided Numerical Analysis of Silicon Solar Cells. Solid-State Electron. *19*, 269–277 (1976).

[4.44] Fossum, J. G., Lee, D. S.: A Physical Model for the Dependence of Carrier Lifetime on Doping Density in Nondegenerate Silicon. Solid-State Electron. *25*, No. 8, 741–747 (1982).

[4.45] Fossum, J. G., Mertens, R. P., Lee, D. S., Nijs, J. F.: Carrier Recombination and Lifetime in Highly Doped Silicon. Solid-State Electron. *26*, No. 6, 569–576 (1983).

[4.46] Gaur, S. P., Navon, D. H.: Two-Dimensional Carrier Flow in a Transistor Structure Under Nonisothermal Conditions. IEEE Trans. Electron Devices *ED-23*, 50–57 (1976).

[4.47] Gaur, S. P.: Performance Limitations of Silicon Bipolar Transistors. IEEE Trans. Electron Devices *ED-26*, 415–421 (1979).

[4.48] Glasbrenner, C. J., Slack, G. A.: Thermal Conductivity of Silicon and Germanium from 3 K to the Melting Point. Physical Review *134*, No. 4A, A1058–A1069 (1964).

[4.49] Grant, W. N.: Electron and Hole Ionization Rates in Epitaxial Silicon at High Electric Fields. Solid-State Electron. *16*, 1189–1203 (1973).

[4.50] Grove, A. S.: Physics and Technology of Semiconductor Devices. New York: Wiley 1967.

[4.51] Hall, R. N.: Electron-Hole Recombination in Germanium. Physical Review *87*, 387 (1952).

[4.52] Haug, A.: Strahlungslose Rekombination in Halbleitern (Theorie). In: Festkörperprobleme XII, pp. 411–447. Braunschweig: Vieweg 1972.

[4.53] Hauser, J. R.: Threshold Energy for Avalanche Multiplication in Semiconductors. J. Appl. Phys. *37*, 507–509 (1966).

[4.54] Heimeier, H. H.: Zweidimensionale numerische Lösung eines nichtlinearen Randwertproblems am Beispiel des Transistors im stationären Zustand. Dissertation, Technische Hochschule Aachen, 1973.

[4.55] Herring, C.: Transport Properties of a Many-Valley Semiconductor. Bell Syst. Techn. J. *34*, No. 2, 237–290 (1955).

[4.56] Hess, K.: Comment on "Effect of Collisional Broadening on Monte Carlo Simulations of High-Field Transport in Semiconductor Devices". IEEE Electron Device Lett. *EDL-2*, No. 11, 297–298 (1981).

[4.57] Heywang, W., Pötzl, H. W.: Bandstruktur und Stromtransport. Berlin-Heidelberg-New York: Springer 1976.

[4.58] Jacoboni, C., Canali, C., Ottaviani, G., Quaranta, A. A.: A Review of Some Charge Transport Properties of Silicon. Solid-State Electron. *20*, 77–89 (1977).

[4.59] Jaggi, R.: High-Field Drift Velocities in Silicon and Germanium. Helv. Phys. Acta *42*, 941–943 (1969).

[4.60] Jaggi, R., Weibel, H.: High-Field Electron Drift Velocities and Current Densities in Silicon. Helv. Phys. Acta *42*, 631–632 (1969).

[4.61] Kireev, P. S.: Semiconductor Physics. Moscow: MIR Publishers 1978.

[4.62] Kotani, N., Kawazu, S.: Computer Analysis of Punch-Through in MOSFET's. Solid-State Electron. *22*, 63–70 (1979).

[4.63] Kotani, N., Kawazu, S.: A Numerical Analysis of Avalanche Breakdown in Short-Channel MOSFET's. Solid-State Electron. *24*, 681–687 (1981).

[4.64] Landsberg, P. T., Robbins, D. J.: The First 70 Semiconductor Auger Processes. Solid-State Electron. *21*, 1289–1294 (1978).

[4.65] Landsberg, P T., Abrahams, M. S.: Surface Recombination Statistics at Traps. Solid-State Electron. *26*, No. 9, 841–849 (1983).

[4.66] Laux, S. E., Lomax, R. J.: Numerical Investigation of Mesh Size Convergence Rate of the Finite Element Method in MESFET Simulation. Solid-State Electron. *24*, 485–493 (1981).

[4.67] Law, M.: Simulation of Recessed Gate FET's. Workshop on CAD of VLSI Processes, Stanford, 1982.

[4.68] Lee, C. A., Logan, R. A., Batdorf, R. L., Kleimack, J. J., Wiegmann, W.: Ionization Rates of Holes and Electrons in Silicon. Physical Review *134*, A761–773 (1964).

[4.69] Li, S. S., Thurber, W. R.: The Dopant Density and Temperature Dependence of Electron Mobility and Resistivity in n-Type Silicon. Solid-State Electron. *20*, 609–616 (1977).

[4.70] Li, S. S.: The Dopant Density and Temperature Dependence of Hole Mobility and Resistivity in Boron Doped Silicon. Solid-State Electron. *21*, 1109–1117 (1978).

[4.71] Müller, W., Risch, L., Schütz, A.: Analysis of Short Channel MOS Transistors in the Avalanche Multiplication Regime. IEEE Trans. Electron Devices *ED-29,* No. 11, 1778 – 1784 (1982).

[4.72] Manck, O.: Numerische Analyse des Schaltverhaltens eines zweidimensionalen bipolaren Transistors. Dissertation, Technische Hochschule Aachen, 1975.

[4.73] Masetti, G., Severi, M., Solmi, S.: Modeling of Carrier Mobility Against Carrier Concentration in Arsenic-, Phosphorus- and Boron-Doped Silicon. IEEE Trans. Electron Devices *ED-30,* No. 7, 764 – 769 (1983).

[4.74] Maycock, P. D.: Thermal Conductivity of Silicon, Germanium, III-V Compounds and III-V Alloys. Solid-State Electron. *10,* 161 – 168 (1967).

[4.75] Miller, S. L.: Ionization Rates for Holes and Electrons in Silicon. Physical Review *105,* 1246 – 1249 (1957).

[4.76] Moll, J. L., VanOverstraeten, R.: Charge Multiplication in Silicon p-n Junctions. Solid-State Electron. *6,* 147 – 157 (1963).

[4.77] Mott, N. F.: Recombination: A Survey. Solid-State Electron. *21,* 1275 – 1280 (1978).

[4.78] Nakagawa, A.: One-Dimensional Device Model of the npn Bipolar Transistor Including Heavy Doping Effects under Fermi Statistics. Solid-State Electron. *22,* 943 – 949 (1979).

[4.79] Navon, D. H., Wang, C. T.: Numerical Modeling of Power MOSFET's. Solid-State Electron. *26,* No. 4, 287 – 290 (1983).

[4.80] Newman, D. S., Ferry, D. K., Sites, J. R.: Measurement and Simulation of GaAs FET's under Electron Beam Irradiation. IEEE Trans. Electron Devices *ED-30,* No. 7, 849 – 855 (1983).

[4.81] Norton, P., Braggins, T., Levinstein, H.: Impurity and Lattice Scattering Parameters as Determined from Hall and Mobility Analysis in n-type Silicon. Phys. Rev. *B8,* No. 12, 5632 – 5653 (1973).

[4.82] Ogawa, T.: Avalanche Breakdown and Multiplication in Silicon *pin* Junctions. Jap. J. Appl. Phys. *4,* 473 – 484 (1965).

[4.83] Oh, S. Y., Ward, D. E., Dutton, R. W.: Transient Analysis of MOS Transistors. IEEE Trans. Electron Devices *ED-27,* 1571 – 1578 (1980).

[4.84] Oka, H., Nishiuchi, K., Nakamura, T., Ishikawa, H.: Two-Dimensional Numerical Analysis of Normally-Off Type Buried Channel MOSFET's. Proc. Int. Electron Devices Meeting, pp. 30 – 33 (1979).

[4.85] Oka, H., Nishiuchi, K., Nakamura, T., Ishikawa, H.: Computer Analysis of a Short-Channel BC MOSFET. IEEE Trans. Electron Devices *ED-27,* 1514 – 1520 (1980).

[4.86] Okuto, Y., Crowell, C. R.: Energy Conservation Considerations in the Characterization of Impact Ionization in Semiconductors. Physical Review *B6,* 3076 – 3081 (1972).

[4.87] Okuto, Y., Crowell, C. R.: Ionization Coefficients in Semiconductors: A Nonlocalized Property. Physical Review *B10,* 4284 – 4296 (1974).

[4.88] Okuto, Y., Crowell, C. R.: Threshold Energy Effect on Avalanche Breakdown Voltage in Semiconductor Junctions. Solid-State Electron. *18,* 161 – 168 (1975).

[4.89] Omura, Y., Sano, E., Ohwada, K.: A Negative Drain Conductance Property in a Super-Thin Film Buried-Channel MOSFET on a Buried Insulator. IEEE Trans. Electron Devices *ED-30,* No. 1, 67 – 73 (1983).

[4.90] Paul, R.: Halbleiterphysik. Heidelberg: Hüthig-Verlag 1975.

[4.91] Plunkett, J. C., Stone, J. L., Leu, A.: A Computer Algorithm for Accurate and Repeatable Profile Analysis Using Anodization and Stripping of Silicon. Solid-State Electron. *20,* 447 – 453 (1977).

[4.92] Queisser, H. J.: Recombination at Deep Traps. Solid-State Electron. *21,* 1495 – 1503 (1978).

[4.93] Reiser, M.: A Two-Dimensional Numerical FET Model for DC, AC, and Large-Signal Analysis. IEEE Trans. Electron Devices *ED-20,* 35 – 44 (1973).

[4.94] Resta, R., Resca, L.: Ionized Impurity Scattering in Semiconductors. Physical Review *B20,* No. 8, 3254 – 3257 (1979).

[4.95] Robinson, J. E., Rodriguez, S.: Ionized Impurity Scattering in Degenerate Many-Valley Semiconductors. Physical Review *135,* No. 3A, A779 – A784 (1964).

[4.96] Roulston, D. J., Arora, N. D., Chamberlain, S. G.: Modeling and Measurement of Minority-Carrier Lifetime versus Doping in Diffused Layers of n^+p Silicon Diodes. IEEE Trans. Electron Devices *ED-29,* No. 2, 284 – 291 (1982).

[4.97] Roychoudhury, D., Basu, P. K.: A New Mobility-Field Expression for the Calculation of Mosfet Characteristics. Solid-State Electron. *9,* 656 – 657 (1976).

[4.98] Ruch, J. G., Kino, G. S.: Measurement of the Velocity-Field Characteristic of Gallium Arsenide. Appl. Phys. Lett. *10,* No. 2, 40–42 (1967).

[4.99] Ruch, J. G., Kino, G. S.: Transport Properties of GaAs. Physical Review *174,* No. 3, 921–931 (1968).

[4.100] Ruch, J. G., Fawcett, W.: Temperature Dependence of the Transport Properties of Gallium Arsenide Determined by a Monte Carlo Method. J. Appl. Phys. *41,* No. 9, 3843–3849 (1970).

[4.101] Sabnis, A. G., Clemens, J. T.: Characterization of the Electron Mobility in the Inverted (100) SI-Surface. Proc. International Electron Devices Meeting, pp. 18–21 (1979).

[4.102] Sah, C. T., Chan, P. C. H., Wang, C.-K., Sah, R. L. Y., Yamakawa, K. A., Lutwack, R.: Effect of Zinc Impurity in Silicon Solar-Cell Efficiency. IEEE Trans. Electron Devices *ED-28,* No. 3, 304–313 (1981).

[4.103] Scarfone, L. M., Richardson, L. M.: Electron Mobilities Based on an Exact Numerical Analysis of the Dielectric Function Dependent Linearized Poisson's Equation for the Potential of Impurity Ions in Semiconductors. Physical Review *B22,* No. 2, 982–990 (1980).

[4.104] Schütz, A., Selberherr, S., Pötzl, H. W.: A Two-Dimensional Model of the Avalanche Effect in MOS Transistors. Solid-State Electron. *25,* 177–183 (1982).

[4.105] Schütz, A., Selberherr, S., Pötzl, H. W.: Analysis of Breakdown Phenomena in MOSFET's. IEEE Trans. Computer-Aided-Design of Integrated Circuits *CAD-1,* 77–85 (1982).

[4.106] Scharfetter, D. L., Gummel, H. K.: Large-Signal Analysis of a Silicon Read Diode Oscillator. IEEE Trans. Electron Devices *ED-16,* 64–77 (1969).

[4.107] Schmid, W.: Experimental Comparison of Localized and Free Carrier Auger Recombination in Silicon. Solid-State Electron. *21,* 1285–1287 (1978).

[4.108] Seeger, K.: Semiconductor Physics. Wien-New York: Springer 1973.

[4.109] Selberherr, S.: Zweidimensionale Modellierung von MOS-Transistoren. Dissertation, Technische Universität Wien, 1981.

[4.110] Selberherr, S., Schütz, A., Pötzl, H.: Two Dimensional MOS-Transistor Modeling. In: Process and Device Simulation for Integrated Circuit Design, pp. 490–581. The Hague: Martinus Nijhoff 1983.

[4.111] Shekhar, C., Khokle, W. S.: Transient Behaviour of Impact Ionization in Silicon. IEEE Trans. Electron Devices *ED-23,* 1109–1110 (1976).

[4.112] Shichijo, H., Hess, K.: Band Structure Dependent Transport and Impact Ionization in GaAs. Physical Review *B23,* No. 8, 4197–4207 (1981).

[4.113] Shichijo, H., Hess, K., Stillman, G. E.: Simulation of High-Field Transport in GaAs Using a Monte Carlo Method and Pseudopotential Band Structures. Appl. Phys. Lett. *38,* 89–91 (1981).

[4.114] Shockley, W.: Hot Electrons in Germanium and Ohm's Law. Bell System Technical J. *30,* 990–1034 (1951).

[4.115] Shockley, W., Read, W. T.: Statistics of the Recombinations of Holes and Electrons. Physical Review *87,* No. 5, 835–842 (1952).

[4.116] Shockley, W.: Problems Related to p-n Junctions in Silicon. Solid-State Electron. *2,* 35–67 (1961).

[4.117] Smith, R. A.: Semiconductors. Cambridge: Cambridge University Press 1978.

[4.118] Spirito, P.: Avalanche Multiplication Factors in Ge and Si Abrupt Junctions. IEEE Trans. Electron Devices *ED-21,* 226–231 (1974).

[4.119] Stone, J. L., Plunkett, J. C.: Ion Implantation Processes in Silicon. In: Impurity Doping Processes in Silicon, pp. 56–146. Amsterdam: North-Holland 1981.

[4.120] Sun, S. C., Plummer, J. D.: Electron Mobility in Inversion and Accumulation Layers on Thermally Oxidized Silicon Surfaces. IEEE Trans. Electron Devices *ED-27,* 1497–1508 (1980).

[4.121] Sutherland, A. D.: An Improved Empirical Fit to Baraff's Universal Curves for the Ionization Coefficients of Electron and Hole Multiplication in Semiconductors. IEEE Trans. Electron Devices *ED-27,* No. 7, 1299–1300 (1980).

[4.122] Sze, S. M., Gibbons, G.: Avalanche Breakdown Voltages of Abrupt and Linearly Graded p-n Junctions in Ge, Si, GaAs, and GaP. Appl. Phys. Lett. *8,* 111–113 (1966).

[4.123] Sze, S. M.: Physics of Semiconductor Devices. New York: Wiley 1969.

[4.124] Tamer, A. A., Rauch, K., Moll, J. L.: Numerical Comparison of DMOS, VMOS and UMOS Power Transistors. IEEE Trans. Electron Devices *ED-30,* No. 1, 73–76 (1983).

[4.125] Tang, J. Y., Shichijo, H., Hess, K., Iafrate, G. J.: Band Structure Dependent Impact Ionization in Silicon and Gallium Arsenide. Journal de Physique *C7,* No. 10, 63 – 69 (1981).

[4.126] Tauber, G. E.: Transport Phenomena in Germanium and Silicon. J. Phys. Chem. Solids *23,* 7 – 18 (1962).

[4.127] Temple, V. A. K., Adler, M. S.: Calculation of the Diffusion Curvature Related Avalanche Breakdown in High-Voltage Planar p-n Junctions. IEEE Trans. Electron Devices *ED-22,* 910 – 916 (1975).

[4.128] Thornber, K. K.: Relation of Drift Velocity to Low-Field Mobility and High-Field Saturation Velocity. J. Appl. Phys. *51,* 2127 – 2136 (1980).

[4.129] Thornber, K. K.: Applications of Scaling to Problems in High-Field Electronic Transport. J. Appl. Phys. *52,* 279 – 290 (1981).

[4.130] Toyabe, T., Asai, S., Yamaguchi, K.: Internal Documents on the CADDET Program. Hitachi, Tokyo, 1979.

[4.131] Tyagi, M. S., VanOverstraeten, R.: Minority Carrier Recombination in Heavily Doped Silicon. Solid-State Electron. *26,* No. 6, 577 – 597 (1983).

[4.132] VanOverstraeten, R., DeMan, H.: Measurement of the Ionization Rates in Diffused Silicon p-n Junctions. Solid-State Electron. *13,* 583 – 608 (1970).

[4.133] Vass, E., Hess, K.: Energy Loss of Warm and Hot Carriers in Surface Inversion Layers of Polar Semiconductors. Z. Physik *B25,* 323 – 325 (1976).

[4.134] Warner, R. M., Ju, D.-H., Grung, B. L.: Electron-Velocity Saturation at a BJT Collector Junction under Low-Level Conditions. IEEE Trans. Electron Devices *ED-30,* No. 3, 230 – 236 (1983).

[4.135] Weast, R. C., Astle, M. J.: CRC Handbook of Chemistry and Physics. Boca Raton, Florida: CRC Press 1981.

[4.136] Weaver, H. T., Nasby, R. D.: Analysis of High Efficiency Silicon Solar Cells. IEEE Trans. Electron Devices *ED-28,* No. 5, 465 – 472 (1981).

[4.137] Wolff, P. A.: Theory of Multiplication in Silicon and Germanium. Physical Review *95,* 1415 – 1420 (1954).

[4.138] Yamaguchi, K.: Field-Dependant Mobility Model for Two-Dimensional Numerical Analysis of MOSFET's. IEEE Trans. Electron Devices *ED-26,* 1068 – 1074 (1979).

[4.139] Yamaguchi, K.: A Mobility Model for Carriers in the MOS Inversion Layer. IEEE Trans. Electron Devices *ED-30,* No. 6, 658 – 663 (1983).

Analytical Investigations About the Basic Semiconductor Equations 5

In this chapter we review some of the existing analytical results which characterize the basic semiconductor equations. Of particular concern will be the questions of existence, uniqueness and structure and smoothness of solutions. These are of importance in both the theoretical context and the practical context, since the knowledge of the structure and smoothness properties of solutions is indeed essential for the selection of appropriate numerical solution procedures. The basic semiconductor equations as given in Chapter 2 are:

$$\operatorname{div}\operatorname{grad}\psi=\frac{q}{\varepsilon}\cdot(n-p-C) \tag{5-1}$$

$$\operatorname{div}\vec{J}_n-q\cdot\frac{\partial n}{\partial t}=q\cdot R(\psi,n,p) \tag{5-2}$$

$$\operatorname{div}\vec{J}_p+q\cdot\frac{\partial p}{\partial t}=-q\cdot R(\psi,n,p) \tag{5-3}$$

$$\vec{J}_n=-q\cdot(\mu_n\cdot n\cdot\operatorname{grad}\psi-D_n\cdot\operatorname{grad}n) \tag{5-4}$$

$$\vec{J}_p=-q\cdot(\mu_p\cdot p\cdot\operatorname{grad}\psi+D_p\cdot\operatorname{grad}p) \tag{5-5}$$

We have omitted in the current relations (5-4) and (5-5) terms which account for current components caused by bandgap narrowing and temperature gradients. All these effects are considered to be only small perturbations which just make the essential analytical results about the basic semiconductor equations less transparent. One should also bear in mind that the current relations will become potentially incorrect if one of the above cited effects would change the equations in a dominating manner (cf. Section 2.3). We shall also ignore the impact of a non-homogeneous temperature distribution on the basic semiconductor equations for the following analytical investigations.

By substituting the current relations (5-4) and (5-5) into the continuity equations (5-2) and (5-3), respectively, we obtain a system of three partial differential equations with the dependent variables ψ, n and p.

$$\operatorname{div}\operatorname{grad}\psi-\frac{q}{\varepsilon}\cdot(n-p-C)=0 \tag{5-6}$$

$$\operatorname{div}(D_n \cdot \operatorname{grad} n - \mu_n \cdot n \cdot \operatorname{grad} \psi) - R(\psi, n, p) = \frac{\partial n}{\partial t} \tag{5-7}$$

$$\operatorname{div}(D_p \cdot \operatorname{grad} p + \mu_p \cdot p \cdot \operatorname{grad} \psi) - R(\psi, n, p) = \frac{\partial p}{\partial t} \tag{5-8}$$

This system will be the basis for all further considerations. For the purpose of analysis it is worthwhile to note that the parameters D_n, D_p, μ_n and μ_p are always positive. Therefore (5-7) and (5-8) can be identified as parabolic partial differential equations under the additional assumption that $R(\psi, n, p)$ does not contain differential spatial second order and temporal first order operators applied to n and p. It is obvious that higher order differential operators than the above given for R are also not allowed. (5-6) represents an elliptic differential equation.
For all mathematical investigations we need a priori information about the domain in which (5-6) to (5-8) hold and the boundary conditions. These questions have not been dealt with in the preceeding chapters, although it is obvious that domain and boundary conditions are subjects to be first investigated by physical reasoning and then classified by mathematical considerations. However, I feel that these two tasks are so intensively connected that they should be carried out together. A minor modification of a boundary condition in the physical sense can change the mathematical problem drastically.

5.1 Domain and Boundary Conditions

The basic semiconductor equations (5-6), (5-7) and (5-8) are posed in a bounded domain $D \in R^n$ $(n=1,2,3)$ representing the device geometry. In principle, all semiconductor devices are three dimensional structures. However, in many cases the device under consideration is intrinsically two or even one dimensional and then one can assume that the partial derivatives of the parameters and of the dependent variables of the basic equations perpendicular to a plane (line) vanish, so reducing the problem by one (two) space dimension(s), and, thus, simplifying the numerical solution drastically. Most of the existing simulation programs are restricted to a rectangular geometry. This is, however, not essential as far as the analysis of the equations is concerned, but it can simplify the numerical solution quite significantly.
The boundary ∂D of D is piecewise smooth for the two or three dimensional problem; it is represented, trivially, by two points for the one dimensional problem. The boundary can be split in principle into two parts.

$$\partial D = \partial D_p \cup \partial D_A \tag{5.1-1}$$

∂D_P represents those parts of the boundary which correspond to real "physical" boundaries like contacts and interfaces to insulating material. ∂D_A consists of artificial boundaries which have to be introduced, for instance, to separate neighboring devices in integrated circuits. This second category of boundaries does, therefore, not correspond to boundaries in the physical sense.

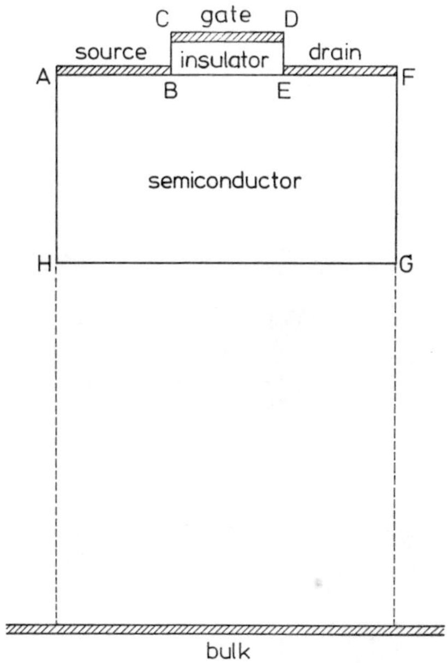

Fig. 5.1-1. Simulation geometry of a planar MOSFET

In order to illustrate this classification refer to Fig. 5.1-1 which shows the idealized two-dimensional simulation geometry for a planar MOS transistor. The total domain for the simulation is bounded by the polygon *A-B-C-D-E-F-G-H-A*. It is to note that the basic equations (5-6) to (5-8) are only posed in the subdomain *A-B-E-F-G-H-A*. For the insulator (subdomain bounded by *B-C-D-E-B*), one usually specifies the Laplace equation for the electrostatic potential, and one neglects the existence of any mobile carriers.

$$\operatorname{div} \operatorname{grad} \psi = 0 \tag{5.1-2}$$

$$n = p = C = 0 \tag{5.1-3}$$

One should be aware that by assuming (5.1-2) and (5.1-3) no gate currents can be calculated and the influence of oxide charges is also neglected. However, physical considerations are beyond the scope of our present discussion.

The boundaries *A-B*, *E-F*, *C-D* and *B-E* can be interpreted as physical boundaries denoting three idealized contacts and an interface between the semiconductor and the insulator, respectively. These boundaries form ∂D_p. As artificial boundaries *A-H*, *B-C*, *D-E*, *F-G* and *G-H* can be identified. These boundaries do not exist in the real device and are only introduced to enable the simulation. It is obvious that these boundaries are not introduced completely arbitrarily. Having a priori information about the functioning of a device it is usually possible to define somewhat natural boundaries which separate the device in a self contained manner from its environment. It should be noted that artificial boundaries are sometimes also introduced to simplify the numerical solution of the basic equations. The boundary

G-H represents such a boundary piece. Considering the real dimensions of a MOS transistor we know that the length of the interface *B-E* is on the order of 1 μm whereas the thickness of the wafer, i.e. the distance between the interface and bulk, is about 500 μm. Thus, the real geometry is a long, small stripe which has a disastrous impact on many classical numerical solution methods for the basic equations. From knowledge about the operation of the MOS transistor we can deduce that by cutting off the simulation geometry at some sufficiently large distance from the interface (*G-H*) we introduce only a small error for most operating conditions of the MOS transistor. In particular, the voltage drop between *G-H* and the bulk contact and the potential distribution along *G-H* has to be known such that the artificial boundary *G-H* is an acceptable simplification [5.31].

∂D_p, the physically motivated part of the boundary, can be roughly split into three classes.

$$\partial D_p = \partial D_O \cup \partial D_S \cup \partial D_I \tag{5.1-4}$$

∂D_O denotes the parts of the boundary corresponding to ohmic contacts; ∂D_S are the parts of the boundary denoting Schottky contacts and ∂D_I are the interfaces to insulating material.

We shall first investigate the ohmic contacts. As a boundary condition for the electrostatic potential we have in general a functional relation between the electrostatic potential and the total current density, which can include first order derivatives with respect to time and the unit vector perpendicular to the boundary and integrals with respect to time and the area of the ohmic contact. We denote this formally with the following implicit relation:

$$g\left(\psi, \frac{\partial \psi}{\partial t}, I, \frac{\partial I}{\partial t}\right) = 0 \tag{5.1-5}$$

with:

$$I = \int\limits_{\partial D_O} (\vec{J}_n + \vec{J}_p) \cdot d\vec{A} \tag{5.1-6}$$

The simplest boundary conditions are obtained for purely voltage or purely current controlled contacts.

For a contact ∂D_O which is voltage controlled (5.1-5) simplifies to an explicit boundary condition for the electrostatic potential.

$$\psi(t) - \psi_b - \psi_D(t) = 0 \tag{5.1-7}$$

$\psi_D(t)$ denotes the externally applied bias and ψ_b represents the built-in potential as defined in Section 2.4. (5.1-7) is a Dirichlet boundary condition for the electrostatic potential.

For a contact ∂D_O which is current controlled in an ideal manner (5.1-5) can be given as:

$$\int\limits_{\partial D_O} (\vec{J}_n + \vec{J}_p) \cdot d\vec{A} - I_D(t) = 0 \tag{5.1-8}$$

$I_D(t)$ is the total current which is forced to flow through the contact. Substituting the current relations (5-4) and (5-5) into the boundary condition (5.1-8) one obtains an integral boundary condition for the electrostatic potential. However, an additional condition is required to avoid ambiguities. One usually assumes that the ohmic contact is ideally conducting which means that there is no voltage drop in the boundary.

$$\psi(t) - \psi_b = \text{const.} \tag{5.1-9}$$

The derivatives with respect to time of the electrostatic potential and the total current enter into the boundary condition for the case of a capacitive, or inductive load. The boundary conditions for a given external load circuit are straightforward to derive with minimal knowledge about circuit theory and, therefore, we skip them. In the mathematical sense one will obtain a fairly complex time dependent boundary condition for the electrostatic potential.

So far we have only a boundary condition for the electrostatic potential at ohmic contacts. We additionally need conditions for the carrier concentrations. It is well established, although a matter open to physical criticism, to assume thermal equilibrium (which corresponds to infinite surface recombination velocities) and vanishing space charge at ohmic contacts.

$$n \cdot p - n_i^2 = 0 \tag{5.1-10}$$

$$n - p - C = 0 \tag{5.1-11}$$

These two conditions can be rearranged into Dirichlet boundary conditions for electrons and holes.

$$n = \frac{\sqrt{C^2 + 4 \cdot n_i^2} + C}{2} \tag{5.1-12}$$

$$p = \frac{\sqrt{C^2 + 4 \cdot n_i^2} - C}{2} \tag{5.1-13}$$

Summarizing, under certain conditions (pure voltage drive), we have at ∂D_O, the ohmic contacts, a Dirichlet boundary condition for the electrostatic potential, and Dirichlet conditions for the carrier concentrations. For more general applications (current drive or external load circuit) the boundary condition for the electrostatic potential is given in integral form.

∂D_S in (5.1-4) denotes, as mentioned, the parts of the boundary corresponding to Schottky contacts. The physics of Schottky contacts is extraordinarily complex and difficult. For the purpose of simulation a highly simplified model is commonly in use. For the electrostatic potential in the case of a voltage drive one can assume the Dirichlet boundary condition:

$$\psi(t) - \psi_b + \psi_s - \psi_D(t) = 0 \tag{5.1-14}$$

$\psi_D(t)$ denotes again the externally applied bias; ψ_s represents the Schottky barrier height, which is a characteristic quantity of the metal and the semiconductor with which the Schottky contact is fabricated. The numerical value of ψ_s is usually on the order of half a volt to one volt. A comprehensive summary can be found in the book by Sze [5.39]. ψ_b is again the built-in potential which in this context is frequently omitted and implicitly accounted for with a properly adjusted value for ψ_s. For the continuity equations it is more difficult to give boundary conditions which are physically reasonable and still sufficiently simple for the purpose of modeling. The carrier concentrations at a Schottky contact depend in general on the current density passing through the contact. Investigations about the interplay of the thermionic emission and diffusion theories have been presented in [5.4], which lead to the following boundary conditions [5.8]:

$$\vec{J}_n \cdot \vec{n} = -q \cdot v_n \cdot \left(n - \frac{\sqrt{C^2 + 4 \cdot n_i^2} + C}{2}\right) \tag{5.1-15}$$

$$\vec{J}_p \cdot \vec{n} = q \cdot v_p \cdot \left(p - \frac{\sqrt{C^2 + 4 \cdot n_i^2} - C}{2}\right) \tag{5.1-16}$$

$\vec{n}$ denotes the unit normal vector on ∂D which exists almost everywhere (that is, everywhere except on a subset of ∂D of Lebesque measure zero). Practically, $\vec{n}$ possibly does not exist on a finite number of edges.

v_n, v_p denote the thermionic recombination velocities for electrons, holes at the contact. Note that (5.1-15), (5.1-16) reduce to (5.1-12), (5.1-13) for infinite thermionic recombination velocities. However, in, e.g. [5.40], it has been assumed for the simulation of FETs that v_n and v_p are zero which is equivalent to assume that no current at all flows through the Schottky contact.

$$\vec{J}_n \cdot \vec{n} = 0 \tag{5.1-17}$$

$$\vec{J}_p \cdot \vec{n} = 0 \tag{5.1-18}$$

This assumption is at first glance reasonable since for most of the practical operating conditions the Schottky contact operates in the reverse biased mode, where the current flow is indeed relatively small. However, one will find an unrealistic decrease of the carrier concentrations in the vicinity of the contact which gives rise to numerical problems. It is obvious that (5.1-15) and (5.1-16) are only applicable for a reverse biased contact. These boundary conditions transform after straightforward calculations into mixed boundary conditions for the carrier concentrations.

In his dissertation Laux [5.16] has used another model for the boundary conditions for the continuity equations.

$$n = \begin{cases} \psi_D \geq 0 & N_c \cdot \exp\left(\dfrac{q \cdot (\psi_D - \psi_s)}{k \cdot T}\right) \\[2ex] \psi_D < 0 & N_c \cdot \exp\left(-\dfrac{q \cdot \psi_s}{k \cdot T}\right) \cdot \dfrac{1}{14 \cdot \sqrt{\dfrac{q \cdot (\psi_s - \psi_D)}{k \cdot T}} - \ln \dfrac{N_c}{N_D} + 1} \end{cases} \tag{5.1-19}$$

$$p = N_v \cdot \exp\left(\frac{q \cdot (\psi_s - U_g)}{k \cdot T}\right) \tag{5.1-20}$$

These conditions are appropriate only, as it has been the intention of the author, for Schottky contacts on n-doped material. It is interesting to note that in this model the hole concentration (the minority carrier concentration) is independent of the applied bias. This corresponds to an infinite thermionic recombination velocity for holes at the Schottky contact.

The last category of boundaries in (5.1-4) are the interfaces ∂D_I between the semiconductor and insulating material. At such interfaces the law of Gauß in differential form must be obeyed.

$$\varepsilon_{\text{sem}} \cdot \left.\frac{\partial \psi}{\partial \vec{n}}\right|_{\text{sem}} - \varepsilon_{\text{ins}} \cdot \left.\frac{\partial \psi}{\partial \vec{n}}\right|_{\text{ins}} = Q_{\text{int}} \tag{5.1-21}$$

ε_{sem} and ε_{ins} denote the permittivity in the semiconductor and the insulator, respectively. Q_{int} represents charges at the interface. For interfaces to a thick insulator, e.g. field oxide, one frequently assumes in the insulator a vanishing electric field component perpendicular to the interface, so that (5.1-21) simplifies to:

$$\varepsilon_{\text{sem}} \cdot \frac{\partial \psi}{\partial \vec{n}} = Q_{\text{int}} \tag{5.1-22}$$

Quite often the existence of interface charges is also neglected. (5.1-22) then reduces to a Neumann boundary condition for the electrostatic potential.

$$\frac{\partial \psi}{\partial \vec{n}} = 0 \tag{5.1-23}$$

In Fig. 5.1-1 *B-E* denotes an interface, as mentioned. In several MOS simulation programs the Laplace equation is not solved explicitly in the insulator. Instead a one dimensional potential drop perpendicular to the interface is assumed which leads to a mixed boundary condition for the electrostatic potential at the interface:

$$\varepsilon_{\text{sem}} \cdot \frac{\partial \psi}{\partial \vec{n}} - \varepsilon_{\text{ins}} \cdot \frac{U_G - \psi}{t_{\text{ins}}} = Q_{\text{int}} \tag{5.1-24}$$

U_g denotes the electrostatic potential at the gate contact *C-D*; t_{ins} is the thickness of the insulator *C-B*. It is obvious that it is much easier to program the mixed boundary condition (5.1-24) instead of solving the Laplace equation in the insulator. However, if the length of the interface *B-E* is not extraordinarily large compared to the thickness of the insulator the error introduced is, from my own personal experience, intolerably large.

For the continuity equations the current components perpendicular to the interface must equal the surface recombination rate R^{SURF} which we have discussed in Section 4.2.

$$\vec{J}_n \cdot \vec{n} = -q \cdot R^{SURF} \tag{5.1-25}$$

$$\vec{J}_p \cdot \vec{n} = q \cdot R^{SURF} \tag{5.1-26}$$

Quite frequently the existence of surface recombination is simply ignored by assuming infinite surface recombination velocities which leads to the boundary conditions (5.1-17), (5.1-18). Under certain circumstances, depending on the specific device and its operating conditions, this can be justified.

The second category we have to deal with are, as outlined, the artificial boundary conditions. Here one assumes either the natural boundary conditions which guarantee that the domain under consideration, i.e. the simulated device area, is self contained (5.1-27) to (5.1-29), or one specifies Dirichlet values for the electrostatic potential and the carrier concentrations, which are a priori estimated (e.g. at *G-H* in Fig. 5.1-1).

$$\frac{\partial \psi}{\partial \vec{n}}=0 \tag{5.1-27}$$

$$\frac{\partial n}{\partial \vec{n}}=0 \tag{5.1-28}$$

$$\frac{\partial p}{\partial \vec{n}}=0 \tag{5.1-29}$$

The applicability of these boundary conditions has to be justified by physical and mathematical reasoning. In Fig. 5.1-1, for instance, the distances *A-B* and *E-F* must be sufficiently large that the error introduced by the artificial boundary conditions at *A-H* and *F-G* is tolerably small. A definite requirement for the applicability of the boundary conditions (5.1-27) to (5.1-29) is that the derivatives of the parameters C, D_n, D_p, μ_n and μ_p with respect to the unit normal vector at the boundary vanish along the artificial boundary.

It is to note that the basic semiconductor equations only constitute a time dependent problem if the boundary condition for the electrostatic potential is time dependent. If the boundary condition for the electrostatic potential is time invariant, the semiconductor equations reduce to a system of three coupled elliptic equations.

5.2 Dependent Variables

For analytical purposes it is often helpful to use dependent variables other than n and p in the basic equations. One set of variables which is frequently employed is (ψ, u, v) which relates to the set (ψ, n, p) by:

$$n=n_i \cdot \exp\left(\frac{\psi}{Ut}\right) \cdot u \tag{5.2-1}$$

$$p=n_i \cdot \exp\left(-\frac{\psi}{Ut}\right) \cdot v \tag{5.2-2}$$

with:

$$Ut=\frac{k \cdot T}{q} \tag{5.2-3}$$

Substituting (5.2-1) and (5.2-2) into (5-6) to (5-8) yields:

$$\operatorname{div} \operatorname{grad} \psi - \frac{q}{\varepsilon} \cdot \left(n_i \cdot \exp\left(\frac{\psi}{Ut}\right) \cdot u - n_i \cdot \exp\left(-\frac{\psi}{Ut}\right) \cdot v - C\right)=0 \tag{5.2-4}$$

$$\operatorname{div}\left(D_n \cdot n_i \cdot \exp\left(\frac{\psi}{Ut}\right) \cdot \operatorname{grad} u\right) - R(\psi, u, v) =$$

$$= n_i \cdot \exp\left(\frac{\psi}{Ut}\right) \cdot \left(\frac{\partial u}{\partial t} + \frac{u}{Ut} \cdot \frac{\partial \psi}{\partial t}\right) \tag{5.2-5}$$

$$\operatorname{div}\left(D_p \cdot n_i \cdot \exp\left(-\frac{\psi}{Ut}\right) \cdot \operatorname{grad} v\right) - R(\psi, u, v) =$$

$$= n_i \cdot \exp\left(-\frac{\psi}{Ut}\right) \cdot \left(\frac{\partial v}{\partial t} - \frac{v}{Ut} \cdot \frac{\partial \psi}{\partial t}\right) \tag{5.2-6}$$

The current relations transform to:

$$\vec{J}_n = \quad q \cdot D_n \cdot n_i \cdot \exp\left(\frac{\psi}{Ut}\right) \cdot \operatorname{grad} u \tag{5.2-7}$$

$$\vec{J}_p = -q \cdot D_p \cdot n_i \cdot \exp\left(-\frac{\psi}{Ut}\right) \cdot \operatorname{grad} v \tag{5.2-8}$$

In (5.2-5) to (5.2-8) we have made use of the simple Einstein relations (5.2-9), (5.2-10) which are consistent with the assumed current relations (cf. Section 2.3).

$$D_n = \mu_n \cdot Ut \tag{5.2-9}$$

$$D_p = \mu_p \cdot Ut \tag{5.2-10}$$

The main advantage of the variables (ψ, u, v) becomes apparent if one considers the static semiconductor equations. Then the right hand sides of (5.2-5) and (5.2-6) are zero and the continuity equations are self-adjoint partial differential equations. The theory and solution methods for systems of self-adjoint partial differential equations have reached at present a very high standard such that a solution of the static semiconductor equations in (ψ, u, v) can be carried out very efficiently. Therefore, the dependent variables (ψ, u, v) have been used quite intensively, e.g. [5.1], [5.3], [5.14], [5.18], [5.36], [5.37]. However, the major drawback of the variables u and v lies in the enormous dynamic range required for real number representation in actual computations. By recalling the definitions (5.2-1) and (5.2-2) we find:

$$u = \frac{n}{n_i} \cdot \exp\left(-\frac{\psi}{Ut}\right) \tag{5.2-11}$$

$$v = \frac{p}{n_i} \cdot \exp\left(\frac{\psi}{Ut}\right) \tag{5.2-12}$$

At 300 K temperature Ut equals 0.02585 V. Thus the exponential terms in (5.2-11), (5.2-12) vary more than 32 orders of magnitude for $\psi \in [-1, 1]$ V. It is therefore obvious that computations with the variables u and v are strictly limited to low voltage applications. However, for analytical investigations the set (ψ, u, v) is in many situations superior to other variables.

Another set of dependent variables, namely $(\psi, \varphi_n, \varphi_p)$, is used as a natural consequence of the derivation of the current relations (cf. Section 2.3). This set relates to (ψ, n, p) through the well known Boltzmann approximations for the carrier concentrations.

$$n = n_i \cdot \exp\left(\frac{\psi - \varphi_n}{Ut}\right) \tag{5.2-13}$$

$$p = n_i \cdot \exp\left(\frac{\varphi_p - \psi}{Ut}\right) \tag{5.2-14}$$

φ_n and φ_p can be interpreted as the quasi-Fermi potentials of electrons and holes, respectively. However, (5.2-13) and (5.2-14) can certainly be regarded purely as a mathematical change of variables, so that the question of the validity of Boltzmann statistics is misplaced at this point. It has to be considered earlier, in the derivation of the current relations. The basic equations in the variables $(\psi, \varphi_n, \varphi_p)$ read:

$$\operatorname{div}\operatorname{grad}\psi - \frac{q}{\varepsilon} \cdot \left(n_i \cdot \exp\left(\frac{\psi - \varphi_n}{Ut}\right) - n_i \cdot \exp\left(\frac{\varphi_p - \psi}{Ut}\right) - C\right) = 0 \tag{5.2-15}$$

$$\operatorname{div}\left(\mu_n \cdot n_i \cdot \exp\left(\frac{\psi - \varphi_n}{Ut}\right) \cdot \operatorname{grad}\varphi_n\right) + R(\psi, \varphi_n, \varphi_p) =$$

$$= \frac{1}{Ut} \cdot n_i \cdot \exp\left(\frac{\psi - \varphi_n}{Ut}\right) \cdot \left(\frac{\partial \varphi_n}{\partial t} - \frac{\partial \psi}{\partial t}\right) \tag{5.2-16}$$

$$\operatorname{div}\left(\mu_p \cdot n_i \cdot \exp\left(\frac{\varphi_p - \psi}{Ut}\right) \cdot \operatorname{grad}\varphi_p\right) - R(\psi, \varphi_n, \varphi_p) =$$

$$= \frac{1}{Ut} \cdot n_i \cdot \exp\left(\frac{\varphi_p - \psi}{Ut}\right) \cdot \left(\frac{\partial \varphi_p}{\partial t} - \frac{\partial \psi}{\partial t}\right) \tag{5.2-17}$$

The current relations transform to:

$$\vec{J}_n = -q \cdot \mu_n \cdot n_i \cdot \exp\left(\frac{\psi - \varphi_n}{Ut}\right) \cdot \operatorname{grad}\varphi_n \tag{5.2-18}$$

$$\vec{J}_p = -q \cdot \mu_p \cdot n_i \cdot \exp\left(\frac{\varphi_p - \psi}{Ut}\right) \cdot \operatorname{grad}\varphi_p \tag{5.2-19}$$

Again we have made use of the Einstein relations (5.2-9) and (5.2-10). An advantage of the set $(\psi, \varphi_n, \varphi_p)$ is that all variables are of the same order of magnitude. Furthermore, negative carrier concentrations n and p are a priori excluded, which may appear due to numerical roundoff errors as undesired non-physical solutions if one uses (ψ, n, p) or (ψ, u, v) as dependent variables. As a drawback of the set $(\psi, \varphi_n, \varphi_p)$, it can be clearly seen that the current relations, and thus the continuity equations, are exponentially nonlinear in φ_n and φ_p. The dependent variables $(\psi, \varphi_n, \varphi_p)$ have been used for numerical computations by, e.g. [5.12], [5.13], [5.16], [5.17]. A comparison of the sets $(\psi, \varphi_n, \varphi_p)$ and (ψ, u, v) for computations has been performed in [5.10] with the anticipated result that for low bias problems the set (ψ, u, v) is preferable. I personally favor the set (ψ, n, p) for computations and I feel that this set should be recommended for that purpose.

We have not transformed the boundary conditions given in Section 5.1 to the above given sets of dependent variables. However, this is only a matter of straightforward calculations. It can therefore be omitted here.
The last set of dependent variables we shall briefly discuss are the stream functions which have been introduced for modeling semiconductor devices by Mock [5.25] using methods from fluid dynamics. The stream function technique has been further developed by Toyabe et al. [5.41] for application to bipolar devices. The basic idea is as follows. Assume we have a device domain D bounded by N smooth ideal ohmic contacts with Dirichlet boundary conditions and N piecewise smooth simplified interfaces (or artificial boundaries) with Neumann boundary conditions.

$$\partial D = \bigcup_{i=1}^{N} \partial D_{Oi} \cup \bigcup_{i=1}^{N} \partial D_{Ii} = \partial D_O \cup \partial D_I \tag{5.2-20}$$

Recalling the basic equations in the variables (ψ, u, v) we may readily split u and v into a recombination and a solenoidal part.

$$u = u_r + u_s \tag{5.2-21}$$

$$v = v_r + v_s \tag{5.2-22}$$

By just considering the static problem (i.e., right hand sides of (5.2-5), (5.2-6) are zero) we may require for u_r, the recombination part of u:

$$\operatorname{div}\left(D_n \cdot n_i \cdot \exp\left(\frac{\psi}{Ut}\right) \cdot \operatorname{grad} u_r\right) - R(\psi, u, v) = 0 \tag{5.2-23}$$

$$u_r|_{\partial D_O} = 0 \tag{5.2-24}$$

$$\vec{n} \cdot \operatorname{grad} u_r|_{\partial D_I} = 0 \tag{5.2-25}$$

and for u_s, the solenoidal part:

$$\operatorname{div}\left(D_n \cdot n_i \cdot \exp\left(\frac{\psi}{Ut}\right) \cdot \operatorname{grad} u_s\right) = 0 \tag{5.2-26}$$

$$u_s|_{\partial D_O} = u_D \tag{5.2-27}$$

$$\vec{n} \cdot \operatorname{grad} u_s|_{\partial D_I} = 0 \tag{5.2-28}$$

Similarly we have for v_r:

$$\operatorname{div}\left(D_p \cdot n_i \cdot \exp\left(-\frac{\psi}{Ut}\right) \cdot \operatorname{grad} v_r\right) - R(\psi, u, v) = 0 \tag{5.2-29}$$

$$v_r|_{\partial D_O} = 0 \tag{5.2-30}$$

$$\vec{n} \cdot \operatorname{grad} v_r|_{\partial D_I} = 0 \tag{5.2-31}$$

and for v_s:

$$\operatorname{div}\left(D_p \cdot n_i \cdot \exp\left(-\frac{\psi}{Ut}\right) \cdot \operatorname{grad} v_s\right) = 0 \tag{5.2-32}$$

$$v_s|_{\partial D_O} = v_D \tag{5.2-33}$$

$$\vec{n} \cdot \operatorname{grad} v_s |_{\partial D_I} = 0 \tag{5.2-34}$$

The solenoidal part of the current densities for a device with N contacts is conveniently split into $N-1$ components.

$$q \cdot D_n \cdot n_i \cdot \exp\left(\frac{\psi}{Ut}\right) \cdot \operatorname{grad} u_s = \sum_{i=1}^{N-1} I_{ni} \cdot \operatorname{grad} x\, \vec{\theta}_{ni} \tag{5.2-35}$$

$$q \cdot D_p \cdot n_i \cdot \exp\left(-\frac{\psi}{Ut}\right) \cdot \operatorname{grad} v_s = \sum_{i=1}^{N-1} I_{pi} \cdot \operatorname{grad} x\, \vec{\theta}_{pi} \tag{5.2-36}$$

The scalars I_{ni} and I_{pi} are the respective electron and hole current components flowing from an arbitrarily chosen ohmic contact ∂D_{Oj}, the source contact, to the $N-1$ other ohmic contacts ∂D_{Om}. $\vec{\theta}_{ni}$ and $\vec{\theta}_{pi}$ are the corresponding normalized stream functions. Fig. 5.2-1 shows an example for the possible arrangement of the source contacts and the current components I_{ni} and I_{pi} in a lateral bipolar transistor. As it is arbitrary, the reference contact has been chosen to be the emitter for the electron current components and the base for the hole current components, a choice which is physically motivated.

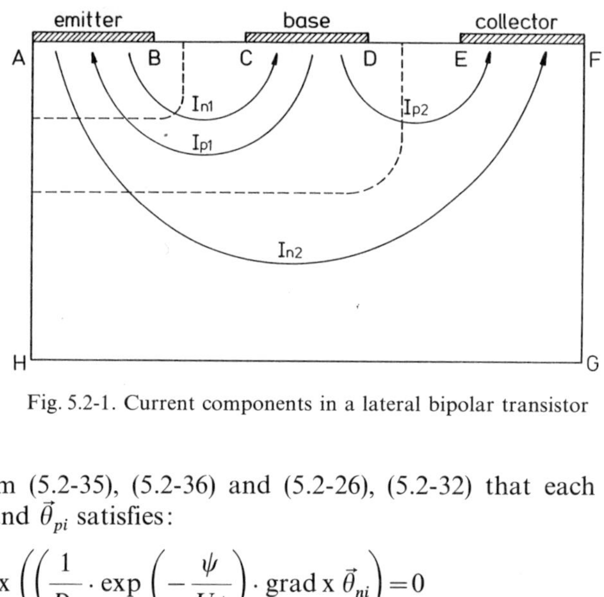

Fig. 5.2-1. Current components in a lateral bipolar transistor

It follows from (5.2-35), (5.2-36) and (5.2-26), (5.2-32) that each of the stream functions $\vec{\theta}_{ni}$ and $\vec{\theta}_{pi}$ satisfies:

$$\operatorname{grad} x \left(\left(\frac{1}{D_n} \cdot \exp\left(-\frac{\psi}{Ut}\right) \cdot \operatorname{grad} x\, \vec{\theta}_{ni}\right) = 0 \tag{5.2-37}$$

$$\operatorname{grad} x \left(\left(\frac{1}{D_p} \cdot \exp\left(\frac{\psi}{Ut}\right) \cdot \operatorname{grad} x\, \vec{\theta}_{pi}\right) = 0 \tag{5.2-38}$$

Furthermore, one has to chose calibration conditions for the stream functions which most conveniently are taken to be:

$$\operatorname{div} \vec{\theta}_{ni} = 0 \tag{5.2-39}$$

$$\operatorname{div} \vec{\theta}_{pi} = 0 \tag{5.2-40}$$

The boundary conditions for the stream functions are very complex, particularly for the three dimensional problem. Therefore, we shall specialize here to the two dimensional problem. In that case the stream function vectors simplify highly as only one component of the full vector remains significant.

$$\vec{\theta}_{ni} = \begin{pmatrix} 0 \\ 0 \\ \theta_{ni} \end{pmatrix} \tag{5.2-41}$$

$$\vec{\theta}_{pi} = \begin{pmatrix} 0 \\ 0 \\ \theta_{pi} \end{pmatrix} \tag{5.2-42}$$

The boundary conditions for the stream functions read then:

$$\vec{n} \cdot \operatorname{grad} \theta_{ni} \,|_{\partial D_o} = 0 \tag{5.2-43}$$

$$\theta_{ni} \,|_{\partial D_I} = 0 \ \text{ or } \ 1 \tag{5.2-44}$$

$$\vec{n} \cdot \operatorname{grad} \theta_{pi} \,|_{\partial D_o} = 0 \tag{5.2-45}$$

$$\theta_{pi} \,|_{\partial D_I} = 0 \ \text{ or } \ 1 \tag{5.2-46}$$

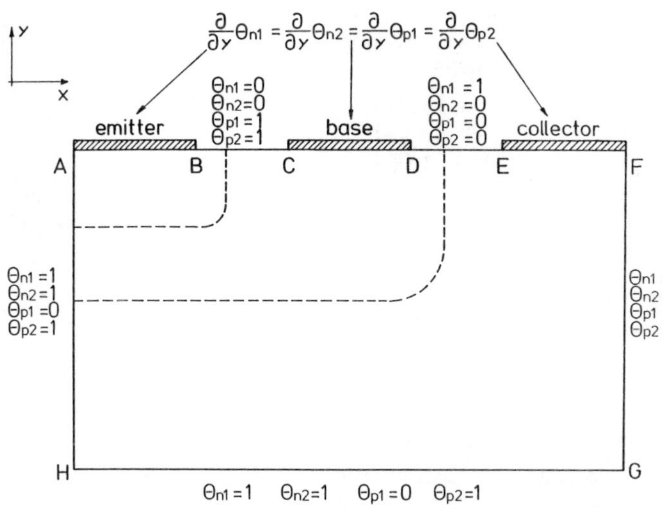

Fig. 5.2-2. Boundary conditions for stream functions in a lateral bipolar transistor

Whether a stream function is 0 or 1 on an insulating boundary segment is most easily explained verbally. If we choose arbitrarily the ohmic contact ∂D_{Oj} as source contact for all current components flowing to the $N-1$ other ohmic contacts ∂D_{Oi}, the stream functions are one (zero) on all insulating segments which are on the right (left) hand side on the path from ∂D_{Oj} to ∂D_{Oi}. Fig. 5.2-2 summarizes the boundary conditions for the example of Fig. 5.2-1. An extensive treatment of the theory of stream functions and the associated problems can be found in the book by Mock [5.29]. From my personal point of view the stream function approach is effective only for fairly simple problems, e.g. two dimensional simulation of MOS transistors where majority carrier current flow is neglected. For general application the complexity of the stream function approach becomes overwhelming and is not competitive with the approach of directly using the carrier concentrations as dependent variables.

5.3 The Existence of Solutions

The question about existence of solutions of the basic semiconductor equations with their associated boundary conditions for a specific device is important in view of mathematical considerations only. Physically, of course, solutions obviously exist. However, since the equations are models they are never exactly correct. How well posed mathematically a model is, gives in some sense a measure of the quality of the model. For instance, if one has a mathematical proof that no solutions exist for some model, this in general indicates that the model is wrong.

For the semiconductor equations it is impossible to prove the existence of solutions without restrictions on the parameters involved. These restrictions apply primarily to the recombination rate and the carrier mobilities. We shall first consider the stationary problem. The boundary conditions for all dependent variables have been assumed in all investigations I am aware of to be piecewise Dirichlet and Neumann conditions. In [5.27] Mock has given an existence proof for zero recombination and constant mobilities. Similar results have been obtained by Bank et al. [5.2], where the mobilities are allowed to be smooth functions of position. Zero recombination, however, was still a requirement. In [5.29] an existence proof has been given for a bounded recombination rate and constant carrier mobilities. Seidman [5.33] has shown the existence of solutions for Shockley-Read-Hall recombination and virtually no restrictions on the carrier mobilities (except for being bounded uniformly away from zero). Existence theorems which account for avalanche generation have not been published so far.

For the transient problem all important results concerning the existence of solutions have been presented by Mock [5.26], [5.29]. Solutions exist when the recombination rate is bounded and independent of the electrostatic potential. Furthermore the carrier mobilities have to be assumed to be smooth functions of position only, and the ratio of electron and hole mobility must be constant. Under these assumptions a solution exists for arbitrarily long time. If the carrier mobilities are constant, the solution is bounded. In case that the carrier mobilities depend on the electric field Mock could only show that solutions exist for $t<t1$, $t1$ sufficiently small.

It is interesting to note that the existence of solutions for the static problem depends primarily upon restrictions on the recombination rate, whereas for the transient problem the restrictions on the carrier mobilities are much stronger.

5.4 Uniqueness or Non-Uniqueness of Solutions

For the static semiconductor equations it cannot be expected that it is possible to prove generally the uniqueness of solutions. Physical mechanisms like avalanche generation and field dependent carrier mobilities are potential sources for multiple stationary solutions, e.g. [5.15], [5.17], [5.32]. Positive results on the uniqueness of a stationary solution can be expected only in a few situations. Mock [5.24] has proved the uniqueness of the equilibrium solution (homogeneous boundary conditions, i.e., no bias). Under the assumption of zero recombination and constant carrier mobilities Mock [5.28] could further show the uniqueness of the solution for sufficiently small bias. This proof has been extended in [5.29] for a bounded recombination rate which is independent of the electrostatic potential and has positive and bounded partial derivatives with respect to the carrier concentrations. Similar results have been obtained by Seidman [5.42]. More results, to my knowledge, are not available.

For the dynamic problem it is much easier to show uniqueness of solutions. It can generally be expected that a solution is unique for specified initial data (ψ, n, p) which fulfill the Poisson equation [5.26], [5.29].

5.5 Scaling

Since the dependent variables (ψ, n, p) in the basic equations (5-6), (5-7) and (5-8) are of greatly different orders of magnitude and show a strongly different behavior in regions with small and large space charge, the first step towards a structural analysis of the basic equations has to be appropriate scaling. A standard way of scaling has been given by DeMari [5.5], [5.6]. The scaling factors are summarized in Table 5.5-1.

Table 5.5-1. *Scaling factors after DeMari*

quantity	symbol	value
$\vec{x}$	x_o	$\sqrt{\varepsilon \cdot k \cdot T/(q^2 \cdot n_i)}$
ψ	ψ_o	$k \cdot T/q$
n, p, C	C_o	n_i
D_n, D_p	D_o	$1\,\text{cm}^2\,\text{s}^{-1}$
μ_n, μ_p		D_o/ψ_o
R		$D_o \cdot C_o/x_o^2$
t		x_o^2/D_o

$\vec{x}$ denotes the independent spatial variables. The scaled basic equations will then read:

$$\operatorname{div}\operatorname{grad}\psi - (n - p - C) = 0 \tag{5.5-1}$$

$$\operatorname{div}(D_n \cdot \operatorname{grad} n - \mu_n \cdot n \cdot \operatorname{grad}\psi) - R(\psi, n, p) = \frac{\partial n}{\partial t} \tag{5.5-2}$$

$$\operatorname{div}(D_p \cdot \operatorname{grad} p + \mu_p \cdot p \cdot \operatorname{grad} \psi) - R(\psi, n, p) = \frac{\partial p}{\partial t} \tag{5.5-3}$$

The scaled current relations are:

$$\vec{J}_n = \mu_n \cdot n \cdot \operatorname{grad} \psi - D_n \cdot \operatorname{grad} n \tag{5.5-4}$$

$$\vec{J}_p = \mu_p \cdot p \cdot \operatorname{grad} \psi + D_p \cdot \operatorname{grad} p \tag{5.5-5}$$

All quantities in (5.5-1) to (5.5-5) are scaled. The differential operators are taken with respect to the scaled independent variables. For the sake of transparency, however, an explicit indication (e.g. an index) has been omitted. It should further be noted that the equations (5.5-1) to (5.5-5) have been multiplied with a combination of the scaling factors, i.e.:

$$\frac{x_o^2}{\psi_o} \quad \text{for (5.5-1)}$$

$$\frac{x_o^2}{D_o \cdot C_o} \quad \text{for (5.5-2), (5.5-3)}$$

$$\frac{x_o}{-q \cdot D_o \cdot C_o} \quad \text{for (5.5-4), (5.5-5)}$$

The scaled equations are very attractive from one computational point of view, namely, there are no operations with constants involved in their evaluation. However, from a mathematical point of view this scaling is not satisfying. The scaled dependent variables are not at all of the same order of magnitude and the scaled space charge may take considerably large values.
A scaling which is more rigorous from the mathematical point of view has been introduced in [5.43], [5.44] and further developed in [5.19], [5.20]. The factors for this scaling are summarized in Table 5.5-2.

Table 5.5-2. "*Better*" *scaling factors*

quantity	symbol	value
$\vec{x}$	x_o	$\max(\vec{x} - \vec{y}),\ \vec{x}, \vec{y} \in D$
ψ	ψ_o	$k \cdot T/q$
n, p, C	C_o	$\max \lvert C(\vec{x}) \rvert,\ \vec{x} \in D$
D_n, D_p	D_o	$\max(D_n(\vec{x}), D_p(\vec{x})),\ \vec{x} \in D$
μ_n, μ_p		D_o/ψ_o
R		$D_o \cdot C_o/x_o^2$
t		x_o^2/D_o

The basic equations will transform with this scaling into:

$$\lambda^2 \cdot \operatorname{div} \operatorname{grad} \psi - (n - p - C) = 0 \tag{5.5-6}$$

$$\operatorname{div}(D_n \cdot \operatorname{grad} n - \mu_n \cdot n \cdot \operatorname{grad} \psi) - R(\psi, n, p) = \frac{\partial n}{\partial t} \tag{5.5-7}$$

$$\operatorname{div}(D_p \cdot \operatorname{grad} p + \mu_p \cdot p \cdot \operatorname{grad} \psi) - R(\psi, n, p) = \frac{\partial p}{\partial t} \tag{5.5-8}$$

with:

$$\lambda^2 = \frac{\psi_o \cdot \varepsilon}{x_o^2 \cdot q \cdot C_o} \tag{5.5-9}$$

The current relations after scaling appear formally identical to (5.5-4), (5.5-5). Analogously, the scaled continuity equations (5.5-7), (5.5-8) are formally identical to (5.5-2), (5.5-3). However, some of the scaling factors differ by orders of magnitude. The equations (5.5-6) to (5.5-8) and the current relations have also been multiplied with combinations of the scaling factors, namely:

$$\frac{\varepsilon}{q \cdot C_o} \quad \text{for (5.5-1)}$$

$$\frac{x_o^2}{D_o \cdot C_o} \quad \text{for (5.5-2), (5.5-3)}$$

$$\frac{x_o}{-q \cdot D_o \cdot C_o} \quad \text{for (5.5-4), (5.5-5)}$$

In order to illustrate the differences of these two scaling approaches a summary of the numerical values of the scaling factors is given in Table 5.5-3. A device with a 30 μm diameter of the simulation geometry, a maximum doping concentration of $10^{20}\,\text{cm}^{-3}$ and a maximum mobility of $1400\,\text{cm}^2\,\text{V}^{-1}\,\text{s}^{-1}$ has been assumed. The scaling factors are calculated for 300 K temperature.

Table 5.5.-3. *Numerical values of the scaling factors*

quantity	Table 5.5-1	Table 5.5-2
$\vec{x}$	$4.09 \cdot 10^{-3}$ cm	$3 \cdot 10^{-3}$ cm
ψ	0.0259 V	0.0259 V
n, p, C	$10^{10}\,\text{cm}^{-3}$	$10^{20}\,\text{cm}^{-3}$
D_n, D_p	$1\,\text{cm}^2\,\text{s}^{-1}$	$36.2\,\text{cm}^2\,\text{s}^{-1}$
μ_n, μ_p	$38.7\,\text{cm}^2\,\text{V}^{-1}\,\text{s}^{-1}$	$1400\,\text{cm}^2\,\text{V}^{-1}\,\text{s}^{-1}$
R	$6 \cdot 10^{14}\,\text{cm}^{-3}\,\text{s}^{-1}$	$4 \cdot 10^{26}\,\text{cm}^{-3}\,\text{s}^{-1}$
t	$1.67 \cdot 10^{-5}$ s	$2.49 \cdot 10^{-7}$ s

Most pronounced is the difference of the scaling factors for the carrier concentrations and the doping concentration. It is to note that in the scaled Poisson equation (5.5-6) the space charge is scaled to unity which allows one to take directly the residuals of (5.5-6) as a measure of the computational accuracy for actual computations. Analogous statements hold for the continuity equations. A comparison of the numerical values of the multiplication factors for the above given example is given in Table 5.5-4.

Table 5.5-4. *Numerical values of multiplying factors*

	Table 5.5-1	Table 5.5-2
Poisson equation	$6.47 \cdot 10^{-4}$	$6.47 \cdot 10^{-14}$
Continuity equations	$1.67 \cdot 10^{-15}$	$2.49 \cdot 10^{-27}$
Current relations	$2.55 \cdot 10^{6}$	$5.17 \cdot 10^{-6}$

The parameter λ^2 in (5.5-6) is a very small constant; its numerical value for the above given example is $1.86 \cdot 10^{-10}$. By physically reasoning λ can be identified as the scaled minimum Debye length in the device. The fact that the Laplacian in the Poisson equation is multiplied with a very small constant enables an asymptotic analysis of the basic semiconductor equations by means of singular perturbation theory. This will be the topic of the next section.

It should be noted that for high voltage problems it is preferable to scale the electrostatic potential with the maximum applied bias instead of the thermal voltage $k \cdot T/q$. Such a modification of the scaling has only a minor impact on the following analysis of the basic equations. However, in view of computational aspects it is advantageous since all scaled dependent variables are then maximally $O(1)$.

5.6 The Singular Perturbation Approach

The scaled basic semiconductor equations (5.5-6) to (5.5-8) constitute a singularly perturbed boundary value problem with perturbation parameter λ [5.21]. This interpretation is attractive because it allows the application of classical strategies for the qualitative analysis of the basic equations, e.g. [5.7], [5.9].

At first it seems intriguing to try to solve the singularly perturbed semiconductor equations by a regular expansion in λ.

$$w(\vec{x},t,\lambda)=\sum_{i=0}^{\infty} \tilde{w}_i(\vec{x},t)\cdot\lambda^i \tag{5.6-1}$$

with:

$$w=\begin{pmatrix}\psi\\ n\\ p\end{pmatrix} \tag{5.6-2}$$

It is to remark that the $\tilde{w}_i(\vec{x},t)$ are independent of λ and shall represent "slowly varying" quantities (slow compared to the variation of the doping profile across *pn*-junctions). Such an expansion implies a smooth dependence of the solution upon λ. By inserting (5.6-1) into equations (5.5-6) to (5.5-8) and setting λ equal to zero we obtain the so-called reduced problem.

$$\tilde{n}_o-\tilde{p}_o-C=0 \tag{5.6-3}$$

$$\operatorname{div}(D_n\cdot\operatorname{grad}\tilde{n}_o-\mu_n\cdot\tilde{n}_o\cdot\operatorname{grad}\tilde{\psi}_o)-R(\tilde{\psi}_o,\tilde{n}_o,\tilde{p}_o)=\frac{\partial\tilde{n}_o}{\partial t} \tag{5.6-4}$$

$$\operatorname{div}(D_p\cdot\operatorname{grad}\tilde{p}_o+\mu_p\cdot\tilde{p}_o\cdot\operatorname{grad}\tilde{\psi}_o)-R(\tilde{\psi}_o,\tilde{n}_o,\tilde{p}_o)=\frac{\partial\tilde{p}_o}{\partial t} \tag{5.6-5}$$

In semiconductor device physics this problem is called the charge-neutral approximation and has proved to be very valuable. Note, that it has been derived here very formally. However, as we have mentioned before the net doping distribution C varies rapidly; actually, it changes almost abruptly at *pn*-junctions. Therefore, the expansion (5.6-1) is insufficient to describe solutions of the basic equations completely. We can see this from relation (5.6-3) which cannot be fulfilled with slowly varying quantities $\tilde{n}_o$, $\tilde{p}_o$ for rapidly changing C.

A *pn*-junction can be described in more mathematical terms as an $n-1$ dimensional manifold Γ (n denotes the number of space dimensions) which splits the domain into two subdomains. If there exists more than one *pn*-junction in a device, which usually is the case, the simulation domain has to be split into subdomains whose "inner" boundaries represent the *pn*-junctions. Since the procedure to be outlined in the following is the same for each of the junctions it is demonstrated for one *pn*-junction only, without loss of generality. By partitioning the simulation geometry along the *pn*-junction into subdomains it is assured that the net doping concentration C does not change sign within each of the subdomains. We, furthermore, split the doping profile within each subdomain into two parts:

$$C = \tilde{C} + \hat{C} \tag{5.6-6}$$

$\tilde{C}$ represents the slowly varying (within the subdomain) contribution to the doping concentration, and $\hat{C}$ is rapidly (typically exponentially) decaying to zero away from the *pn*-junction. For an ideal abrupt junction $\hat{C}$ would be zero. We replace now the total doping concentration C in (5.4-3) by the smooth part $\tilde{C}$.

$$\tilde{n}_o - \tilde{p}_o - \tilde{C} = 0 \tag{5.6-7}$$

As the expansion (5.6-1) is inappropriate to describe the full solution in the vicinity of a *pn*-junction Γ it has to be supplemented by "layer" terms according to singular perturbation theory.

$$w(\vec{x}, t, \lambda) = \sum_{i=0}^{\infty} \tilde{w}_i(\vec{x}, t) \cdot \lambda^i + \sum_{i=0}^{\infty} \hat{w}_i\left(r, \frac{s}{\lambda}, t\right) \cdot \lambda^i \tag{5.6-8}$$

The following coordinates have been employed. $r = r(\vec{x})$ is the point on the *pn*-junction Γ closest to $\vec{x}$; $s = s(\vec{x})$ denotes the distance between $\vec{x}$ and Γ. These coordinates are illustrated in Fig. 5.6-1.

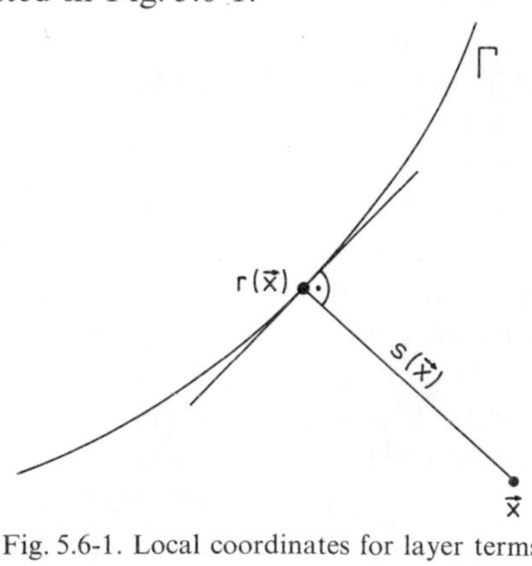

Fig. 5.6-1. Local coordinates for layer terms

If one evaluates the scaled continuity equations (5.5-7) and (5.5-8) after substituting (5.6-8) and carrying out the differentiation, one obtains, by comparing coefficients of order $O(\lambda^{-1})$, a system of ordinary differential equations in $(\hat{\psi}_o, \hat{n}_1, \hat{p}_1)$ (cf. [5.23]). The quantities $\hat{n}_o$, $\hat{p}_o$ vanish which means that no zero order internal layers exist. However, my concern here is only the reduced problem. In order to define this problem completely, four "interface conditions" have to be additionally imposed on the two second order equations (5.6-4) and (5.6-5) at the *pn*-junction Γ. These are obtained by matching the full solution (5.6-8) at the *pn*-junction.

$$\tilde{n}_o \cdot \exp\left(\frac{-\tilde{\psi}_o}{Ut}\right)\bigg|_{\vec{x}-} = \tilde{n}_o \cdot \exp\left(\frac{-\tilde{\psi}_o}{Ut}\right)\bigg|_{\vec{x}+} \tag{5.6-9}$$

$$\tilde{p}_o \cdot \exp\left(\frac{\tilde{\psi}_o}{Ut}\right)\bigg|_{\vec{x}-} = \tilde{p}_o \cdot \exp\left(\frac{\tilde{\psi}_o}{Ut}\right)\bigg|_{\vec{x}+} \tag{5.6-10}$$

$$\vec{n} \cdot \vec{J}_{\tilde{n}_o}|_{\vec{x}-} = \vec{n} \cdot \vec{J}_{\tilde{n}_o}|_{\vec{x}+} \tag{5.6-11}$$

$$\vec{n} \cdot \vec{J}_{\tilde{p}_o}|_{\vec{x}-} = \vec{n} \cdot \vec{J}_{\tilde{p}_o}|_{\vec{x}+} \tag{5.6-12}$$

$w|_{\vec{x}-}$ and $w|_{\vec{x}+}$ denote the onesided limits of w as $\vec{x}$ tends to Γ from each side of the junction. $\vec{n}$ denotes the unit normal vector on Γ. $\vec{J}_{\tilde{n}_o}$ and $\vec{J}_{\tilde{p}_o}$ are the zeroth order terms of the slowly varying parts of the scaled electron and hole current densities.

$$\vec{J}_{\tilde{n}_o} = \mu_n \cdot \tilde{n}_o \cdot \operatorname{grad} \tilde{\psi}_o - D_n \cdot \operatorname{grad} \tilde{n}_o \tag{5.6-13}$$

$$\vec{J}_{\tilde{p}_o} = \mu_p \cdot \tilde{p}_o \cdot \operatorname{grad} \tilde{\psi}_o + D_p \cdot \operatorname{grad} \tilde{p}_o \tag{5.6-14}$$

Equations (5.6-7) and (5.6-4), (5.6-5) together with the interface conditions (5.6-9) to (5.6-12) and the boundary conditions defined in Section 5.1 represent the final reduced problem whose solution is an $O(\lambda)$ approximation to the full solution away from Γ and away from the boundary segments ∂D_S, (∂D_I) as will be explained subsequently [5.19], [5.20], [5.38]. This reduced problem is a valuable tool for the development and analysis of numerical solution methods for the full problem, since the reduced problem, particularly conditions (5.6-9) to (5.6-12), has to be solved implicitly by any discretization method which should deliver sufficient accuracy for a reasonable number of grid points [5.30], [5.34], [5.35].

One crucial point in these considerations is the behavior of the solution of the full problem and of the expansion (5.6-8) close to the boundary ∂D. By setting λ equal to zero we lose one degree of freedom in imposing boundary conditions. Therefore, the reduced solution will only be close to the full solution (i.e., the solution of the singular perturbation problem (5.4-14)) near those parts of the boundary ∂D where the boundary conditions do not contradict the algebraic equation (5.6-7) [5.23]. This is the case for ideal ohmic contacts, but not for Schottky contacts and interfaces where large normal components of the electric field occur, which lead to boundary layers [5.22]. However, one can overcome this problem by adding to the boundary layers at Schottky contacts and interfaces layer terms in (5.6-8) in a similar manner as has been done for the internal layers at *pn*-junctions [5.23]. Anyway, the reduced solution will for many applications (numerical computations) constitute an excellent initial guess for the full solution, as has been confirmed by numerical experiments [5.11].

5.7 References

[5.1] Adachi, T., Yoshii, A., Sudo, T.: Two-Dimensional Semiconductor Analysis Using Finite-Element Method. IEEE Trans. Electron Devices *ED-26,* 1026 – 1031 (1979).

[5.2] Bank, R. E., Jerome, J. W., Rose, D. J.: Analytical and Numerical Aspects of Semiconductor Device Modeling. Report 82-11274-2, Bell Laboratories (1982).

[5.3] Chryssafis, A., Love, W.: A Computer-Aided Analysis of One-Dimensional Thermal Transients in n-p-n Power Transistors. Solid-State Electron. *22,* 249 – 256 (1979).

[5.4] Crowell, C. R., Beguwala, M.: Recombination Velocity Effects on Current Diffusion and IMREF in Schottky Barriers. Solid-State Electron. *4,* No. 11, 1149 – 1157 (1971).

[5.5] DeMari, A.: An Accurate Numerical Steady-State One-Dimensional Solution of the P-N Junction. Solid-State Electron. *11,* 33 – 58 (1968).

[5.6] DeMari, A.: An Accurate Numerical One-Dimensional Solution of the P-N Junction under Arbitrary Transient Conditions. Solid-State Electron. *11,* 1021 – 2053 (1968).

[5.7] Eckhaus, W.: Asymptotic Analysis of Singular Perturbations. Amsterdam: North-Holland 1979.

[5.8] Engl, W. L., Dirks, H. K., Meinerzhagen, B.: Device Modeling. Proc. IEEE *71,* No. 1, 10 – 33 (1983).

[5.9] Fife, P. C.: Semilinear Elliptic Boundary Value Problems with Small Parameters. Arch. Rat. Mech. Anal. *52,* 205 – 232 (1973).

[5.10] Fontana, T. P., McGregor, D. M., Lowther, R. P.: DEFINES: A Semiconductor Device Finite Element Simulation. Electrocon International Inc. 1982.

[5.11] Franz, A. F., Franz, G. A., Selberherr, S., Ringhofer, C., Markowich, P.: Finite Boxes – A Generalization of the Finite Difference Method Suitable for Semiconductor Device Simulation. IEEE Trans. Electron Devices *ED-30,* No. 9, 1070 – 1082 (1983).

[5.12] Hachtel, G. D., Mack, M. H., O'Brien, R. R., Speelpennig, B.: Semiconductor Analysis Using Finite Elements – Part 1: Computational Aspects. IBM J. Res. Dev. *25,* 232 – 245 (1981).

[5.13] Hachtel, G. D., Mack, M. H., O'Brien, R. R.: Semiconductor Analysis Using Finite Elements – Part 2: IGFET and BJT Case Studies. IBM J. Res. Dev. *25,* 246 – 260 (1981).

[5.14] Heimeier, H. H.: A Two-Dimensional Numerical Analysis of a Silicon N-P-N Transistor. IEEE Trans. Electron Devices *ED-20,* 708 – 714 (1973).

[5.15] Kurata, M.: Numerical Analysis for Semiconductor Devices. Lexington, Mass.: Lexington Press 1982.

[5.16] Laux, S. E.: Two-Dimensional Simulation of Gallium-Arsenide MESFET's Using the Finite-Element Method. Dissertation, University of Michigan, 1981.

[5.17] Laux, S. E., Lomax, R. J.: Effect of Mesh Spacing on Static Negative Resistance in GaAs MESFET Simulation. IEEE Trans. Electron Devices *ED-28,* No. 1, 120 – 122 (1981).

[5.18] Manck, O., Heimeier, H. H., Engl, W. L.: High Injection in a Two-Dimensional Transistor. IEEE Trans. Electron Devices *ED-21,* 403 – 409 (1974).

[5.19] Markowich, P. A., Ringhofer, C. A., Selberherr, S., Langer, E.: A Singularly Perturbed Boundary Value Problem Modelling a Semiconductor Device. Report 2388, MRC, University of Wisconsin, 1982.

[5.20] Markowich, P. A., Ringhofer, C. A., Langer, E., Selberherr, S.: An Asymptotic Analysis of Single-Junction Semiconductor Devices. Report 2527, MRC, University of Wisconsin, 1983.

[5.21] Markowich, P. A., Ringhofer, C. A., Selberherr, S.: A Singular Perturbation Approach for the Analysis of the Fundamental Semiconductor Equations. Report 2482, MRC, University of Wisconsin, 1983.

[5.22] Markowich, P. A.: A Singular Perturbation Analysis of the Fundamental Semiconductor Device Equations. Habilitation, Technische Universität Wien, 1983.

[5.23] Markowich, P. A.: A Qualitative Analysis of the Fundamental Semiconductor Device Equations. COMPEL *2,* No. 3, 97 – 115 (1983).

[5.24] Mock, M. S.: On Equations Describing Steady-State Carrier Distributions in a Semiconductor Device. Comm. Pure and Appl. Math. *25,* 781 – 792 (1972).

[5.25] Mock, M. S.: A Two-Dimensional Mathematical Model of the Insulated-Gate Field-Effect Transistor. Solid-State Electron. *16,* 601 – 609 (1973).

[5.26] Mock, M. S.: An Initial Value Problem from Semiconductor Device Theory. SIAM J. Math. Anal. *5,* No. 4, 597 – 612 (1974).

[5.27] Mock, M. S.: Asymptotic Behaviour of Solutions of Transport Equations for Semiconductor Devices. J. Math. Anal. Appl. *49,* 215 – 255 (1975).

[5.28] Mock, M. S.: An Example of Nonuniqueness of Stationary Solutions in Semiconductor Device Models. COMPEL *1,* No. 3, 165 – 174 (1982).

[5.29] Mock, M. S.: Analysis of Mathematical Models of Semiconductor Devices. Dublin: Boole Press 1983.

[5.30] Ringhofer, C., Selberherr, S.: Implications of Analytical Investigations about the Semiconductor Equations on Device Modeling Programs. Report 2513, MRC, University of Wisconsin, 1983.

[5.31] Schütz, A., Selberherr, S., Pötzl, H. W.: A Two-Dimensional Model of the Avalanche Effect in MOS Transistors. Solid-State Electron. *25,* 177 – 183 (1982).

[5.32] Schütz, A., Selberherr, S., Pötzl, H. W.: Analysis of Breakdown Phenomena in MOSFET's. IEEE Trans. Computer-Aided-Design of Integrated Circuits *CAD-1,* 77 – 85 (1982).

[5.33] Seidman, T. I.: Steady State Solutions of Diffusion-Reaction Systems with Electrostatic Convection. Nonlinear Analysis, Theory, Methods Appl. *4,* No. 3, 623 – 637 (1980).

[5.34] Selberherr, S., Ringhofer, C.: Discretization Methods for the Semiconductor Equations. Proc. NASECODE III Conf., pp. 31 – 45. Dublin: Boole Press 1983.

[5.35] Selberherr, S., Ringhofer, C.: Implications of Analytical Investigations about the Semiconductor Equations on Device Modeling Programs. IEEE Trans. Computer-Aided Design of Integrated Circuits *CAD-3,* No. 1, 52 – 64 (1984).

[5.36] Slotboom, J. W.: Iterative Scheme for 1- and 2-Dimensional D.C.-Transistor Simulation. Electron. Lett. *5,* 677 – 678 (1969).

[5.37] Slotboom, J. W.: Computer-Aided Two-Dimensional Analysis of Bipolar Transistors. IEEE Trans. Electron Devices *ED-20,* 669 – 679 (1973).

[5.38] Smith, D. R.: On a Singularly Perturbed Boundary Value Problem Arising in the Physical Theory of Semiconductors. Report 10-80-M21-200/1-FMA, Technische Universität München, 1980.

[5.39] Sze, S. M.: Physics of Semiconductor Devices. New York: Wiley 1969.

[5.40] Toyabe, T., Asai, S., Yamaguchi, K.: Internal Documents on the CADDET Program. Hitachi, Tokyo, 1979.

[5.41] Toyabe, T., Mock, M. S., Okabe, T., Ujiie, K., Nagata, M.: A Two-Dimensional Analysis of I2L with Multi-Stream Function Technique. Proc. NASECODE I Conf., pp. 290 – 292. Dublin: Boole Press 1979.

[5.42] VanDell, W. R.: Accuracy and Efficiency in High Power Semiconductor Device Modeling. Proc. NASECODE III Conf., pp. 281 – 286. Dublin: Boole Press 1983.

[5.43] Vasil'eva, A. B., Stel'makh, V. G.: Singularly Disturbed Systems of the Theory of Semiconductor Devices. Math. Fiz. *17,* No. 2, 339 – 348 (1977).

[5.44] Vasil'eva, A. B., Butuzov, V. F.: Singularly Perturbed Equations in the Critical Case. Translated Report 2039, MRC, University of Wisconsin, 1978.

6 The Discretization of the Basic Semiconductor Equations

The system of partial differential equations which forms the basic semiconductor equations together with appropriate boundary conditions has been investigated and characterized analytically in the previous chapter. This system cannot be solved explicitly in general. Therefore, the solution must be calculated by means of numerical approaches. We shall consider in this chapter such solution procedures for the scaled equations which read:

$$\lambda^2 \cdot \operatorname{div} \operatorname{grad} \psi - (n - p - C) = 0 \tag{6-1}$$

$$\operatorname{div} (D_n \cdot \operatorname{grad} n - \mu_n \cdot n \cdot \operatorname{grad} \psi) - R(\psi, n, p) = \frac{\partial n}{\partial t} \tag{6-2}$$

$$\operatorname{div} (D_p \cdot \operatorname{grad} p + \mu_p \cdot p \cdot \operatorname{grad} \psi) - R(\psi, n, p) = \frac{\partial p}{\partial t} \tag{6-3}$$

Any numerical approach for the solution of such a system consists essentially of three tasks. First, the domain, i.e. the simulation geometry of the device, has to be partitioned into a finite number of subdomains, in which the solution can be approximated easily with a desired accuracy. Secondly, the differential equations have to be approximated in each of the subdomains by algebraic equations which involve only values of the continuous dependent variables at discrete points in the domain and knowledge of the structure of the chosen functions which approximate the dependent variables within each of the subdomains. In that way one obtains a fairly large system of, in general nonlinear, algebraic equations with unknowns comprised of approximations of the continuous dependent variables at discrete points. The solution of this system is the final third task to be carried out. As this problem can be viewed rather independently it will be treated separately in Chapter 7.

It should be noted a priori that with the above outlined procedure it is impossible to obtain an exact solution of the analytically formulated problem. Instead one can obtain in the best case an exact solution of the nonlinear algebraic equations which form the discrete problem. Such a solution represents a good approximation to the solution of the analytically formulated problem depending upon the fineness of the partitions of the simulation subdomains and the suitability of the approximating functions for the dependent variables within the subdomains.

There are many classical methods which propose constructive possibilities for the partitioning of the domain (discretization) and the choice of functions to approximate the dependent variables within the subdomains. However, many techniques have been developed especially for the semiconductor equations. These techniques are probably not essential from a mathematical analysis point of view, but they allow an enormous gain in computational efficiency, which for the purpose of engineering can be indeed essential.

For the derivation of the discrete problem we shall discuss in this chapter the "finite difference method", the "finite box method" which indeed is just a more general finite difference method and the "finite element method". We shall concentrate somewhat more on the finite difference method as it has proved to be in some sense advantageous for the solution of the semiconductor equations compared to the finite element method. However, this statement is not rigorous in a mathematical sense; a serious mathematical preference for one method or the other cannot be given. The finite difference method and the finite element method are frequently considered mutually independent from the very beginning. However, it is often a matter of interpretation only to obtain exactly the same discrete equations from either a finite difference approach or a finite element approach [6.51].

For the sake of simplicity in nomenclature we shall primarily consider the case of two space dimensions in this chapter. Most results can be generalized to three space dimensions in a straightforward manner.

We shall first consider the solution of the static semiconductor equations which are obtained automatically if the boundary conditions for the electrostatic potential are time invariant. The partial derivatives of the electron and hole concentration with respect to time vanish identically in that case. Then we have to deal with an elliptic system of partial differential equations which reads in scaled form:

$$\lambda^2 \cdot \operatorname{div} \operatorname{grad} \psi - n + p + C = 0 \tag{6-4}$$

$$\operatorname{div}(D_n \cdot \operatorname{grad} n - \mu_n \cdot n \cdot \operatorname{grad} \psi) - R(\psi, n, p) = 0 \tag{6-5}$$

$$\operatorname{div}(D_p \cdot \operatorname{grad} p + \mu_p \cdot p \cdot \operatorname{grad} \psi) - R(\psi, n, p) = 0 \tag{6-6}$$

These static semiconductor equations are considered in Section 6.1, Section 6.2 and Section 6.3 for a specialized finite difference method, a finite box method and the finite element method, respectively. The transient problem given by (6-1), (6-2) and (6-3) is dealt with in Section 6.4. The heat flow equation is not considered explicitly, since its treatment is analogous in concept. Some constructive remarks for the design of appropriate discretization meshes are given in Section 6.5.

6.1 Finite Differences

In the classical method of finite differences the domain, in which a solution of a differential equation is sought, is partitioned into subregions by a mesh which is a set of meshlines parallel to the coordinate axes. This task is most easily accomplished for a rectangular domain because the boundaries of the domain are then straight lines parallel to the coordinate axes too; they coincide therefore with mesh lines. We put NX lines parallel to the y-axis and NY lines parallel to the x-axis through a rectangular domain. The first and the last line coincide with the boundaries. We have $NX \cdot NY$ intersection points of these lines on which an approximate solution for the differential equations (6-4), (6-5) and (6-6) is sought. An example of such a mesh is given in Fig. 6.1-1. In this example $NX=41$ and $NY=22$. The x-axis lies parallel to the surface with positive direction from emitter to collector and the y-axis is perpendicular to the surface with negative direction into the bulk. The total number of points is 902; 122 points are on the boundaries; 6 points lie on the emitter $A-B$; 9 points on the base $C-D$ and 6 points on the collector $E-F$. In the following we shall derive the algebraic equations for each of the meshpoints.

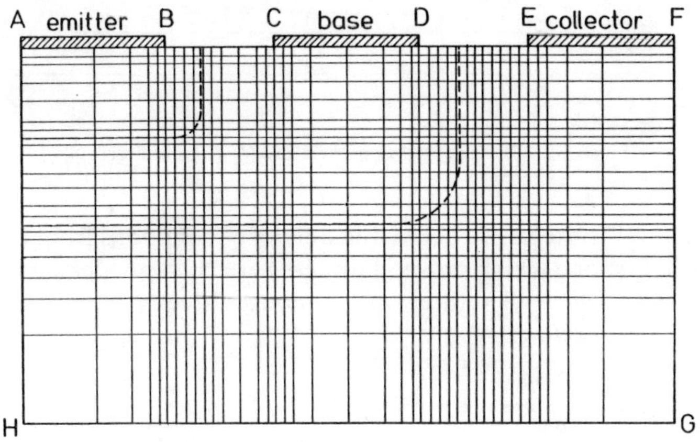

Fig. 6.1-1. Finite difference mesh for a lateral bipolar transistor

At first we replace the differential equations at the inner points by difference equations where only the nearest neighboring points for each of the inner points are invoked. We restrict ourselves here to the classical five point discretization. The adopted nomenclature is shown in Fig. 6.1-2. x_i (y_j) is assumed to be the distance from the origin to the i-th (j-th) mesh line parallel to the y-axis (x-axis). We shall use the following abbreviations:

$$h_i = x_{i+1} - x_i, \quad i = 1, NX-1 \tag{6.1-1}$$

$$k_j = y_{j+1} - y_j, \quad j = 1, NY-1 \tag{6.1-2}$$

To simplify the notation we further use:

$$u_{i,j} = u(x_i, y_j), \qquad i = 1, NX, \qquad j = 1, NY \tag{6.1-3}$$

$$u_{i+1/2,j} = u\left(\frac{x_i + x_{i+1}}{2}, y_j\right), \quad i = 1, NX-1, \; j = 1, NY \tag{6.1-4}$$

$$u_{i,j+1/2} = u\left(x_i, \frac{y_j + y_{j+1}}{2}\right), \quad i = 1, NX, \qquad j = 1, NY-1 \tag{6.1-5}$$

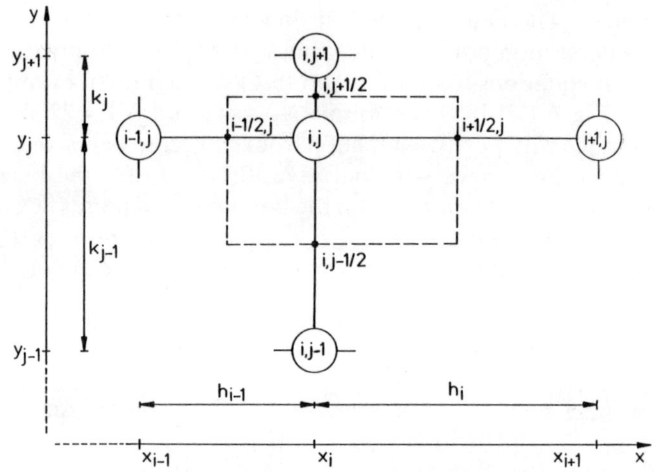

Fig. 6.1-2. The adopted nomenclature for finite differences

There exist several strategies to derive the difference equations from the differential equations. The first one which we shall discuss here is, in my personal opinion, the simplest one where the differential operators are directly replaced by difference operators. We first write explicitly "div" and "grad" in the basic equations.

$$\lambda^2 \cdot \left(\frac{\partial^2 \psi}{\partial x^2} + \frac{\partial^2 \psi}{\partial y^2}\right) - n + p + C = 0 \tag{6.1-6}$$

$$\frac{\partial}{\partial x}\left(D_n \cdot \frac{\partial n}{\partial x} - \mu_n \cdot n \cdot \frac{\partial \psi}{\partial x}\right) + \frac{\partial}{\partial y}\left(D_n \cdot \frac{\partial n}{\partial y} - \mu_n \cdot n \cdot \frac{\partial \psi}{\partial y}\right) - R(\psi, n, p) = 0 \tag{6.1-7}$$

$$\frac{\partial}{\partial x}\left(D_p \cdot \frac{\partial p}{\partial x} + \mu_p \cdot p \cdot \frac{\partial \psi}{\partial x}\right) + \frac{\partial}{\partial y}\left(D_p \cdot \frac{\partial p}{\partial y} + \mu_p \cdot p \cdot \frac{\partial \psi}{\partial y}\right) - R(\psi, n, p) = 0 \tag{6.1-8}$$

We have assumed, as already mentioned at the outset, that the problem is intrinsically two-dimensional which means that the partial derivatives of all parameters with respect to the third independent space variable are zero.
Assuming that u is three times continuously differentiable we replace all first order partial derivatives by:

$$\left.\frac{\partial u}{\partial x}\right|_{i,j} = \frac{u_{i+1/2,j} - u_{i-1/2,j}}{\frac{h_i + h_{i-1}}{2}} + \frac{h_{i-1} - h_i}{4} \cdot \left.\frac{\partial^2 u}{\partial x^2}\right|_{i,j} + O\left(\frac{h_i^3 + h_{i-1}{}^3}{h_i + h_{i-1}}\right) \tag{6.1-9}$$

$$\left.\frac{\partial u}{\partial y}\right|_{i,j} = \frac{u_{i,j+1/2} - u_{i,j-1/2}}{\frac{k_j + k_{j-1}}{2}} + \frac{k_{j-1} - k_j}{4} \cdot \left.\frac{\partial^2 u}{\partial y^2}\right|_{i,j} + O\left(\frac{k_j^3 + k_{j-1}{}^3}{k_j + k_{j-1}}\right) \tag{6.1-10}$$

We shall use in the following a pessimistic estimation of local truncation errors, which greatly simplifies the notation. The local truncation error is the residuum which occurs when inserting the solution of the continuous problem into the discrete scheme. We define:

$$h = \max(h_i; i=1, NX-1) \tag{6.1-11}$$

$$k = \max(k_j; j=1, NY-1) \tag{6.1-12}$$

We write $O(h)$, $O(h^2)$ etc. for the quantities in the truncation error which are controlled by linear, quadratic etc. expressions in the local x-mesh spacing and similarly $O(k)$, $O(k^2)$ etc. The local truncation error in (6.1-9), (6.1-10) for an equidistant mesh or even for a quasi-uniform mesh is of second order in the mesh spacing weighted with the third partial derivatives. A quasi-uniform mesh is a mesh for which (6.1-13), (6.1-14) holds:

$$h_{i+1} = h_i \cdot (1 + O(h_i)), \quad i=1, NX-2 \tag{6.1-13}$$

$$k_{j+1} = k_j \cdot (1 + O(k_j)), \quad j=1, NY-2 \tag{6.1-14}$$

However, as outlined in the previous chapter, the solution of the semiconductor equations exhibits a smooth behavior in some regions of the device domain whereas in others it varies rapidly. Thus a strongly non-uniform mesh is often mandatory. We can therefore only expect, pessimistically, a truncation error of first order in the mesh spacing and we shall assume throughout the following a non-uniform mesh. With the difference approximations (6.1-9) and (6.1-10) we rewrite the basic equations (6.1-6) to (6.1-8) at all inner points $1<i<NX$, $1<j<NY$. We obtain for the Poisson equation:

$$\lambda^2 \cdot \left(\frac{\left.\frac{\partial \psi}{\partial x}\right|_{i+1/2,j} - \left.\frac{\partial \psi}{\partial x}\right|_{i-1/2,j}}{\frac{h_i + h_{i-1}}{2}} + O(h) \cdot \left.\frac{\partial^3 \psi}{\partial x^3}\right|_{i,j} + \right.$$
$$\left. + \frac{\left.\frac{\partial \psi}{\partial y}\right|_{i,j+1/2} - \left.\frac{\partial \psi}{\partial y}\right|_{i,j-1/2}}{\frac{k_j + k_{j-1}}{2}} + O(k) \cdot \left.\frac{\partial^3 \psi}{\partial y^3}\right|_{i,j} \right) -$$
$$- n_{i,j} + p_{i,j} + C_{i,j} = 0 \tag{6.1-15}$$

The continuity equation for electrons reads:

$$\frac{(-J_{nx})|_{i+1/2,j} - (-J_{nx})|_{i-1/2,j}}{\frac{h_i + h_{i-1}}{2}} + O(h) \cdot \left.\frac{\partial^2}{\partial x^2} J_{nx}\right|_{i,j} +$$
$$+ \frac{(-J_{ny})|_{i,j+1/2} - (-J_{ny})|_{i,j-1/2}}{\frac{k_j + k_{j-1}}{2}} + O(k) \cdot \left.\frac{\partial^2}{\partial y^2} J_{ny}\right|_{i,j} -$$
$$- R(\psi, n, p)|_{i,j} = 0 \tag{6.1-16}$$

J_{nx} and J_{ny} are the scaled electron current density components in x and y direction, respectively.

$$J_{nx}=\mu_n\cdot n\cdot\frac{\partial\psi}{\partial x}-D_n\cdot\frac{\partial n}{\partial x} \tag{6.1-17}$$

$$J_{ny}=\mu_n\cdot n\cdot\frac{\partial\psi}{\partial y}-D_n\cdot\frac{\partial n}{\partial y} \tag{6.1-18}$$

Similarly, the continuity equation for holes reads:

$$\frac{J_{px}|_{i+1/2,j}-J_{px}|_{i-1/2,j}}{\frac{h_i+h_{i-1}}{2}}+O(h)\cdot\frac{\partial^2}{\partial x^2}J_{px}\bigg|_{i,j}+$$
$$+\frac{J_{py}|_{i,j+1/2}-J_{py}|_{i,j-1/2}}{\frac{k_j+k_{j-1}}{2}}+O(k)\cdot\frac{\partial^2}{\partial y^2}J_{py}\bigg|_{i,j}-$$
$$-R(\psi,n,p)|_{i,j}=0 \tag{6.1-19}$$

J_{px} and J_{py} are the scaled hole current density components in x and y direction, respectively.

$$J_{px}=\mu_p\cdot p\cdot\frac{\partial\psi}{\partial x}+D_p\cdot\frac{\partial p}{\partial x} \tag{6.1-20}$$

$$J_{py}=\mu_p\cdot p\cdot\frac{\partial\psi}{\partial y}+D_p\cdot\frac{\partial p}{\partial y} \tag{6.1-21}$$

The next step is to replace the midinterval values of the quantities $\partial\psi/\partial x$, $\partial\psi/\partial y$, J_{nx}, J_{ny}, J_{px}, J_{py} with an appropriate difference approximation. We use the assumption that these quantities are constant within each interval. We obtain for the partial derivatives of the electrostatic potential:

$$\frac{\partial\psi}{\partial x}\bigg|_{i+1/2,j}=\frac{\psi_{i+1,j}-\psi_{i,j}}{h_i}+O(h^2)\cdot\frac{\partial^3\psi}{\partial x^3}\bigg|_{i+1/2,j} \tag{6.1-22}$$

$$\frac{\partial\psi}{\partial y}\bigg|_{i,j+1/2}=\frac{\psi_{i,j+1}-\psi_{i,j}}{k_j}+O(k^2)\cdot\frac{\partial^3\psi}{\partial y^3}\bigg|_{i,j+1/2} \tag{6.1-23}$$

These approximations substituted into (6.1-15) do not increase the order of the local truncation error. We obtain after the substitution:

$$\lambda^2\cdot\left(\frac{\frac{\psi_{i+1,j}-\psi_{i,j}}{h_i}-\frac{\psi_{i,j}-\psi_{i-1,j}}{h_{i-1}}}{\frac{h_i+h_{i-1}}{2}}+\frac{\frac{\psi_{i,j+1}-\psi_{i,j}}{k_j}-\frac{\psi_{i,j}-\psi_{i,j-1}}{k_{j-1}}}{\frac{k_j+k_{j-1}}{2}}\right)-n_{i,j}+p_{i,j}+C_{i,j}=0 \tag{6.1-24}$$

The discrete Poisson equation (6.1-24) has a local truncation error linearly proportional to the mesh spacing and the third partial derivatives of the electrostatic potential for a non-uniform mesh. As mentioned earlier this statement might be too pessimistic for some applications. We have assumed that the electrostatic potential varies linearly on the paths to the four nearest neighboring mesh points. This implies that the electric field components on these paths are constant. It is to remark explicitly that we have no information about the variation of the electrostatic potential off these paths.
The discretization of the continuity equations is much more crucial. First, we write the approximations (6.1-25) to (6.1-28).

$$J_{nx}(x\in[x_i,x_{i+1}],y_j)=J_{nx}|_{i+1/2,j}+\left(x-x_i-\frac{h_i}{2}\right)\cdot\frac{\partial}{\partial x}J_{nx}\bigg|_{i+1/2,j}+ \\ +O(h^2)\cdot\frac{\partial^2}{\partial x^2}J_{nx}\bigg|_{i+1/2,j} \tag{6.1-25}$$

$$J_{ny}(x_i,y\in[y_j,y_{j+1}])=J_{ny}|_{i,j+1/2}+\left(y-y_j-\frac{k_j}{2}\right)\cdot\frac{\partial}{\partial y}J_{ny}\bigg|_{i,j+1/2}+ \\ +O(k^2)\cdot\frac{\partial^2}{\partial y^2}J_{ny}\bigg|_{i,j+1/2} \tag{6.1-26}$$

$$J_{px}(x\in[x_i,x_{i+1}],y_j)=J_{px}|_{i+1/2,j}+\left(x-x_i-\frac{h_i}{2}\right)\cdot\frac{\partial}{\partial x}J_{px}\bigg|_{i+1/2,j}+ \\ +O(h^2)\cdot\frac{\partial^2}{\partial x^2}J_{px}\bigg|_{i+1/2,j} \tag{6.1-27}$$

$$J_{py}(x_i,y\in[y_j,y_{j+1}])=J_{py}|_{i,j+1/2}+\left(y-y_j-\frac{k_j}{2}\right)\cdot\frac{\partial}{\partial y}J_{py}\bigg|_{i,j+1/2}+ \\ +O(k^2)\cdot\frac{\partial^2}{\partial y^2}J_{py}\bigg|_{i,j+1/2} \tag{6.1-28}$$

We obtain differential equations for the carrier concentrations n and p for each meshinterval by ignoring the $O(h^2)$, $O(k^2)$ terms. For instance, for the interval $[x_i, x_{i+1}]$ we have for the electron concentration:

$$\mu_n\cdot n\frac{\partial\psi}{\partial x}-D_n\cdot\frac{\partial n}{\partial x}=J_{nx}|_{i+1/2,j}+\left(x-x_i-\frac{h_i}{2}\right)\cdot\frac{\partial}{\partial x}J_{nx}\bigg|_{i+1/2,j} \tag{6.1-29}$$

$$n(x_i,y_j)=n_{i,j},\quad n(x_{i+1},y_j)=n_{i+1,j} \tag{6.1-30}$$

This equation is solved to determine the variation of the electron concentration along the path $(x\in[x_i,x_{i+1}],y_j)$. We have to assume that the partial derivative of the electrostatic potential is constant on the path under consideration, which is the assumption we have already invoked for the Poisson equation (cf. (6.1-22)). Furthermore, a scaled Einstein relation is assumed to hold for the scaled carrier diffusivities and mobilities; both quantities are assumed to be constant on the integration path. (6.1-29) represents a first order differential equation with one parameter ($J_{nx}|_{i+1/2,j}$) subject to two boundary conditions (6.1-30). The solution to this problem (the boundary conditions are not matched so far) reads:

$$n(x\in[x_i,x_{i+1}],y_j)=C\cdot\exp\left(\frac{\psi(x,y_j)}{Ut}\right)+h_i\cdot\frac{J_{nx}|_{i+1/2,j}}{\mu_n|_{i+1/2,j}}\cdot\frac{1-\exp\left(\dfrac{\psi(x,y_j)}{Ut}\right)}{\psi_{i+1,j}-\psi_{i,j}}+$$
$$+h_i^2\cdot\frac{\dfrac{\partial}{\partial x}J_{nx}\Big|_{i+1/2,j}}{\mu_n|_{i+1/2,j}}\cdot\frac{\left(\dfrac{Ut}{\psi_{i+1,j}-\psi_{i,j}}-\dfrac{1}{2}\right)\cdot\left(1-\exp\left(\dfrac{\psi(x,y_j)}{Ut}\right)\right)+\dfrac{x-x_i}{h_i}}{\psi_{i+1,j}-\psi_{i,j}} \tag{6.1-31}$$

Note that Ut in (6.1-31) represents the scaled thermal voltage. This quantity is, obviously, one if the electrostatic potential is scaled with the thermal voltage. However, as it is sometimes advantageous to scale the electrostatic potential differently (cf. Section 5.5) I shall not make use of such a simplification here. Assuming that $|\psi_{i+1,j}-\psi_{i,j}|=O(h)$ a laborious but simple calculation shows that the last term in (6.1-31) is $O(h^3)$. By ignoring this term and matching the boundary conditions (6.1-30) we obtain:

$$n(x\in[x_i,x_{i+1}],y_j)=(1-g_{i,j}^x(x,\psi))\cdot n_{i,j}+g_{i,j}^x(x,\psi)\cdot n_{i+1,j} \tag{6.1-32}$$

with:

$$g_{i,j}^x(x,\psi)=\frac{1-\exp\left(\dfrac{\psi_{i+1,j}-\psi_{i,j}}{Ut}\cdot\dfrac{x-x_i}{h_i}\right)}{1-\exp\left(\dfrac{\psi_{i+1,j}-\psi_{i,j}}{Ut}\right)} \tag{6.1-33}$$

Note that the growth function (6.1-33) degenerates (as expected) for $\psi_{i+1,j}=\psi_{i,j}$ to a linear function.

$$g_{i,j}^x(x,\psi)=\frac{x-x_i}{h_i} \tag{6.1-34}$$

Fully analogously we obtain from (6.1-26), (6.1-27) and (6.1-28) the following expressions:

$$n(x_i,y\in[y_j,y_{j+1}])=(1-g_{i,j}^y(y,\psi))\cdot n_{i,j}+g_{i,j}^y(y,\psi)\cdot n_{i,j+1} \tag{6.1-35}$$

with:

$$g_{i,j}^y(y,\psi)=\frac{1-\exp\left(\dfrac{\psi_{i,j+1}-\psi_{i,j}}{Ut}\cdot\dfrac{y-y_j}{k_j}\right)}{1-\exp\left(\dfrac{\psi_{i,j+1}-\psi_{i,j}}{Ut}\right)} \tag{6.1-36}$$

$$p(x\in[x_i,x_{i+1}],y_j)=(1-g_{i,j}^x(x,-\psi))\cdot p_{i,j}+g_{i,j}^x(x,-\psi)\cdot p_{i+1,j} \tag{6.1-37}$$

$$p(x_i,y\in[y_j,y_{j+1}])=(1-g_{i,j}^y(y,-\psi))\cdot p_{i,j}+g_{i,j}^y(y,-\psi)\cdot p_{i,j+1} \tag{6.1-38}$$

The growth function (6.1-33) is shown in Fig. 6.1-3 in a normalized interval with $(\psi_{i+1,j}-\psi_{i,j})/Ut$ as parameter.

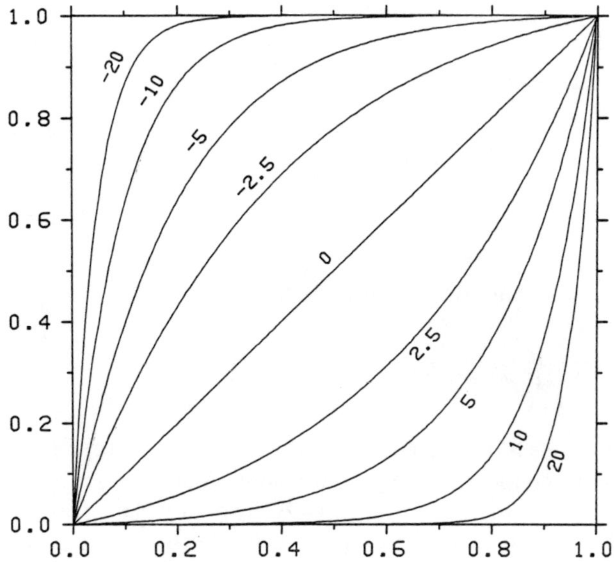

Fig. 6.1-3. Growth function of carrier concentrations in a finite difference interval

The required current density components of each of the intervals, which are the parameters of the respective differential equations for the carrier concentrations, are evaluated to (including the last term in, e.g., (6.1-31)):

$$J_{nx}\big|_{i+1/2,j} = D_n\big|_{i+1/2,j} \cdot \frac{B\left(\dfrac{\psi_{i,j}-\psi_{i+1,j}}{Ut}\right)\cdot n_{i,j} - B\left(\dfrac{\psi_{i+1,j}-\psi_{i,j}}{Ut}\right)\cdot n_{i+1,j}}{h_i} + \\ + h_i \cdot \left(\frac{1}{2}\cdot \coth\left(\frac{\psi_{i+1,j}-\psi_{i,j}}{2\cdot Ut}\right) - \frac{Ut}{\psi_{i+1,j}-\psi_{i,j}}\right)\cdot \frac{\partial}{\partial x} J_{nx}\bigg|_{i+1/2,j} \tag{6.1-39}$$

$$J_{ny}\big|_{i,j+1/2} = D_n\big|_{i,j+1/2} \cdot \frac{B\left(\dfrac{\psi_{i,j}-\psi_{i,j+1}}{Ut}\right)\cdot n_{i,j} - B\left(\dfrac{\psi_{i,j+1}-\psi_{i,j}}{Ut}\right)\cdot n_{i,j+1}}{k_j} + \\ + k_j \cdot \left(\frac{1}{2}\cdot \coth\left(\frac{\psi_{i,j+1}-\psi_{i,j}}{2\cdot Ut}\right) - \frac{Ut}{\psi_{i,j+1}-\psi_{i,j}}\right)\cdot \frac{\partial}{\partial y} J_{ny}\bigg|_{i,j+1/2} \tag{6.1-40}$$

$$J_{px}\big|_{i+1/2,j} = D_p\big|_{i+1/2,j} \cdot \frac{B\left(\dfrac{\psi_{i,j}-\psi_{i+1,j}}{Ut}\right)\cdot p_{i+1,j} - B\left(\dfrac{\psi_{i+1,j}-\psi_{i,j}}{Ut}\right)\cdot p_{i,j}}{h_i} + \\ + h_i \cdot \left(\frac{1}{2}\cdot \coth\left(\frac{\psi_{i,j}-\psi_{i+1,j}}{2\cdot Ut}\right) - \frac{Ut}{\psi_{i,j}-\psi_{i+1,j}}\right)\cdot \frac{\partial}{\partial x} J_{px}\bigg|_{i+1/2,j} \tag{6.1-41}$$

$$J_{py}|_{i,j+1/2} = D_p|_{i,j+1/2} \cdot \frac{B\left(\dfrac{\psi_{i,j}-\psi_{i,j+1}}{Ut}\right)\cdot p_{i,j+1} - B\left(\dfrac{\psi_{i,j+1}-\psi_{i,j}}{Ut}\right)\cdot p_{i,j}}{k_j} +$$
$$+ k_j \cdot \left(\frac{1}{2}\cdot \coth\left(\frac{\psi_{i,j}-\psi_{i,j+1}}{2\cdot Ut}\right) - \frac{Ut}{\psi_{i,j}-\psi_{i,j+1}}\right)\cdot \frac{\partial}{\partial y} J_{py}\bigg|_{i,j+1/2} \tag{6.1-42}$$

$B(x)$ is the Bernoulli function which is defined as:

$$B(x) = \frac{x}{e^x - 1} \tag{6.1-43}$$

These approximations have been first suggested by Scharfetter and Gummel [6.48]; they have been derived more heuristically though. Again, under the assumption that $\psi_{i,j}$ only differs by $O(h)$ and $O(k)$, respectively, from any of its nearest neighbors, one can easily show that the last terms in (6.1-39) to (6.1-42) are $O(h^2)$ and $O(k^2)$, respectively. By ignoring these terms and by substituting (6.1-39) and (6.1-40) into (6.1-16) we obtain the discrete form of the continuity equation for electrons.

$$D_n|_{i+1/2,j} \cdot \frac{B\left(\dfrac{\psi_{i+1,j}-\psi_{i,j}}{Ut}\right)\cdot n_{i+1,j} - B\left(\dfrac{\psi_{i,j}-\psi_{i+1,j}}{Ut}\right)\cdot n_{i,j}}{h_i \cdot \dfrac{h_i+h_{i-1}}{2}} -$$

$$-D_n|_{i-1/2,j} \cdot \frac{B\left(\dfrac{\psi_{i,j}-\psi_{i-1,j}}{Ut}\right)\cdot n_{i,j} - B\left(\dfrac{\psi_{i-1,j}-\psi_{i,j}}{Ut}\right)\cdot n_{i-1,j}}{h_{i-1} \cdot \dfrac{h_i+h_{i-1}}{2}} +$$

$$+D_n|_{i,j+1/2} \cdot \frac{B\left(\dfrac{\psi_{i,j+1}-\psi_{i,j}}{Ut}\right)\cdot n_{i,j+1} - B\left(\dfrac{\psi_{i,j}-\psi_{i,j+1}}{Ut}\right)\cdot n_{i,j}}{k_j \cdot \dfrac{k_j+k_{j-1}}{2}} -$$

$$-D_n|_{i,j-1/2} \cdot \frac{B\left(\dfrac{\psi_{i,j}-\psi_{i,j-1}}{Ut}\right)\cdot n_{i,j} - B\left(\dfrac{\psi_{i,j-1}-\psi_{i,j}}{Ut}\right)\cdot n_{i,j-1}}{k_{j-1} \cdot \dfrac{k_j+k_{j-1}}{2}} -$$

$$-R(\psi, n, p)|_{i,j} = 0 \tag{6.1-44}$$

Fully analogously we obtain the discretized continuity equation for holes.

$$D_p|_{i+1/2,j} \cdot \frac{B\left(\frac{\psi_{i,j}-\psi_{i+1,j}}{Ut}\right) \cdot p_{i+1,j} - B\left(\frac{\psi_{i+1,j}-\psi_{i,j}}{Ut}\right) \cdot p_{i,j}}{h_i \cdot \frac{h_i+h_{i-1}}{2}} -$$

$$-D_p|_{i-1/2,j} \cdot \frac{B\left(\frac{\psi_{i-1,j}-\psi_{i,j}}{Ut}\right) \cdot p_{i,j} - B\left(\frac{\psi_{i,j}-\psi_{i-1,j}}{Ut}\right) \cdot p_{i-1,j}}{h_{i-1} \cdot \frac{h_i+h_{i-1}}{2}} +$$

$$+D_p|_{i,j+1/2} \cdot \frac{B\left(\frac{\psi_{i,j}-\psi_{i,j+1}}{Ut}\right) \cdot p_{i,j+1} - B\left(\frac{\psi_{i,j+1}-\psi_{i,j}}{Ut}\right) \cdot p_{i,j}}{k_j \cdot \frac{k_j+k_{j-1}}{2}} -$$

$$-D_p|_{i,j-1/2} \cdot \frac{B\left(\frac{\psi_{i,j-1}-\psi_{i,j}}{Ut}\right) \cdot p_{i,j} - B\left(\frac{\psi_{i,j}-\psi_{i,j-1}}{Ut}\right) \cdot p_{i,j-1}}{k_{j-1} \cdot \frac{k_j+k_{j-1}}{2}} -$$

$$-R(\psi, n, p)|_{i,j}=0 \tag{6.1-45}$$

The discrete continuity equations (6.1-44), (6.1-45) have a local truncation error linearly proportional to the mesh spacing and the sum of the first and second partial derivatives of the respective current density components, provided the electrostatic potential is resolved by the mesh to first order accuracy. However, if the mesh is quasi-uniform (cf. (6.1-13)), the assumption on the resolution of the electrostatic potential can be waved. This can be seen by considering, e.g., (6.1-16), (6.1-39) and (6.1-40). Such discretization schemes which exponentially fit the dependent variables have been investigated also by, e.g., Kellogg [6.22], Kellogg et al. [6.23] and Doolan et al. [6.10].

The local truncation errors for all three equations are summarized in (6.1-46) to (6.1-48). The index denotes the respective equation.

$$T_\psi < O(h) \cdot \left|\frac{\partial^3 \psi}{\partial x^3}\right| + O(k) \cdot \left|\frac{\partial^3 \psi}{\partial y^3}\right| \tag{6.1-46}$$

$$T_n < O(h) \cdot \left|\frac{\partial J_{nx}}{\partial x}\right| + O(k) \cdot \left|\frac{\partial J_{ny}}{\partial y}\right| + O(h) \cdot \left|\frac{\partial^2 J_{nx}}{\partial x^2}\right| + O(k) \cdot \left|\frac{\partial^2 J_{ny}}{\partial y^2}\right| \tag{6.1-47}$$

$$T_p < O(h) \cdot \left|\frac{\partial J_{px}}{\partial x}\right| + O(k) \cdot \left|\frac{\partial J_{py}}{\partial y}\right| + O(h) \cdot \left|\frac{\partial^2 J_{px}}{\partial x^2}\right| + O(k) \cdot \left|\frac{\partial^2 J_{py}}{\partial y^2}\right| \tag{6.1-48}$$

A thorough investigation of these errors and the associated convergence properties, which is beyond the scope of this more towards engineering oriented text, can be found in e.g. [6.32], [6.33].
Another approach for deriving difference approximations to differential equations is the box integration method [6.13]. The differential equation is integrated over each of the following subdomains:

$$D_{i,j}=\left\{x_i-\frac{h_i-1}{2}\leq x\leq x_i+\frac{h_i}{2},\ y_j-\frac{k_{j-1}}{2}\leq y\leq y_j+\frac{k_j}{2}\right\} \tag{6.1-49}$$

These subdomains partition the interior of the domain without overlap or exclusion. For the Poisson equation we obtain:

$$\begin{aligned}&\int\limits_{x_i-\frac{h_{i-1}}{2}}^{x_i+\frac{h_i}{2}}\int\limits_{y_j-\frac{k_{j-1}}{2}}^{y_j+\frac{k_j}{2}}\lambda^2\cdot\operatorname{div}\operatorname{grad}\psi\cdot dy\cdot dx-\\&-\int\limits_{x_i-\frac{h_{i-1}}{2}}^{x_i+\frac{h_i}{2}}\int\limits_{y_j-\frac{k_{j-1}}{2}}^{y_j+\frac{k_j}{2}}(n-p-C)\cdot dy\cdot dx=0\end{aligned} \tag{6.1-50}$$

The first integral in (6.1-50) is transformed into a boundary integral by using one of Green's theorems:

$$\iint\limits_{D_{i,j}}\operatorname{div}(P\cdot\operatorname{grad}u)\cdot dx\cdot dy=\int\limits_{\partial D_{i,j}}\left(P\cdot\frac{\partial u}{\partial x}\cdot dy-P\cdot\frac{\partial u}{\partial y}\cdot dx\right) \tag{6.1-51}$$

The right hand side integral has to be split into four parts because of the discontinuous edges of the domain $D_{i,j}$. We obtain for (6.1-51) in more explicit form:

$$\begin{aligned}&\int\limits_{x_i-\frac{h_{i-1}}{2}}^{x_i+\frac{h_i}{2}}\int\limits_{y_j-\frac{k_{j-1}}{2}}^{y_j+\frac{k_j}{2}}\operatorname{div}(P(x,y)\cdot\operatorname{grad}u(x,y))\cdot dy\cdot dx=\\&=\int\limits_{y_j-\frac{k_{j-1}}{2}}^{y_j+\frac{k_j}{2}}P\left(x_i+\frac{h_i}{2},y\right)\cdot\frac{\partial}{\partial x}u\left(x_i+\frac{h_i}{2},y\right)\cdot dy+\\&+\int\limits_{x_i+\frac{h_i}{2}}^{x_i-\frac{h_{i-1}}{2}}P\left(x,y_j+\frac{k_j}{2}\right)\cdot\frac{\partial}{\partial y}u\left(x,y_j+\frac{k_j}{2}\right)\cdot dx-\end{aligned}$$

$$-\int_{y_j+\frac{k_j}{2}}^{y_j-\frac{k_{j-1}}{2}} P\left(x_i-\frac{h_{i-1}}{2}, y\right)\cdot\frac{\partial}{\partial x}u\left(x_i-\frac{h_{i-1}}{2}, y\right)\cdot dy-$$

$$-\int_{x_i-\frac{h_{i-1}}{2}}^{x_i+\frac{h_i}{2}} P\left(x, y_j-\frac{k_{j-1}}{2}\right)\cdot\frac{\partial}{\partial y}u\left(x, y_j-\frac{k_{j-1}}{2}\right)\cdot dx \qquad (6.1\text{-}52)$$

The integrals on the right hand side of (6.1-52) are now approximated with finite differences. This is done frequently under the following assumptions:

$$P\left(x_i+\frac{h_i}{2}, y\right)\cdot\frac{\partial}{\partial x}u\left(x_i+\frac{h_i}{2}, y\right)=P_{i+1/2,j}\cdot\frac{u_{i+1,j}-u_{i,j}}{h_i} \qquad (6.1\text{-}53)$$

$$P\left(x, y_j+\frac{k_j}{2}\right)\cdot\frac{\partial}{\partial y}u\left(x, y_j+\frac{k_j}{2}\right)=P_{i,j+1/2}\cdot\frac{u_{i,j+1}-u_{i,j}}{k_j} \qquad (6.1\text{-}54)$$

$$P\left(x_i-\frac{h_{i-1}}{2}, y\right)\cdot\frac{\partial}{\partial x}u\left(x_i-\frac{h_{i-1}}{2}, y\right)=P_{i-1/2,j}\cdot\frac{u_{i,j}-u_{i-1,j}}{h_{i-1}} \qquad (6.1\text{-}55)$$

$$P\left(x, y_j-\frac{k_{j-1}}{2}\right)\cdot\frac{\partial}{\partial y}u\left(x, y_j-\frac{k_{j-1}}{2}\right)=P_{i,j-1/2}\cdot\frac{u_{i,j}-u_{i,j-1}}{k_{j-1}} \qquad (6.1\text{-}56)$$

However, it is to remark that these approximations have to be used with care. For instance, for the continuity equations we shall find that they are absolutely inappropriate.

By substituting (6.1-53) to (6.1-56) into (6.1-52) we obtain:

$$\int_{x_i-\frac{h_{i-1}}{2}}^{x_i+\frac{h_i}{2}}\int_{y_j-\frac{k_{j-1}}{2}}^{y_j+\frac{k_j}{2}} \operatorname{div}\left(P(x,y)\cdot\operatorname{grad}u(x,y)\right)\cdot dy\cdot dx\cong$$

$$\begin{aligned}&\cong P_{i+1/2,j}\cdot\frac{u_{i+1,j}-u_{i,j}}{h_i}\cdot\frac{k_{j-1}+k_j}{2}+\\&+P_{i,j+1/2}\cdot\frac{u_{i,j+1}-u_{i,j}}{k_j}\cdot\frac{h_{i-1}+h_i}{2}-\\&-P_{i-1/2,j}\cdot\frac{u_{i,j}-u_{i-1,j}}{h_{i-1}}\cdot\frac{k_{j-1}+k_j}{2}-\\&-P_{i,j-1/2}\cdot\frac{u_{i,j}-u_{i,j-1}}{k_{j-1}}\cdot\frac{h_{i-1}+h_i}{2}\end{aligned} \qquad (6.1\text{-}57)$$

In case of the Poisson equation we have $P(x,y)=1$ which liberates us from the problem of approximating midpoint values. The second integral in (6.1-50) is quite a difficult problem to be solved elegantly. The most trivial approach is to use the assumption that the carrier concentrations and the net doping concentration are constant within the integration domain which yields:

$$\int_{x_i-\frac{h_{i-1}}{2}}^{x_i+\frac{h_i}{2}} \int_{y_j-\frac{k_{j-1}}{2}}^{y_j+\frac{k_j}{2}} (n-p-C)\cdot dy\cdot dx \cong (n_{i,j}-p_{i,j}-C_{i,j})\cdot\frac{k_{j-1}+k_j}{2}\cdot\frac{h_{i-1}+h_i}{2} \tag{6.1-58}$$

This approximation is rather poor since we know that the quantities n, p and C can vary rapidly. However, for the two-dimensional problem I am not aware of any better approach. For the one-dimensional problem a better approximation and its derivation can be found in the book by Mock [6.43].

By substituting (6.1-57) and (6.1-58) into (6.1-50) we obtain the final discrete form of the Poisson equation derived by box integration.

$$\begin{aligned}\lambda^2\cdot\Bigg(&\frac{\psi_{i+1,j}-\psi_{i,j}}{h_i}\cdot\frac{k_{j-1}+k_j}{2}+\frac{\psi_{i,j+1}-\psi_{i,j}}{k_j}\cdot\frac{h_{i-1}+h_i}{2}+\\ &+\frac{\psi_{i-1,j}-\psi_{i,j}}{h_{i-1}}\cdot\frac{k_{j-1}+k_j}{2}+\frac{\psi_{i,j-1}-\psi_{i,j}}{k_{j-1}}\cdot\frac{h_{i-1}+h_i}{2}\Bigg)-\\ &-(n_{i,j}-p_{i,j}-C_{i,j})\cdot\frac{k_{j-1}+k_j}{2}\cdot\frac{h_{i-1}+h_i}{2}\end{aligned} \tag{6.1-59}$$

It is a trivial exercise to show that the previously derived discrete equation (6.1-24) is equivalent to (6.1-59).

For the continuity equations the situation is again, as could have been expected, more difficult. In a formal manner we may integrate the continuity equation for electrons over a box.

$$\begin{aligned}&\int_{x_i-\frac{h_{i-1}}{2}}^{x_i+\frac{h_i}{2}} \int_{y_j-\frac{k_{j-1}}{2}}^{y_j+\frac{k_j}{2}} \operatorname{div}(D_n\cdot\operatorname{grad} n)\cdot dy\cdot dx-\\ -&\int_{x_i-\frac{h_{i-1}}{2}}^{x_i+\frac{h_i}{2}} \int_{y_j-\frac{k_{j-1}}{2}}^{y_j+\frac{k_j}{2}} \operatorname{div}(\mu_n\cdot n\cdot\operatorname{grad}\psi)\cdot dy\cdot dx-\\ -&\int_{x_i-\frac{h_{i-1}}{2}}^{x_i+\frac{h_i}{2}} \int_{y_j-\frac{k_{j-1}}{2}}^{y_j+\frac{k_j}{2}} R(\psi,n,p)\cdot dy\cdot dx=0\end{aligned} \tag{6.1-60}$$

Fully analogous is the treatment of the continuity equation for holes.

$$\int_{x_i-\frac{h_{i-1}}{2}}^{x_i+\frac{h_i}{2}} \int_{y_j-\frac{k_{j-1}}{2}}^{y_j+\frac{k_j}{2}} \operatorname{div}(D_p \cdot \operatorname{grad} p) \cdot dy \cdot dx +$$

$$+ \int_{x_i-\frac{h_{i-1}}{2}}^{x_i+\frac{h_i}{2}} \int_{y_j-\frac{k_{j-1}}{2}}^{y_j+\frac{k_j}{2}} \operatorname{div}(\mu_p \cdot p \cdot \operatorname{grad} \psi) \cdot dy \cdot dx -$$

$$- \int_{x_i-\frac{h_{i-1}}{2}}^{x_i+\frac{h_i}{2}} \int_{y_j-\frac{k_{j-1}}{2}}^{y_j+\frac{k_j}{2}} R(\psi, n, p) \cdot dy \cdot dx = 0 \tag{6.1-61}$$

The evaluation of the integrals, particularly the first two, in (6.1-60), (6.1-61) has to be done very carefully. The approximation (6.1-57) is inappropriate for the first integral in (6.1-60), (6.1-61) as it assumes a linear behavior of u (which corresponds to n and p here) perpendicular to the four integration paths and a constant normal derivative along the integration paths. One may keep the assumption that the partial derivatives of n and p perpendicular to the integration paths are constant along the integration paths as I do not know a better choice. However, we have shown that the carrier concentrations exhibit in general an exponential behavior between neighboring mesh points (cf. (6.1-32) to (6.1-38)) depending upon the electrostatic potential. Recalling these results it is straightforward to calculate the various required partial derivatives and midinterval values of the carrier concentrations. We obtain for the electron concentration:

$$n_{i+1/2,j} = \frac{n_{i,j}}{1+\exp\left(\dfrac{\psi_{i,j}-\psi_{i+1,j}}{2 \cdot Ut}\right)} + \frac{n_{i+1,j}}{1+\exp\left(\dfrac{\psi_{i+1,j}-\psi_{i,j}}{2 \cdot Ut}\right)} \tag{6.1-62}$$

$$n_{i,j+1/2} = \frac{n_{i,j}}{1+\exp\left(\dfrac{\psi_{i,j}-\psi_{i,j+1}}{2 \cdot Ut}\right)} + \frac{n_{i,j+1}}{1+\exp\left(\dfrac{\psi_{i,j+1}-\psi_{i,j}}{2 \cdot Ut}\right)} \tag{6.1-63}$$

$$\left.\frac{\partial n}{\partial x}\right|_{i+1/2,j} = \frac{\dfrac{\psi_{i+1,j}-\psi_{i,j}}{Ut}}{\exp\left(\dfrac{\psi_{i+1,j}-\psi_{i,j}}{2 \cdot Ut}\right) - \exp\left(\dfrac{\psi_{i,j}-\psi_{i+1,j}}{2 \cdot Ut}\right)} \cdot \frac{n_{i+1,j}-n_{i,j}}{h_i} \tag{6.1-64}$$

$$\left.\frac{\partial n}{\partial x}\right|_{i,j+1/2} = \frac{\dfrac{\psi_{i,j+1}-\psi_{i,j}}{Ut}}{\exp\left(\dfrac{\psi_{i,j+1}-\psi_{i,j}}{2\cdot Ut}\right)-\exp\left(\dfrac{\psi_{i,j}-\psi_{i,j+1}}{2\cdot Ut}\right)} \cdot \frac{n_{i,j+1}-n_{i,j}}{k_j} \tag{6.1-65}$$

We have similar results for the hole concentration:

$$p_{i+1/2,j} = \frac{p_{i,j}}{1+\exp\left(\dfrac{\psi_{i+1,j}-\psi_{i,j}}{2\cdot Ut}\right)} + \frac{p_{i+1,j}}{1+\exp\left(\dfrac{\psi_{i,j}-\psi_{i+1,j}}{2\cdot Ut}\right)} \tag{6.1-66}$$

$$p_{i,j+1/2} = \frac{p_{i,j}}{1+\exp\left(\dfrac{\psi_{i,j+1}-\psi_{i,j}}{2\cdot Ut}\right)} + \frac{p_{i,j+1}}{1+\exp\left(\dfrac{\psi_{i,j}-\psi_{i,j+1}}{2\cdot Ut}\right)} \tag{6.1-67}$$

$$\left.\frac{\partial p}{\partial x}\right|_{i+1/2,j} = \frac{\dfrac{\psi_{i+1,j}-\psi_{i,j}}{Ut}}{\exp\left(\dfrac{\psi_{i+1,j}-\psi_{i,j}}{2\cdot Ut}\right)-\exp\left(\dfrac{\psi_{i,j}-\psi_{i+1,j}}{2\cdot Ut}\right)} \cdot \frac{p_{i+1,j}-p_{i,j}}{h_i} \tag{6.1-68}$$

$$\left.\frac{\partial p}{\partial x}\right|_{i,j+1/2} = \frac{\dfrac{\psi_{i,j+1}-\psi_{i,j}}{Ut}}{\exp\left(\dfrac{\psi_{i,j+1}-\psi_{i,j}}{2\cdot Ut}\right)-\exp\left(\dfrac{\psi_{i,j}-\psi_{i,j+1}}{2\cdot Ut}\right)} \cdot \frac{p_{i,j+1}-p_{i,j}}{k_j} \tag{6.1-69}$$

Note, that the weights for the partial derivatives of the electron concentration in (6.1-64), (6.1-65) are identical to the corresponding weights for the partial derivatives of the hole concentration in (6.1-68), (6.1-69).
With the approximations (6.1-62) to (6.1-69) it is straightforward to evaluate the integrals (6.1-60), (6.1-61). Under the additional assumptions that the carrier mobilities (diffusivities) are constant along the integration paths and that the recombination rate is constant within the integration domain we obtain for the continuity equation for electrons:

$$D_n|_{i+1/2,j} \cdot \frac{\dfrac{\psi_{i+1,j}-\psi_{i,j}}{Ut}}{\exp\left(\dfrac{\psi_{i+1,j}-\psi_{i,j}}{2\cdot Ut}\right)-\exp\left(\dfrac{\psi_{i,j}-\psi_{i+1,j}}{2\cdot Ut}\right)} \cdot \frac{n_{i+1,j}-n_{i,j}}{h_i} \cdot \frac{k_{j-1}+k_j}{2} +$$

$$+D_n|_{i,j+1/2}\cdot\frac{\dfrac{\psi_{i,j+1}-\psi_{i,j}}{Ut}}{\exp\left(\dfrac{\psi_{i,j+1}-\psi_{i,j}}{2\cdot Ut}\right)-\exp\left(\dfrac{\psi_{i,j}-\psi_{i,j+1}}{2\cdot Ut}\right)}\cdot\frac{n_{i,j+1}-n_{i,j}}{k_j}\cdot\frac{h_{i-1}+h_i}{2}+$$

$$+D_n|_{i-1/2,j}\cdot\frac{\dfrac{\psi_{i-1,j}-\psi_{i,j}}{Ut}}{\exp\left(\dfrac{\psi_{i-1,j}-\psi_{i,j}}{2\cdot Ut}\right)-\exp\left(\dfrac{\psi_{i,j}-\psi_{i-1,j}}{2\cdot Ut}\right)}\cdot\frac{n_{i-1,j}-n_{i,j}}{h_{i-1}}\cdot\frac{k_{j-1}+k_j}{2}+$$

$$+D_n|_{i,j-1/2}\cdot\frac{\dfrac{\psi_{i,j-1}-\psi_{i,j}}{Ut}}{\exp\left(\dfrac{\psi_{i,j-1}-\psi_{i,j}}{2\cdot Ut}\right)-\exp\left(\dfrac{\psi_{i,j}-\psi_{i,j-1}}{2\cdot Ut}\right)}\cdot\frac{n_{i,j-1}-n_{i,j}}{k_{j-1}}\cdot\frac{h_{i-1}+h_i}{2}-$$

$$-\mu_n|_{i+1/2,j}\cdot\left(\frac{n_{i+1,j}}{1+\exp\left(\dfrac{\psi_{i+1,j}-\psi_{i,j}}{2\cdot Ut}\right)}+\frac{n_{i,j}}{1+\exp\left(\dfrac{\psi_{i,j}-\psi_{i+1,j}}{2\cdot Ut}\right)}\right)\cdot$$

$$\cdot\frac{\psi_{i+1,j}-\psi_{i,j}}{h_i}\cdot\frac{k_{j-1}+k_j}{2}-$$

$$-\mu_n|_{i,j+1/2}\cdot\left(\frac{n_{i,j+1}}{1+\exp\left(\dfrac{\psi_{i,j+1}-\psi_{i,j}}{2\cdot Ut}\right)}+\frac{n_{i,j}}{1+\exp\left(\dfrac{\psi_{i,j}-\psi_{i,j+1}}{2\cdot Ut}\right)}\right)\cdot$$

$$\cdot\frac{\psi_{i,j+1}-\psi_{i,j}}{k_j}\cdot\frac{h_i+h_{i-1}}{2}-$$

$$-\mu_n|_{i-1/2,j}\cdot\left(\frac{n_{i-1,j}}{1+\exp\left(\dfrac{\psi_{i-1,j}-\psi_{i,j}}{2\cdot Ut}\right)}+\frac{n_{i,j}}{1+\exp\left(\dfrac{\psi_{i,j}-\psi_{i-1,j}}{2\cdot Ut}\right)}\right)\cdot$$

$$\cdot\frac{\psi_{i-1,j}-\psi_{i,j}}{h_{i-1}}\cdot\frac{k_j+k_{j-1}}{2}-$$

$$-\mu_n|_{i,j-1/2}\cdot\left(\frac{n_{i,j-1}}{1+\exp\left(\dfrac{\psi_{i,j-1}-\psi_{i,j}}{2\cdot Ut}\right)}+\frac{n_{i,j}}{1+\exp\left(\dfrac{\psi_{i,j}-\psi_{i,j-1}}{2\cdot Ut}\right)}\right)\cdot$$

$$\cdot\frac{\psi_{i,j-1}-\psi_{i,j}}{k_{j-1}}\cdot\frac{h_i+h_{i-1}}{2}-$$

$$-R_{i,j}\cdot\frac{h_i+h_{i-1}}{2}\cdot\frac{k_j+k_{j-1}}{2}=0 \qquad (6.1\text{-}70)$$

Fully analogously reads the discretized continuity equation for holes:

$$
\begin{aligned}
& D_p|_{i+1/2,j} \cdot \frac{\dfrac{\psi_{i+1,j}-\psi_{i,j}}{Ut}}{\exp\left(\dfrac{\psi_{i+1,j}-\psi_{i,j}}{2\cdot Ut}\right)-\exp\left(\dfrac{\psi_{i,j}-\psi_{i+1,j}}{2\cdot Ut}\right)} \cdot \frac{p_{i+1,j}-p_{i,j}}{h_i} \cdot \frac{k_{j-1}+k_j}{2} + \\
& + D_p|_{i,j+1/2} \cdot \frac{\dfrac{\psi_{i,j+1}-\psi_{i,j}}{Ut}}{\exp\left(\dfrac{\psi_{i,j+1}-\psi_{i,j}}{2\cdot Ut}\right)-\exp\left(\dfrac{\psi_{i,j}-\psi_{i,j+1}}{2\cdot Ut}\right)} \cdot \frac{p_{i,j+1}-p_{i,j}}{k_j} \cdot \frac{h_{i-1}+h_i}{2} + \\
& + D_p|_{i-1/2,j} \cdot \frac{\dfrac{\psi_{i-1,j}-\psi_{i,j}}{Ut}}{\exp\left(\dfrac{\psi_{i-1,j}-\psi_{i,j}}{2\cdot Ut}\right)-\exp\left(\dfrac{\psi_{i,j}-\psi_{i-1,j}}{2\cdot Ut}\right)} \cdot \frac{p_{i-1,j}-p_{i,j}}{h_{i-1}} \cdot \frac{k_{j-1}+k_j}{2} + \\
& + D_p|_{i,j-1/2} \cdot \frac{\dfrac{\psi_{i,j-1}-\psi_{i,j}}{Ut}}{\exp\left(\dfrac{\psi_{i,j-1}-\psi_{i,j}}{2\cdot Ut}\right)-\exp\left(\dfrac{\psi_{i,j}-\psi_{i,j-1}}{2\cdot Ut}\right)} \cdot \frac{p_{i,j-1}-p_{i,j}}{k_{j-1}} \cdot \frac{h_{i-1}+h_i}{2} + \\
& + \mu_p|_{i+1/2,j} \cdot \left(\frac{p_{i+1,j}}{1+\exp\left(\dfrac{\psi_{i,j}-\psi_{i+1,j}}{2\cdot Ut}\right)} + \frac{p_{i,j}}{1+\exp\left(\dfrac{\psi_{i+1,j}-\psi_{i,j}}{2\cdot Ut}\right)} \right) \cdot \\
& \quad \cdot \frac{\psi_{i+1,j}-\psi_{i,j}}{h_i} \cdot \frac{k_{j-1}+k_j}{2} + \\
& + \mu_p|_{i,j+1/2} \cdot \left(\frac{p_{i,j+1}}{1+\exp\left(\dfrac{\psi_{i,j}-\psi_{i,j+1}}{2\cdot Ut}\right)} + \frac{p_{i,j}}{1+\exp\left(\dfrac{\psi_{i,j+1}-\psi_{i,j}}{2\cdot Ut}\right)} \right) \cdot \\
& \quad \cdot \frac{\psi_{i,j+1}-\psi_{i,j}}{k_j} \cdot \frac{h_i+h_{i-1}}{2} + \\
& + \mu_p|_{i-1/2,j} \cdot \left(\frac{p_{i-1,j}}{1+\exp\left(\dfrac{\psi_{i,j}-\psi_{i-1,j}}{2\cdot Ut}\right)} + \frac{p_{i,j}}{1+\exp\left(\dfrac{\psi_{i-1,j}-\psi_{i,j}}{2\cdot Ut}\right)} \right) \cdot \\
& \quad \cdot \frac{\psi_{i-1,j}-\psi_{i,j}}{h_{i-1}} \cdot \frac{k_j+k_{j-1}}{2} + \\
& + \mu_p|_{i,j-1/2} \cdot \left(\frac{p_{i,j-1}}{1+\exp\left(\dfrac{\psi_{i,j}-\psi_{i,j-1}}{2\cdot Ut}\right)} + \frac{p_{i,j}}{1+\exp\left(\dfrac{\psi_{i,j-1}-\psi_{i,j}}{2\cdot Ut}\right)} \right) \cdot \\
& \quad \cdot \frac{\psi_{i,j-1}-\psi_{i,j}}{k_{j-1}} \cdot \frac{h_i+h_{i-1}}{2} - \\
& - R_{i,j} \cdot \frac{h_i+h_{i-1}}{2} \cdot \frac{k_j+k_{j-1}}{2} = 0 \qquad (6.1\text{-}71)
\end{aligned}
$$

The assumption that the recombination rate is constant within the integration domain is indeed a very poor one. However, a better treatment is only possible for the one-dimensional problem as has been demonstrated in [6.11].
It is a recommended exercise to show that the difference schemes (6.1-70), (6.1-71) are fully equivalent to (6.1-44), (6.1-45).
In the following we summarize the discretization schemes for the Poisson and the continuity equation in a form which is more suitable for computer implementation. We obtain, after collecting all factors for the discrete dependent variables, the following three equations:

$$\begin{aligned}
&\psi_{i,j-1}\cdot\lambda^2\cdot\frac{h_{i-1}+h_i}{2\cdot k_{j-1}}+\\
&+\psi_{i-1,j}\cdot\lambda^2\cdot\frac{k_{j-1}+k_j}{2\cdot h_{i-1}}-\\
&-\psi_{i,j}\cdot\lambda^2\cdot\left(\frac{h_{i-1}+h_i}{2\cdot k_{j-1}}+\frac{k_{j-1}+k_j}{2\cdot h_{i-1}}+\frac{k_{j-1}+k_j}{2\cdot h_i}+\frac{h_{i-1}+h_i}{2\cdot k_j}\right)+\\
&+\psi_{i+1,j}\cdot\lambda^2\cdot\frac{k_{j-1}+k_j}{2\cdot h_i}+\\
&+\psi_{i,j+1}\cdot\lambda^2\cdot\frac{h_{i-1}+h_i}{2\cdot k_j}-\\
&-(n_{i,j}-p_{i,j}-C_{i,j})\cdot\frac{h_{i-1}+h_i}{2}\cdot\frac{k_{j-1}+k_j}{2}=0
\end{aligned}\tag{6.1-72}$$

$$\begin{aligned}
&n_{i,j-1}\cdot D_n|_{i,j-1/2}\cdot B\left(\frac{\psi_{i,j-1}-\psi_{i,j}}{Ut}\right)\cdot\frac{h_{i-1}+h_i}{2\cdot k_{j-1}}+\\
&+n_{i-1,j}\cdot D_n|_{i-1/2,j}\cdot B\left(\frac{\psi_{i-1,j}-\psi_{i,j}}{Ut}\right)\cdot\frac{k_{j-1}+k_j}{2\cdot h_{i-1}}-\\
&-n_{i,j}\cdot\left(D_n|_{i,j-1/2}\cdot B\left(\frac{\psi_{i,j}-\psi_{i,j-1}}{Ut}\right)\cdot\frac{h_{i-1}+h_i}{2\cdot k_{j-1}}+\right.\\
&\qquad+D_n|_{i-1/2,j}\cdot B\left(\frac{\psi_{i,j}-\psi_{i-1,j}}{Ut}\right)\cdot\frac{k_{j-1}+k_j}{2\cdot h_{i-1}}+\\
&\qquad+D_n|_{i+1/2,j}\cdot B\left(\frac{\psi_{i,j}-\psi_{i+1,j}}{Ut}\right)\cdot\frac{k_{j-1}+k_j}{2\cdot h_i}+\\
&\qquad\left.+D_n|_{i,j+1/2}\cdot B\left(\frac{\psi_{i,j}-\psi_{i,j+1}}{Ut}\right)\cdot\frac{h_{i-1}+h_i}{2\cdot k_j}\right)+\\
&+n_{i+1,j}\cdot D_n|_{i+1/2,j}\cdot B\left(\frac{\psi_{i+1,j}-\psi_{i,j}}{Ut}\right)\cdot\frac{k_{j-1}+k_j}{2\cdot h_i}+\\
&+n_{i,j+1}\cdot D_n|_{i,j+1/2}\cdot B\left(\frac{\psi_{i,j+1}-\psi_{i,j}}{Ut}\right)\cdot\frac{h_{i-1}+h_i}{2\cdot k_j}-\\
&-R_{i,j}\cdot\frac{h_{i-1}+h_i}{2}\cdot\frac{k_{j-1}+k_j}{2}=0
\end{aligned}\tag{6.1-73}$$

$$p_{i,j-1} \cdot D_p|_{i,j-1/2} \cdot B\left(\frac{\psi_{i,j}-\psi_{i,j-1}}{Ut}\right) \cdot \frac{h_{i-1}+h_i}{2 \cdot k_{j-1}} +$$

$$+p_{i-1,j} \cdot D_p|_{i-1/2,j} \cdot B\left(\frac{\psi_{i,j}-\psi_{i-1,j}}{Ut}\right) \cdot \frac{k_{j-1}+k_j}{2 \cdot h_{i-1}} -$$

$$-p_{i,j} \cdot \left(D_p|_{i,j-1/2} \cdot B\left(\frac{\psi_{i,j-1}-\psi_{i,j}}{Ut}\right) \cdot \frac{h_{i-1}+h_i}{2 \cdot k_{j-1}} +\right.$$

$$+D_p|_{i-1/2,j} \cdot B\left(\frac{\psi_{i-1,j}-\psi_{i,j}}{Ut}\right) \cdot \frac{k_{j-1}+k_j}{2 \cdot h_{i-1}} +$$

$$+D_p|_{i+1/2,j} \cdot B\left(\frac{\psi_{i+1,j}-\psi_{i,j}}{Ut}\right) \cdot \frac{k_{j-1}+k_j}{2 \cdot h_i} +$$

$$\left.+D_p|_{i,j+1/2} \cdot B\left(\frac{\psi_{i,j+1}-\psi_{i,j}}{Ut}\right) \cdot \frac{h_{i-1}+h_i}{2 \cdot k_j}\right) +$$

$$+p_{i+1,j} \cdot D_p|_{i+1/2,j} \cdot B\left(\frac{\psi_{i,j}-\psi_{i+1,j}}{Ut}\right) \cdot \frac{k_{j-1}+k_j}{2 \cdot h_i} +$$

$$+p_{i,j+1} \cdot D_p|_{i,j+1/2} \cdot B\left(\frac{\psi_{i,j}-\psi_{i,j+1}}{Ut}\right) \cdot \frac{h_{i-1}+h_i}{2 \cdot k_j} -$$

$$-R_{i,j} \cdot \frac{h_{i-1}+h_i}{2} \cdot \frac{k_{j-1}+k_j}{2} = 0 \tag{6.1-74}$$

The midinterval values of the carrier diffusivities and the carrier mobilities can be obtained, if they are not explicitly available, by a simple interpolation. One may use, for instance, linear interpolation, e.g.:

$$D_n|_{i+1/2,j} = \frac{D_n|_{i,j} + D_n|_{i+1,j}}{2} \tag{6.1-75}$$

It might be more physically motivated to assume that the relaxation times, which are proportional to the reciprocals of the carrier diffusivities, are linear functions between neighboring mesh points which leads to the following interpolation scheme, e.g.:

$$D_n|_{i+1/2,j} = \frac{2}{\dfrac{1}{D_n|_{i,j}} + \dfrac{1}{D_n|_{i+1,j}}} \tag{6.1-76}$$

The impact of any of these interpolation schemes on the final solution, however, is small.
The Bernoulli function (6.1-43) has to be implemented very carefully for numerical computations. A rational approximation can be used for the most efficient implementation on a particular computer [6.19]. If, however, the easier way of implementation via the exponential function of a supplied runtime library and basic

arithmetic is chosen, some effort has to be spent to avoid potential underflow and overflow traps. It can be suggested to implement the Bernoulli function as follows:

$$B(x)=\begin{cases} x\leq x1 & -x \\ x1<x<x2 & \dfrac{x}{\exp(x)-1} \\ x2\leq x\leq x3 & 1-\dfrac{x}{2} \\ x3<x<x4 & \dfrac{x\cdot\exp(-x)}{1-\exp(-x)} \\ x4\leq x<x5 & x\cdot\exp(-x) \\ x5\leq x & 0 \end{cases} \tag{6.1-77}$$

The constants $x1$ to $x5$ depend on the individual computer hardware. They are defined by:

$$\text{`}\exp(x1)\text{'}\ \text{`}-\text{'}\ 1\ \text{`}=\text{'}\ -1 \tag{6.1-78}$$

$$x2\,\text{`}/\text{'}\,(\text{`}\exp(x2)\text{'}\ \text{`}-\text{'}\ 1)\ \text{`}=\text{'}\ 1\,\text{`}-\text{'}\,(x2\,\text{`}/\text{'}\,2)\wedge x2<0 \tag{6.1-79}$$

$$1\,\text{`}-\text{'}\,(x3\,\text{`}/\text{'}\,2)\ \text{`}=\text{'}\ x3\,\text{`}\cdot\text{'}\ \text{`}\exp(\text{`}-\text{'}\,x3)\text{'}\ \text{`}/\text{'}\,(1\,\text{`}-\text{'}\ \text{`}\exp(\text{`}-\text{'}\,x3)\text{'})\wedge x3>0 \tag{6.1-80}$$

$$1\,\text{`}-\text{'}\ \text{`}\exp(\text{`}-\text{'}\,x4)\text{'}\ \text{`}=\text{'}\ 1 \tag{6.1-81}$$

$$\text{`}\exp(\text{`}-\text{'}\,x5)\text{'}\ \text{`}=\text{'}\ 0 \tag{6.1-82}$$

The quoted operators and the quoted exponential function denote the finite and discrete implementation of operators and exponential function on a digital computer. It is an easy and straightforward task to calculate these constants once on a given computer.

So far we have derived the discrete equations at all inner points of a given mesh. In the following we shall discuss the discretization of boundary conditions. We recall for that purpose the mesh shown in Fig. 6.1-1, which was described in the very beginning of this section. Let us assume that the emitter $A-B$, base $C-D$ and collector $E-F$ represent ideal ohmic contacts. The boundary pieces $B-C$ and $D-E$ which separate the contacts are assumed to be simplified interfaces to insulating material (cf. Section 5.1). The remaining boundary pieces $F-G$, $G-H$ and $H-A$ represent artificial boundaries.

For the ohmic contacts it depends upon the application which boundary condition has to be satisfied. We shall consider here purely voltage controlled and purely current controlled ohmic contacts. For the former boundary condition we have:

$$\psi=\psi_b+\psi_D \tag{6.1-83}$$

$$n=\frac{\sqrt{C^2+4\cdot n_i^2}+C}{2} \tag{6.1-84}$$

$$p=\frac{\sqrt{C^2+4\cdot n_i^2}-C}{2} \tag{6.1-85}$$

A discretization of these equations is trivial as only point values are involved. We obtain directly:

$$\psi_{i,j}=\psi_b|_{i,j}+\psi_D|_{i,j} \tag{6.1-86}$$

$$n_{i,j}=\frac{\sqrt{C_{i,j}^2+4\cdot n_i^2}+C_{i,j}}{2} \tag{6.1-87}$$

$$p_{i,j}=\frac{\sqrt{C_{i,j}^2+4\cdot n_i^2}-C_{i,j}}{2} \tag{6.1-88}$$

In the specific example in Fig. 6.1-1 we have the following subset of points for the three contacts:

emitter $1\leq i\leq 6 \quad j=22 \quad \psi_D|_{i,j}=U_e$ (6.1-89)

base $16\leq i\leq 24 \quad j=22 \quad \psi_D|_{i,j}=U_b$ (6.1-90)

collector $36\leq i\leq 41 \quad j=22 \quad \psi_D|_{i,j}=U_c$ (6.1-91)

It should be noted that (6.1-87), (6.1-88) should not be evaluated directly because of inherent problems with cancellation. One should preferably use:

if $C_{i,j}>0$ then

$$n_{i,j}=\frac{\sqrt{C_{i,j}^2+4\cdot n_i^2}+C_{i,j}}{2}$$

$$p_{i,j}=\frac{n_i^2}{n_{i,j}}$$

else (6.1-92)

$$p_{i,j}=\frac{\sqrt{C_{i,j}^2+4\cdot n_i^2}-C_{i,j}}{2}$$

$$n_{i,j}=\frac{n_i^2}{p_{i,j}}$$

In the case of an ideal current controlled contact (6.1-83) is replaced by:

$$\int\limits_{\partial D_o}(\vec{J}_n+\vec{J}_p)\cdot d\vec{A}-I_D=0 \tag{6.1-93}$$

$$(\psi-\psi_b)|_{\partial D_o}=\text{const.} \tag{6.1-94}$$

The boundary conditions (6.1-84), (6.1-85) for the carrier concentrations are not altered for the case of an ideal contact. The integral (6.1-93) can be approximated by, for instance, the midpoint integration rule. To outline the procedure it is most convenient to take an example. Let us assume that the base contact C-D in the example of Fig. 6.1-1 is ideally current controlled. We then obtain for the discretized equation at this boundary:

$$(J_{nx}|_{15+1/2,22}+J_{px}|_{15+1/2,22})\cdot\frac{k_{21}}{2}+$$

$$+\sum_{m=16}^{24}\left((J_{ny}|_{m,22-1/2}+J_{py}|_{m,22-1/2})\cdot\frac{h_m+h_{m-1}}{2}\right)-$$

$$-(J_{nx}|_{24+1/2,22}+J_{px}|_{24+1/2,22})\cdot\frac{k_{21}}{2}-$$

$$-\frac{I_D}{W}=0 \tag{6.1-95}$$

$$\psi_{h+1,22}-\psi_b|_{h+1,22}-\psi_{h,22}+\psi_b|_{h,22}=0, \quad h=(16,17,\ldots 23) \tag{6.1-96}$$

W in (6.1-95) denotes the width of the contact, which is the artificial extent of the contact in the third dimension. (6.1-95) and (6.1-96) form nine equations for the electrostatic potential at the nine boundary points. The current density components at the midinterval points are defined in (6.1-39) to (6.1-42). It should be noted that in the discrete equation (6.1-95) the dependent variables at all mesh points which surround the contact are involved. This fact complicates the numerical solution of the final nonlinear algebraic system as the regular structure which is obtained by the discretization at all inner points is disturbed. This regular structure arises since only four neighboring meshpoints are involved in the discrete equations at a specific meshpoint.

The next category of boundary conditions to discretize are the interfaces. In our example these are, as mentioned, the boundary pieces $B-C$ and $D-E$. We shall first consider the (physically) simplified boundary conditions which then read:

$$\frac{\partial\psi}{\partial\vec{n}}=0 \tag{6.1-97}$$

$$\vec{J}_n\cdot\vec{n}=0 \tag{6.1-98}$$

$$\vec{J}_p\cdot\vec{n}=0 \tag{6.1-99}$$

In our example these equations evaluate for the boundary pieces $B-C$ and $D-E$ to:

$$\left.\frac{\partial\psi}{\partial y}\right|_{i,j}=0 \tag{6.1-100}$$

$$J_{ny}|_{i,j}=0 \tag{6.1-101}$$

$$J_{py}|_{i,j}=0 \tag{6.1-102}$$

These equations hold for the following subset of points:

$$B-C \quad 7\le i\le 15 \quad j=22 \tag{6.1-103}$$

$$D-E \quad 25\le i\le 35 \quad j=22 \tag{6.1-104}$$

The discretization of the Neumann boundary conditions (6.1-100) to (6.1-102) is elegantly performed by so called "mirror imaging". Without any loss of generality we may write the linear interpolation scheme for an equidistant grid ($k_j = k_{j-1}$):

$$u_{i,j} = \frac{u_{i,j+1/2} + u_{i,j-1/2}}{2} + O(k^2) \cdot \left. \frac{\partial^2 u}{\partial y^2} \right|_{i,j} \tag{6.1-105}$$

If we substitute successively the quantities $\partial\psi/\partial y$, J_{ny} and J_{py} for u into (6.1-105), we obtain, by remembering the boundary conditions (6.1-100) to (6.1-102):

$$\left. \frac{\partial \psi}{\partial y} \right|_{i,j+1/2} = - \left. \frac{\partial \psi}{\partial y} \right|_{i,j-1/2} + O(k^2) \cdot \left. \frac{\partial^3 \psi}{\partial y^3} \right|_{i,j} \tag{6.1-106}$$

$$J_{ny}|_{i,j+1/2} = -J_{ny}|_{i,j-1/2} + O(k^2) \cdot \left. \frac{\partial^2}{\partial y^2} J_{ny} \right|_{i,j} \tag{6.1-107}$$

$$J_{py}|_{i,j+1/2} = -J_{py}|_{i,j-1/2} + O(k^2) \cdot \left. \frac{\partial^2}{\partial y^2} J_{py} \right|_{i,j} \tag{6.1-108}$$

The quantities defined by (6.1-106) to (6.1-108) do not really exist; they represent artificial images which implicitly resolve the boundary conditions (6.1-100) to (6.1-102) with a local truncation error of the same order as we have for the midinterval values at inner points (cf. (6.1-22), (6.1-25) with (6.1-39), (6.1-27) with (6.1-41)). By substituting (6.1-106) to (6.1-108) into (6.1-15), (6.1-16) and (6.1-19), respectively, we obtain the discrete Poisson equation and continuity equations at the boundary.

$$\lambda^2 \cdot \left(\frac{\left. \frac{\partial \psi}{\partial x} \right|_{i+1/2,j} - \left. \frac{\partial \psi}{\partial x} \right|_{i-1/2,j}}{\frac{h_i + h_{i-1}}{2}} + \frac{-2 \cdot \left. \frac{\partial \psi}{\partial y} \right|_{i,j-1/2}}{k_{j-1}} \right) - $$

$$- n_{i,j} + p_{i,j} + C_{i,j} = 0 \tag{6.1-109}$$

$$\frac{(-J_{nx})|_{i+1/2,j} - (-J_{nx})|_{i-1/2,j}}{\frac{h_i + h_{i-1}}{2}} + \frac{2 \cdot J_{ny}|_{i,j-1/2}}{k_{j-1}} - $$

$$- R(\psi, n, p)|_{i,j} = 0 \tag{6.1-110}$$

$$\frac{J_{px}|_{i+1/2,j} - J_{px}|_{i-1/2,j}}{\frac{h_i + h_{i-1}}{2}} + \frac{-2 \cdot J_{py}|_{i,j-1/2}}{k_{j-1}} - $$

$$- R(\psi, n, p)|_{i,j} = 0 \tag{6.1-111}$$

In case of non-negligible interface charge and surface recombination (6.1-97) to (6.1-99) are replaced by:

$$\frac{\partial \psi}{\partial \vec{n}} - Q_{\text{int}} = 0 \tag{6.1-112}$$

$$\vec{J}_n \cdot \vec{n} - R^{SURF} = 0 \tag{6.1-113}$$

$$\vec{J}_p \cdot \vec{n} + R^{SURF} = 0 \tag{6.1-114}$$

In our example these equations read explicitly:

$$\left.\frac{\partial \psi}{\partial y}\right|_{i,j} - Q_{\text{int}}|_{i,j} = 0 \tag{6.1-115}$$

$$J_{ny}|_{i,j} - R^{SURF}|_{i,j} = 0 \tag{6.1-116}$$

$$J_{py}|_{i,j} + R^{SURF}|_{i,j} = 0 \tag{6.1-117}$$

We may again use the interpolation scheme (6.1-105) which yields for the artificial quantities:

$$\left.\frac{\partial \psi}{\partial y}\right|_{i,j+1/2} = -\left.\frac{\partial \psi}{\partial y}\right|_{i,j-1/2} + 2 \cdot Q_{\text{int}}|_{i,j} + O(k^2) \cdot \left.\frac{\partial^3 \psi}{\partial y^3}\right|_{i,j} \tag{6.1-118}$$

$$J_{ny}|_{i,j+1/2} = -J_{ny}|_{i,j-1/2} + 2 \cdot R^{SURF}|_{i,j} + O(k^2) \cdot \left.\frac{\partial^2}{\partial y^2} J_{ny}\right|_{i,j} \tag{6.1-119}$$

$$J_{p,y}|_{i,j+1/2} = -J_{py}|_{i,j-1/2} - 2 \cdot R^{SURF}|_{i,j} + O(k^2) \cdot \left.\frac{\partial^2}{\partial y^2} J_{py}\right|_{i,j} \tag{6.1-120}$$

The remaining steps are fully analogous as for the homogeneous boundary conditions. We obtain for the discrete equations after straightforward calculations:

$$\lambda^2 \cdot \left(\frac{\left.\frac{\partial \psi}{\partial x}\right|_{i+1/2,j} - \left.\frac{\partial \psi}{\partial x}\right|_{i-1/2,j}}{\frac{h_i + h_{i-1}}{2}} + \frac{-2 \cdot \left.\frac{\partial \psi}{\partial y}\right|_{i,j-1/2} + 2 \cdot Q_{\text{int}}\Big|_{i,j}}{k_{j-1}} \right) - \\ - n_{i,j} + p_{i,j} + C_{i,j} = 0 \tag{6.1-121}$$

$$\frac{(-J_{nx})|_{i+1/2,j} - (-J_{nx})|_{i-1/2,j}}{\frac{h_i + h_{i-1}}{2}} + \frac{2 \cdot J_{ny}|_{i,j-1/2} - 2 \cdot R^{SURF}|_{i,j}}{k_{j-1}} - \\ - R(\psi, n, p)|_{i,j} = 0 \tag{6.1-122}$$

$$\frac{J_{px}|_{i+1/2,j} - J_{px}|_{i-1/2,j}}{\frac{h_i + h_{i-1}}{2}} + \frac{-2 \cdot J_{py}|_{i,j-1/2} - 2 \cdot R^{SURF}|_{i,j}}{k_{j-1}} - \\ - R(\psi, n, p)|_{i,j} = 0 \tag{6.1-123}$$

If the electric field in an insulator cannot be assumed to be negligibly small, we have to account for Gauß's law in differential form:

$$\varepsilon_{\text{sem}} \cdot \left.\frac{\partial \psi}{\partial \vec{n}}\right|_{\text{sem}} - \varepsilon_{\text{ins}} \cdot \left.\frac{\partial \psi}{\partial \vec{n}}\right|_{\text{ins}} - Q_{\text{int}} = 0 \tag{6.1-124}$$

We must also account for that insulator as an explicit part of the simulation. Note that the Poisson equation is only valid in the semiconductor whereas in the insulator the Laplace equation holds. In the following we assume to have an interface at $y=y_m$. A meshline must be put exactly on that location. $y<y_m$ denotes the semiconductor; $y>y_m$ is the insulator. Under this assumption (6.1-124) gives:

$$\varepsilon_{\text{sem}} \cdot \left. \frac{\partial \psi}{\partial y} \right|_{\text{sem}} - \varepsilon_{\text{ins}} \cdot \left. \frac{\partial \psi}{\partial y} \right|_{\text{ins}} - Q_{\text{int}} = 0 \tag{6.1-125}$$

The partial derivatives can be replaced by the following, simple difference expressions:

$$\left. \left. \frac{\partial \psi}{\partial y} \right|_{\text{sem}} \right|_{i,m} = \frac{\psi_{i,m} - \psi_{i,m-1}}{k_{m-1}} + \frac{k_{m-1}}{2} \cdot \left. \left. \frac{\partial^2 \psi}{\partial y^2} \right|_{\text{sem}} \right|_{i,m} + O(k^2) \tag{6.1-126}$$

$$\left. \left. \frac{\partial \psi}{\partial y} \right|_{\text{ins}} \right|_{i,m} = \frac{\psi_{i,m+1} - \psi_{i,m}}{k_m} - \frac{k_m}{2} \cdot \left. \left. \frac{\partial^2 \psi}{\partial y^2} \right|_{\text{ins}} \right|_{i,m} + O(k^2) \tag{6.1-127}$$

These expressions can be proved by straightforward Taylor series expansions. In principle, it would suffice to substitute (6.1-126) and (6.1-127) into (6.1-125) to have a discrete boundary condition for the electrostatic potential at meshpoint (x_i, y_m). However, it is not attractive, although consistent for the boundary condition, that the second order partial derivative of the electrostatic potential is in the truncation error of (6.1-126), (6.1-127). This can be overcome by substituting the differential equations, the Poisson equation and the Laplace equation, into (6.1-126) and (6.1-127), respectively. We can see immediately that we then obtain a local truncation error of second order if the discrete approximations for the differential equations to be substituted have a local truncation error of second order. This is the case, as mentioned, for an equidistant or a quasi-uniform mesh. By recalling results which we have obtained earlier in this section we may write:

$$\left. \left. \frac{\partial^2 \psi}{\partial y^2} \right|_{\text{sem}} \right|_{i,m} = - \frac{\dfrac{\psi_{i+1,m} - \psi_{i,m}}{h_i} - \dfrac{\psi_{i,m} - \psi_{i-1,m}}{h_{i-1}}}{\dfrac{h_i + h_{i-1}}{2}} + \frac{1}{\lambda^2} \cdot (n_{i,m} - p_{i,m} - C_{i,m}) \tag{6.1-128}$$

$$\left. \left. \frac{\partial^2 \psi}{\partial y^2} \right|_{\text{ins}} \right|_{i,m} = - \frac{\dfrac{\psi_{i+1,m} - \psi_{i,m}}{h_i} - \dfrac{\psi_{i,m} - \psi_{i-1,m}}{h_{i-1}}}{\dfrac{h_i + h_{i-1}}{2}} \tag{6.1-129}$$

Substituting (6.1-128) and (6.1-129) into (6.1-126) and (6.1-127), respectively, and the results into (6.1-125) it is a matter of straightforward calulations to obtain the final discrete interface condition.

$$\lambda^2 \cdot \left(\frac{\dfrac{\psi_{i+1,m}-\psi_{i,m}}{h_i} - \dfrac{\psi_{i,m}-\psi_{i-1,m}}{h_{i-1}}}{\dfrac{h_i+h_{i-1}}{2}} + \right.$$

$$\left. + \frac{\varepsilon_{ins} \cdot \dfrac{\psi_{i,m+1}-\psi_{i,m}}{k_m} - \varepsilon_{sem} \cdot \dfrac{\psi_{i,m}-\psi_{i,m-1}}{k_{m-1}} + Q_{int}|_{i,m}}{\dfrac{\varepsilon_{ins} \cdot k_m + \varepsilon_{sem} \cdot k_{m-1}}{2}} \right) -$$

$$- \frac{\varepsilon_{sem} \cdot k_{m-1}}{\varepsilon_{ins} \cdot k_m + \varepsilon_{sem} \cdot k_{m-1}} \cdot (n_{i,m} - p_{i,m} - C_{i,m}) = 0 \qquad (6.1\text{-}130)$$

As I have demonstrated it is quite easy to derive the discrete interface condition. It is obvious that in using a different discretization approach the result obtained must be identical up to higher order terms of the truncation error. Applying the box integration approach yields results fully identical to (6.1-130) (cf. [6.13]). However, this almost trivial task of discretizing an interface condition must exhibit some intransparent spots (which hopefully have become clear now) because the interface condition has been dealt with even in the recent literature, e.g. [6.52], [6.53], although the basic strategy has been known for more than 100 years!

The method of improving the local truncation error of a "naively" discretized boundary condition with the aid of the differential equation can certainly be used for any boundary condition which involves normal derivatives. In fact, this approach is the basis for the mirror imaging method which has been outlined above for the discretization of simplified interfaces.

The last category of boundaries we have to consider are the artificial boundaries. There we have in general Neumann boundary conditions which are exactly identical with (6.1-97) to (6.1-99) for the simplified interfaces, the treatment of which has already been outlined.

For curved boundaries it is not possible to give a generally applicable procedure for the discretization of boundary conditions. A five-point scheme which accounts for the boundary condition as well as for the differential equation exists only in rare cases [6.34]. If the boundary condition is a Dirichlet condition, the discretization of differential equations at points near the boundary is straightforward; it can be found in, for instance, [6.50]. However, finite difference formulae accounting for normal derivatives at a curved boundary are extremely awkward. Details on that subject can be found in [6.20]. A discretization of interface conditions at beveled boundaries has been presented in, e.g., [6.27], [6.28] for a special case though.

6.2 Finite Boxes

In the classical finite difference approach for the solution of partial differential equations the meshlines introduced to partition the simulation domain start out at the boundary of the domain and are continued throughout the domain up to the opposite part of the boundary. As already stressed previously, a rapidly varying behavior of the solution of the basic semiconductor equations is, in case of many devices, confined to small regions of the simulation domain. This can result in an enormous number of superfluous points located in regions where the solution exhibits a smooth behavior, thus, wasting computer storage and time. Adler [6.1], [6.2] has introduced the (theoretically already known [6.13]) method of terminating meshlines in the finite difference approach in order to avoid this problem. Adler restricted himself to allow terminating lines only in one coordinate direction. An example of such a mesh is shown in Fig. 6.2-1. The simulation domain and the point allocation is the same as in the previously shown finite difference mesh (Fig. 6.1-1). The total number of meshpoints is reduced by the terminating lines from 902 to 687. 88 points lie on the boundary and 34 points are at terminations of meshlines in the interior of the simulation domain.

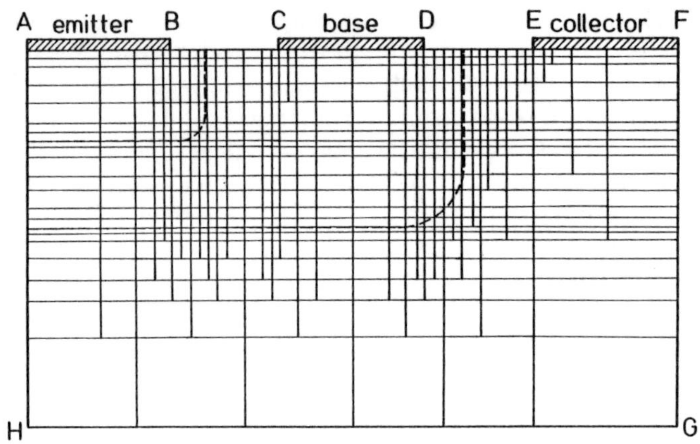

Fig. 6.2-1. Terminating line mesh for a lateral bipolar transistor

This approach has been further generalized, as Adler speculated in [6.2], by Franz et al. [6.15] to the concept of finite boxes. There meshlines are allowed to start and to terminate in either direction. Fig. 6.2-2 shows an example of a finite boxes mesh in which points have been removed compared to Fig. 6.2-1. The total number of meshpoints is thereby decreased to 583. 64 points are located at the boundary and 66 points are at terminations of meshlines in the interior of the domain.

It is obvious that the advantage obtained through the reduction of meshpoints by the terminating lines or finite boxes approach has to be paid for at a different place. In the case of classical finite differences it is a trivial task to find the location of quantities at neighboring points if the quantities are stored by columns or rows, because the number of points in a column or row is a mesh specific constant. In the case of terminating lines one needs four additional variables for each point, i.e., two indices for the neighboring points in the direction perpendicular to the direction in which mesh lines may terminate and the space coordinates. For finite boxes six variables per meshpoint are required to fully describe the mesh. These are the four indices for all neighboring points and the space coordinates.

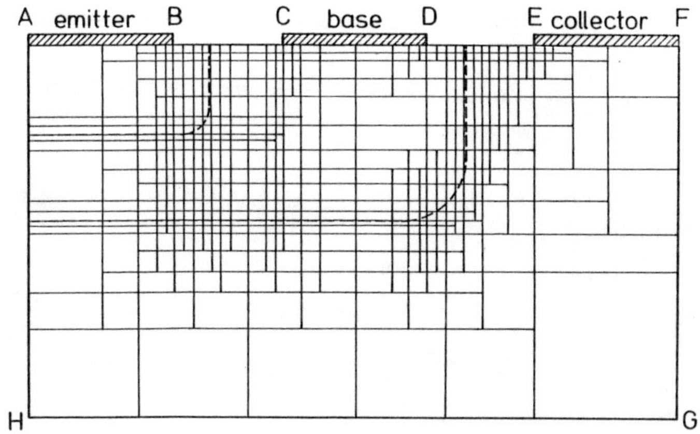

Fig. 6.2-2. Finite box mesh for a lateral bipolar transistor

In the following we shall derive the difference approximations for the basic semiconductor equations at the termination point of a meshline. We shall use the nomenclature adopted in Fig. 6.2-3. In this example a meshline terminates in the x-direction. Thus, the endpoint (x_i, y_j) does not have a neighboring point at (x_{i+1}, y_j). It should be noted that one has to prohibit two neighboring meshlines terminating such that the termination points are nearest neighboring points. This would complicate the derivation of consistent difference schemes significantly.
In the previous section we have reduced the problem of finding difference approximations into supplying appropriate midinterval quantities (cf. (6.1-15), (6.1-16) and (6.1-17)). In our example (Fig. 6.2-3) we have to derive proper expressions for $\partial\psi/\partial x$, J_{nx} and J_{px} at location $(x_i + h_i/2, y_j)$. All other midinterval quantities can be calculated from the results of the previous section.

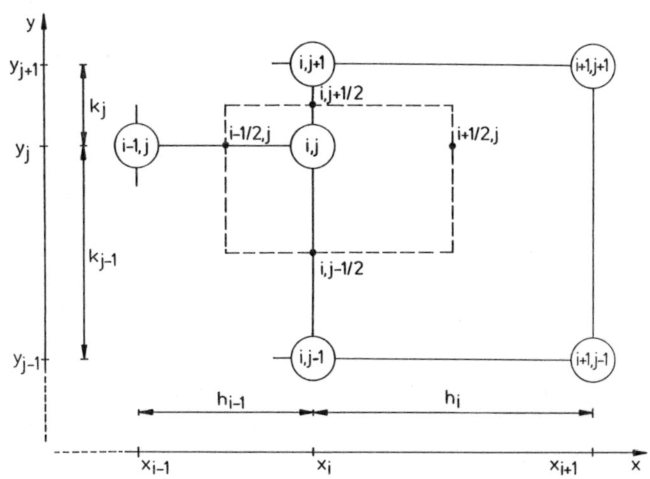

Fig. 6.2-3. The adopted nomenclature for terminating lines

We first recall the simple formula for linear interpolation and write:

$$u_{i+1/2,j}=\frac{k_{j-1}}{k_{j-1}+k_j}\cdot u_{i+1/2,j+1}+\frac{k_j}{k_{j-1}+k_j}\cdot u_{i+1/2,j-1}+O(k^2)\cdot\left.\frac{\partial^2 u}{\partial y^2}\right|_{i+1/2,j} \tag{6.2-1}$$

Substituting successively $\partial\psi/\partial x$, J_{nx} and J_{px} into (6.2-1) we obtain the required midinterval quantities.

$$\left.\frac{\partial\psi}{\partial x}\right|_{i+1/2,j}=\frac{k_{j-1}\cdot\left.\frac{\partial\psi}{\partial x}\right|_{i+1/2,j+1}+k_j\cdot\left.\frac{\partial\psi}{\partial x}\right|_{i+1/2,j-1}}{k_{j-1}+k_j}+O(k^2)\cdot\left.\frac{\partial^2}{\partial y^2}\frac{\partial\psi}{\partial x}\right|_{i+1/2,j} \tag{6.2-2}$$

$$J_{nx}|_{i+1/2,j}=\frac{k_{j-1}\cdot J_{nx}|_{i+1/2,j+1}+k_j\cdot J_{nx}|_{i+1/2,j-1}}{k_{j-1}+k_j}+O(k^2)\cdot\left.\frac{\partial^2}{\partial y^2}J_{nx}\right|_{i+1/2,j} \tag{6.2-3}$$

$$J_{px}|_{i+1/2,j}=\frac{k_{j-1}\cdot J_{px}|_{i+1/2,j+1}+k_j\cdot J_{px}|_{i+1/2,j-1}}{k_{j-1}+k_j}+O(k^2)\cdot\left.\frac{\partial^2}{\partial y^2}J_{px}\right|_{i+1/2,j} \tag{6.2-4}$$

In (6.2-2) to (6.2-4) we have introduced two midinterval quantities which have not been required so far, but which can be approximated with (6.1-22), (6.1-39) and (6.1-41), respectively. For instance, for (6.2-2) we obtain finally:

$$\begin{aligned}\left.\frac{\partial\psi}{\partial x}\right|_{i+1/2,j}&=\frac{k_{j-1}\cdot\frac{\psi_{i+1,j+1}-\psi_{i,j+1}}{h_i}+k_j\cdot\frac{\psi_{i+1,j-1}-\psi_{i,j-1}}{h_i}}{k_{j-1}+k_j}+\\&+O(h^2)\cdot\left.\frac{\partial^3\psi}{\partial x^3}\right|_{i+1/2,j+1}+O(h^2)\cdot\left.\frac{\partial^3\psi}{\partial x^3}\right|_{i+1/2,j-1}+O(k^2)\cdot\left.\frac{\partial^3\psi}{\partial x\cdot y^2}\right|_{i+1,2/j}\end{aligned} \tag{6.2-5}$$

The discrete approximation for the Poisson equation can then be expressed after some algebraic manipulation as:

$$\begin{aligned}&\psi_{i,j-1}\cdot\lambda^2\cdot\left(\frac{h_{i-1}+h_i}{2\cdot k_{j-1}}-\frac{k_j}{2\cdot h_i}\right)+\\&+\psi_{i+1,j-1}\cdot\lambda^2\cdot\frac{k_j}{2\cdot h_i}+\\&+\psi_{i-1,j}\cdot\lambda^2\cdot\frac{k_{j-1}+k_j}{2\cdot h_{i-1}}-\\&-\psi_{i,j}\cdot\lambda^2\cdot\left(\frac{h_{i-1}+h_i}{2\cdot k_{j-1}}+\frac{k_{j-1}+k_j}{2\cdot h_{i-1}}+\frac{h_{i-1}+h_i}{2\cdot k_j}\right)+\\&+\psi_{i+1,j}\cdot\lambda^2\cdot\frac{k_{j-1}+k_j}{2\cdot h_i}+\end{aligned}$$

$$+\psi_{i,j+1}\cdot\lambda^2\cdot\left(\frac{h_{i-1}+h_i}{2\cdot k_j}-\frac{k_{j-1}}{2\cdot h_i}\right)+$$

$$+\psi_{i+1,j-1}\cdot\lambda^2\cdot\frac{k_{j-1}}{2\cdot h_i}-$$

$$-(n_{i,j}-p_{i,j}-C_{i,j})\cdot\frac{h_{i-1}+h_i}{2}\cdot\frac{k_{j-1}+k_j}{2}=0 \tag{6.2-6}$$

The treatment of the continuity equations is fully analogous. First, one has to substitute (6.2-3) and (6.2-4) into (6.1-16) and (6.1-17), respectively, and then (6.1-39), (6.1-40) and (6.1-41), (6.1-42) into the intermediate results.

If a meshline terminates in any other direction, the treatment for the required midinterval quantities is absolutely equivalent as in our example; it needs, therefore, not to be dealt with here explicitly.

One drawback of the terminating lines and the finite boxes approach becomes directly visible in (6.2-6). For a point which is not the termination point of a meshline quantities at four neighboring points are involved in the discrete approximation of the differential equation (cf. 6.1-72) whereas for a point at the termination of the meshline quantities at five neighboring points are involved. Furthermore, the influence of points is not "reciprocal". For instance, in equation (6.2-6) for $\psi_{i,j}$ the quantities $\psi_{i+1,j-1}$ and $\psi_{i+1,j+1}$ have non-zero entrics whereas in the respective equation for $\psi_{i+1,j-1}$ and $\psi_{i+1,j+1}$ the quantity $\psi_{i,j}$ does not enter. This means that the system of equations comprised of the individual equations at all mesh points is structurally unsymmetric which complicates the solution.

The local truncation errors for the three equations are altered compared to (6.1-46), (6.1-47) and (6.1-48). For a point at a terminating meshline in x-direction we have:

$$T_\psi < O(h)\cdot\left|\frac{\partial^3\psi}{\partial x^3}\right|+O(k)\cdot\left|\frac{\partial^3\psi}{\partial y^3}\right|+O\left(\frac{k^2}{h}\right)\cdot\left|\frac{\partial^3\psi}{\partial x\cdot y^2}\right| \tag{6.2-7}$$

$$T_n < O(h)\cdot\left|\frac{\partial J_{nx}}{\partial x}\right|+O(k)\cdot\left|\frac{\partial J_{ny}}{\partial y}\right|+O(h)\cdot\left|\frac{\partial^2 J_{nx}}{\partial x^2}\right|+O(k)\cdot\left|\frac{\partial^2 J_{ny}}{\partial y^2}\right|+$$

$$+O\left(\frac{k^2}{h}\right)\cdot\left|\frac{\partial^2 J_{nx}}{\partial y^2}\right| \tag{6.2-8}$$

$$T_p < O(h)\cdot\left|\frac{\partial J_{px}}{\partial x}\right|+O(k)\cdot\left|\frac{\partial J_{py}}{\partial y}\right|+O(h)\cdot\left|\frac{\partial^2 J_{px}}{\partial x^2}\right|+O(k)\cdot\left|\frac{\partial^2 J_{py}}{\partial y^2}\right|+$$

$$+O\left(\frac{k^2}{h}\right)\cdot\left|\frac{\partial^2 J_{px}}{\partial y^2}\right| \tag{6.2-9}$$

Similarly, we obtain for a point at a terminating meshline in y-direction:

$$T_\psi < O(h)\cdot\left|\frac{\partial^3\psi}{\partial x^3}\right|+O(k)\cdot\left|\frac{\partial^3\psi}{\partial y^3}\right|+O\left(\frac{h^2}{k}\right)\cdot\left|\frac{\partial^3\psi}{\partial x^2\cdot y}\right| \tag{6.2-10}$$

$$T_n < O(h) \cdot \left|\frac{\partial J_{nx}}{\partial x}\right| + O(k) \cdot \left|\frac{\partial J_{ny}}{\partial y}\right| + O(h) \cdot \left|\frac{\partial^2 J_{nx}}{\partial x^2}\right| + O(k) \cdot \left|\frac{\partial^2 J_{ny}}{\partial y^2}\right| + $$
$$+ O\left(\frac{h^2}{k}\right) \cdot \left|\frac{\partial^2 J_{ny}}{\partial x^2}\right| \tag{6.2-11}$$

$$T_p < O(h) \cdot \left|\frac{\partial J_{px}}{\partial x}\right| + O(k) \cdot \left|\frac{\partial J_{py}}{\partial y}\right| + O(h) \cdot \left|\frac{\partial^2 J_{px}}{\partial x^2}\right| + O(k) \cdot \left|\frac{\partial^2 J_{py}}{\partial y^2}\right| + $$
$$+ O\left(\frac{h^2}{k}\right) \cdot \left|\frac{\partial^2 J_{py}}{\partial x^2}\right| \tag{6.2-12}$$

From these expressions we can directly deduce that we should introduce a terminating meshline only where the respective mixed derivatives are small or, in other words, where the solution is almost constant in the direction perpendicular to the terminating meshline. In order to get consistency between the discrete approximations at regular points and those at terminations it has been shown in [6.32] to be necessary to assume mesh-spacing ratio restrictions at terminating meshlines. For a terminating meshline in the x-direction it is required that:

$$\frac{k_j}{h_i} < c, \quad \frac{k_{j-1}}{k_j} < c \tag{6.2-13}$$

Analogously, the prerequisite conditions for a terminating meshline in y-direction are:

$$\frac{h_i}{k_j} < c, \quad \frac{h_{i-1}}{h_i} < c \tag{6.2-14}$$

c is a moderate constant. As an additional more qualitative criterion one may prohibit meshlines from terminating inside of layer regions (space charge regions, inversion layers etc.).

Finally, it should be noted that it is completely incorrect to directly interpolate the dependent variable in order to get a value for a missing neighbor. Recalling (6.2-1) we would obtain, for instance for the electrostatic potential, at the missing point (x_{i+1}, y_j):

$$\psi_{i+1,j} = \frac{k_{j-1}}{k_{j-1}+k_j} \cdot \psi_{i+1,j+1} + \frac{k_j}{k_{j-1}+k_j} \cdot \psi_{i+1,j-1} + O(k^2) \cdot \left.\frac{\partial^2 \psi}{\partial y^2}\right|_{i+1,j} \tag{6.2-15}$$

The interpolation error in (6.2-15) is proportional to the product of the second partial derivative of the electrostatic potential and the square of the mesh spacing. It is incorrect to use an interpolation for the approximation of a point value in a finite difference approximation of a second order differential equation, which includes an error term of second order.

6.3 Finite Elements

The finite element method is a very new method. The modern use of finite elements started in the early 1940's with attempts by Hrenikoff [6.21] and McHenry [6.35] in the field of structural engineering. The term "finite element" was introduced by Clough [6.7] in 1960 in view of the direct analogy to engineering. Since then the finite element method has developed enormously and it can be seen as "a general discretization procedure of continuum problems posed by mathematically defined statements" [6.55]. A further historical introduction can be found in, e.g., [6.9].

In order to apply the finite element method we have, similarly to the finite difference method, to partition the simulation domain without overlap or exclusion into a finite number of subregions, or finite elements. Probably the most frequently chosen shape for the individual element is the triangle. Fig. 6.3-1 shows the partitioning of the simulation geometry of a lateral bipolar transistor. The domain is the same as already chosen for the mesh example in the finite difference section and the finite boxes section of this chapter. The mesh point allocation is identical to the one for the finite boxes method shown in Fig. 6.2-2. The number of elements (i.e. triangles) is 968 in this example.

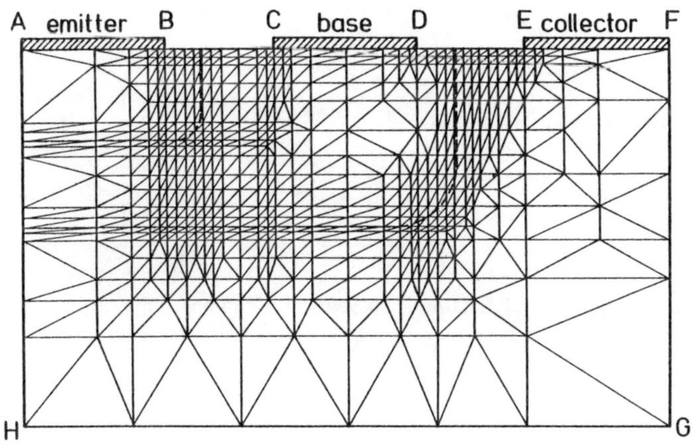

Fig. 6.3-1. Finite element mesh (triangular elements) for a lateral bipolar transistor

The finite element method seeks an approximation u^a to the exact solution u in each of the elements. Thus, in the i-th element an approximation (partial solution) is sought in such a manner that outside the element the contribution to the total approximate solution is zero. The total approximate solution is then simply the sum of the partial solutions over all elements.

$$u^a = \sum_{j=1}^{n} u_j^a \tag{6.3-1}$$

Next one has to choose a representation of the element approximation u_j^a. The most common form is the polynomial approximation probably due to the fact that

polynomials are easy to manipulate algebraically and computationally, and, most importantly, smooth functions can be approximated arbitrarily accurately by an appropriate polynomial. We drop the subscript j in the following since we consider at first only a single element.
The simplest polynomial suitable for triangular elements is:

$$u^a(x,y)=a_0+a_1\cdot x+a_2\cdot y \tag{6.3-2}$$

(6.3-2) is a linear approximation to the solution within the j-th element. To determine uniquely the coefficients a_i of this approximation three values of the approximate solution are required. These are in general taken at the nodes of the triangle (cf. Fig. 6.3-2).

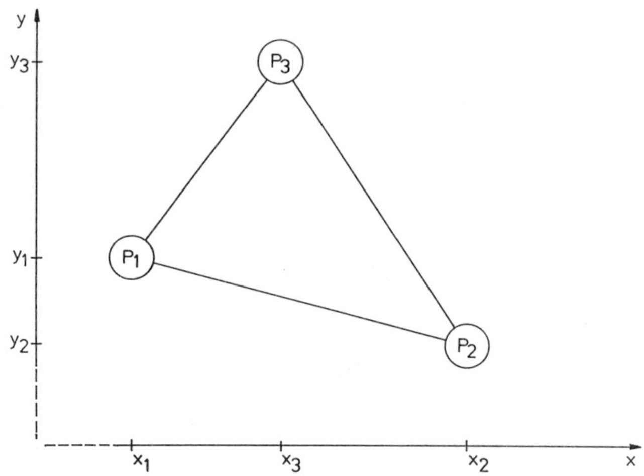

Fig. 6.3-2. Nomenclature for triangular element with linear approximation function

A more accurate approximation gives a complete quadratic polynomial.

$$u^a(x,y)=a_0+a_1\cdot x+a_2\cdot y+a_3\cdot x^2+a_4\cdot x\cdot y+a_5\cdot y^2 \tag{6.3-3}$$

Six values of the approximate solution are required to fully determine the coefficients a_i. As shown in Fig. 6.3-3, these may be taken at the nodes and in the middle of the edges of the triangle.
It can be imagined that the degree of the polynomial can be further increased, almost arbitrarily high. Incomplete polynomials may be used too. One may essentially prescribe an almost arbitrary functional behavior; only a few restrictions must be obeyed. There should be no preference for either the x- or the y-direction; the approximation should have geometrical invariance. Furthermore, the approximation must be continuous and must be capable of assuming an arbitrary linear shape [6.9], [6.55]. Lastly, the approximation should be conformal, i.e., the approximate solution should be continuous at the transition between adjacent elements. This is not a necessary condition; however, it significantly eases efforts towards a convergence and consistency analysis [6.49].

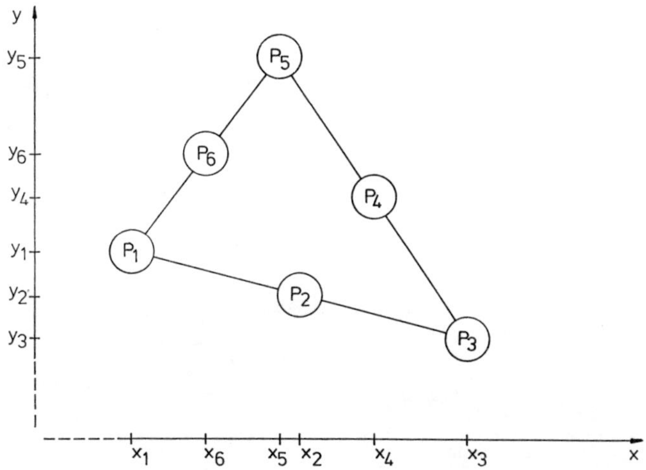

Fig. 6.3-3. Nomenclature for triangular element with quadratic approximation function

Any triangle with nodes $P_1(x_1, y_1)$, $P_2(x_2, y_2)$ and $P_3(x_3, y_3)$ which are numbered counter-clockwise can be transformed with the linear transformation (6.3-4) into a right unit triangle.

$$\begin{aligned} x &= x_1 + (x_2 - x_1) \cdot \xi + (x_3 - x_1) \cdot \eta \\ y &= y_1 + (y_2 - y_1) \cdot \xi + (y_3 - y_1) \cdot \eta \end{aligned} \tag{6.3-4}$$

We shall therefore consider in the following the unit triangle which enables a significantly simpler notation. The linear approximation (6.3-2) transforms in the coordinates (ξ, η) to:

$$u^a(\xi, \eta) = u_1 + (u_2 - u_1) \cdot \xi + (u_3 - u_1) \cdot \eta \tag{6.3-5}$$

u_i, $i = 1, 3$ are the values of the approximate solution at the respective nodes of the triangle.
Analogously the quadratic approximation (6.3-3) transforms in the coordinates (ξ, η) to:

$$\begin{aligned} u^a(\xi, \eta) = u_1 &+ (4 \cdot u_2 - 3 \cdot u_1 - u_3) \cdot \xi + (4 \cdot u_6 - 3 \cdot u_1 - u_5) \cdot \eta + \\ &+ (2 \cdot u_1 + 2 \cdot u_3 - 4 \cdot u_2) \cdot \xi^2 + 4 \cdot (u_1 - u_2 + u_4 - u_6) \cdot \xi \cdot \eta + \\ &+ (2 \cdot u_1 + 2 \cdot u_5 - 4 \cdot u_6) \cdot \eta^2 \end{aligned} \tag{6.3-6}$$

Particularly for the semiconductor equations rectangular elements have proved to be suitable. The solution is most simply approximated in a rectangle with a bilinear form.

$$u^a(x, y) = a_0 + a_1 \cdot x + a_2 \cdot y + a_3 \cdot x \cdot y \tag{6.3-7}$$

By numbering the nodes of the general rectangular element counter-clockwise (cf. Fig. 6.3-4) we may transform it with (6.3-8) into the unit square.

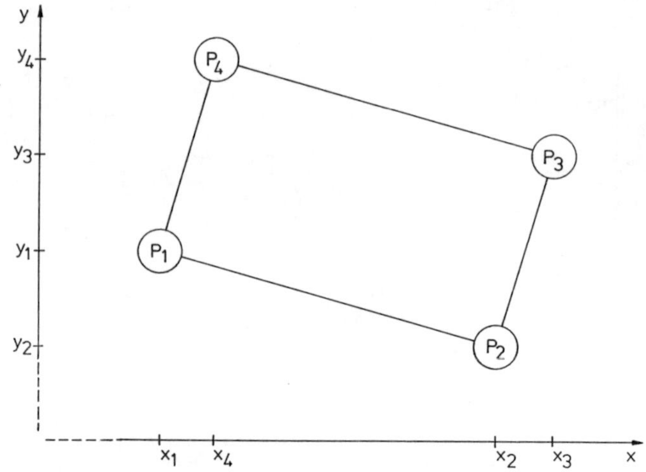

Fig. 6.3-4. Nomenclature for rectangular element with bilinear approximation function

$$
\begin{aligned}
x &= x_1 + (x_2 - x_1) \cdot \xi + (x_4 - x_1) \cdot \eta \\
y &= y_1 + (y_2 - y_1) \cdot \xi + (y_4 - y_1) \cdot \eta
\end{aligned}
\tag{6.3-8}
$$

(6.3-7) evaluates after transformation to:

$$
u^a(\xi, \eta) = u_1 + (u_2 - u_1) \cdot \xi + (u_4 - u_1) \cdot \eta + (u_1 - u_2 + u_3 - u_4) \cdot \xi \cdot \eta \tag{6.3-9}
$$

The coefficients a_i, $i = 1,4$ have been replaced by the proper combination of the nodal values of the approximate solution. The rectangular elements with bilinear approximation and the triangular elements with linear approximation can be very efficiently combined to minimize the number of elements which essentially determines the overhead costs of a computer program. Such an approach has been investigated in [6.31]. An example of a finite element mesh with both triangular and rectangular elements is shown in Fig. 6.3-5. Compared to the triangular mesh in

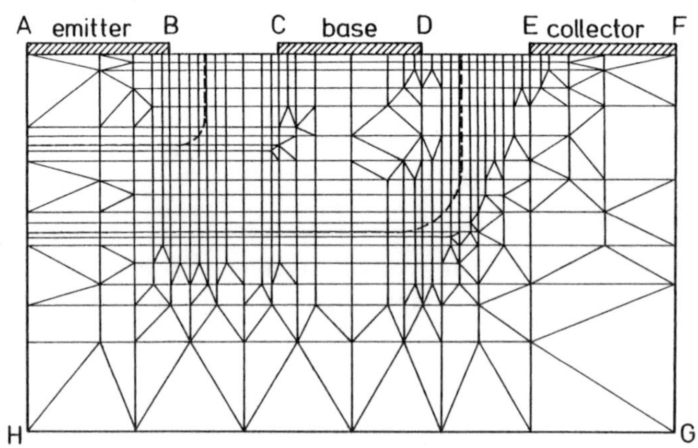

Fig. 6.3-5. Finite element mesh (rectangular and triangular elements) for a lateral bipolar transistor

Fig. 6.3-1 all pairs of triangular elements which form a rectangle parallel to the coordinate axes have been replaced by a rectangular element. The total number of elements has been reduced thereby from 968 triangles to 451 rectangles plus 198 triangles, or, 649 elements in total. Note that the allocation of the nodes has not been changed. Higher order approximations for rectangular elements certainly exist (cf. [6.49], [6.55]). However, in view of our application to the semiconductor equations I speculate that these will bring about only an insignificant improvement. Some investigations on the impact of the various possible approximations upon the semiconductor equations has been presented in, e.g., [6.5], [6.17], [6.18], [6.29]. A rigorous judgement of the results of these investigations is quite difficult, if at all possible.

We have given examples of approximations to the solution in triangular elements (6.3-5), (6.3-6) and rectangular elements (6.3-9). These approximations are more attractively formulated by the aid of so called shape functions $\varphi_i(\xi,\eta)$ in such a manner that the approximations transform into:

$$u^a(\xi,\eta)=\sum_{i=1}^{m} u_1\cdot\varphi_i(\xi,\eta) \tag{6.3-10}$$

m is the number of variable degrees of freedom in an element. The shape functions for our examples can be calculated with only small effort. For the linear approximation in triangles (6.3-5) they evaluate to:

$$\begin{aligned}\varphi_1(\xi,\eta)&=1-\xi-\eta\\ \varphi_2(\xi,\eta)&=\xi\\ \varphi_3(\xi,\eta)&=\eta\end{aligned} \tag{6.3-11}$$

For the quadratic approximation in triangles (6.3-6) the shape functions read:

$$\begin{aligned}\varphi_1(\xi,\eta)&=(1-2\cdot\xi-2\cdot\eta)\cdot(1-\xi-\eta)\\ \varphi_2(\xi,\eta)&=4\cdot\xi\cdot(1-\xi-\eta)\\ \varphi_3(\xi,\eta)&=\xi\cdot(2\cdot\xi-1)\\ \varphi_4(\xi,\eta)&=4\cdot\xi\cdot\eta\\ \varphi_5(\xi,\eta)&=\eta\cdot(2\cdot\eta-1)\\ \varphi_6(\xi,\eta)&=4\cdot\eta\cdot(1-\xi-\eta)\end{aligned} \tag{6.3-12}$$

For the bilinear approximation in rectangular elements (6.3-9) the shape functions are:

$$\begin{aligned}\varphi_1(\xi,\eta)&=(1-\xi)\cdot(1-\eta)\\ \varphi_2(\xi,\eta)&=\xi\cdot(1-\eta)\\ \varphi_3(\xi,\eta)&=\xi\cdot\eta\\ \varphi_4(\xi,\eta)&=(1-\xi)\cdot\eta\end{aligned} \tag{6.3-13}$$

Note that the shape functions outside of their specified element are identically zero. The shape functions are only non-vanishing within the element for which they are defined.

In order to proceed with our discussion of the finite element method we first shall note a few results from functional analysis. We define $F(u)$ to be a system of k differential equations.

$$F(u)=\begin{pmatrix} F_1(u) \\ F_2(u) \\ \vdots \\ F_k(u) \end{pmatrix}=0 \tag{6.3-14}$$

u is a vector function defining k components (dependent variables) for the k differential equations.

$$u=\begin{pmatrix} u_1 \\ u_2 \\ \vdots \\ u_k \end{pmatrix}=0 \tag{6.3-15}$$

This system is posed in a domain D subject to boundary conditions:

$$f(u)=\begin{pmatrix} f_1(u) \\ f_2(u) \\ \vdots \\ f_k(u) \end{pmatrix}\Big|_{\partial D}=0 \tag{6.3-16}$$

As the system (6.3-14) has to be satisfied in the domain D and simultaneously the boundary condition (6.3-16) has to be fulfilled on ∂D, it follows that:

$$\int\limits_{D} V^T \cdot F(u) \cdot d\vec{x} + \int\limits_{\partial D} v^T \cdot f(u) \cdot d\vec{A}=0 \tag{6.3-17}$$

V and v are vectors of arbitrary functions with $\operatorname{rank}(V)=\operatorname{rank}(v)=k$.

$$V=\begin{pmatrix} V_1 \\ V_2 \\ \vdots \\ V_k \end{pmatrix}, \quad v=\begin{pmatrix} v_1 \\ v_2 \\ \vdots \\ v_k \end{pmatrix} \tag{6.3-18}$$

However, in general one has to restrict the possible families of functions in (6.3-18) such that the integrals in (6.3-17) can be evaluated. This integral formulation is termed the "weak formulation" of the system of differential equations (6.3-14), (6.3-16).

With (6.3-1) we have introduced an approximation u^a for the exact solution u.

$$u \cong u^a = \sum_{j=1}^{n} u_j^a \tag{6.3-19}$$

We have furthermore described the solution u_j^a within each finite element as a sum of nodal values times shape functions. We obtain therefore for the total approximate solution:

$$u^a = \sum_{j=1}^{n} \left(\sum_{i=1}^{m} u_i \cdot \varphi_i \right)_j \tag{6.3-20}$$

The integral equation (6.3-17) is clearly never satisfied with the approximate solution (6.3-20) for arbitrary functions V and v. However, we may use, instead of "arbitrary" function V and v, a finite set of prescribed linearly independent functions.

$$V = V_\nu, \quad v = v_\nu, \quad \nu = 1, N \tag{6.3-21}$$

N denotes the total number of unknowns in (6.3-20), i.e. the total number of nodal values, involved in the approximation of the solution. (6.3-17) thus gives a system of N equations in N unknowns.

$$\begin{aligned} &\int_D V_\nu^T \cdot F\left(\sum_{j=1}^{n} \left(\sum_{i=1}^{m} u_i \cdot \varphi_i \right)_j \right) \cdot d\vec{x} + \\ &+ \int_{\partial D} v_\nu^T \cdot f\left(\sum_{j=1}^{n} \left(\sum_{i=1}^{m} u_i \cdot \varphi_i \right)_j \right) \cdot d\vec{A} = 0 \end{aligned} \tag{6.3-22}$$

This approach is called the "method of weighted residuals" because $F(u^a)$ represents the residual obtained by substituting the approximate solution u^a into the differential equation; $f(u^a)$ is the residual of the boundary condition. V_ν and v_ν are called weight functions. Depending on the choice of weight functions various names are associated with the method of weighted residuals [6.55]. Frequently the shape functions $\varphi_{i,j}$ are used directly to determine the weight functions. This approach is then called the Galerkin method. In detail the components of V_ν and v_ν are composed of the union of all elemental shape functions of node ν.

(6.3-22) may now be directly applied to the basic semiconductor equations. We choose shape functions for the dependent variables (ψ, n, p):

$$\psi^a = \sum_{j=1}^{n} \left(\sum_{i=1}^{m} \psi_i \cdot \varphi_i^\psi \right)_j \tag{6.3-23}$$

$$n^a = \sum_{j=1}^{n} \left(\sum_{i=1}^{m} n_i \cdot \varphi_i^n \right)_j \tag{6.3-24}$$

$$p^a = \sum_{j=1}^{n} \left(\sum_{i=1}^{m} p_i \cdot \varphi_i^p \right)_j \tag{6.3-25}$$

The Poisson equation substituted into (6.3-22) gives:

$$\begin{aligned} &\lambda^2 \cdot \int_D V_\nu \cdot \operatorname{div}\operatorname{grad} \sum_{j=1}^{n} \left(\sum_{i=1}^{m} \psi_i \cdot \varphi_i^\psi \right)_j \cdot d\vec{x} - \\ &- \int_D \cdot V_\nu \cdot \left(\sum_{j=1}^{n} \left(\sum_{i=1}^{m} n_i \cdot \varphi_i^n - p_i \cdot \varphi_i^p \right)_j - C \right) \cdot d\vec{x} + \\ &+ \int_{\partial D} v_\nu \cdot f^\psi(\psi^a, n^a, p^a) \cdot d\vec{A} = 0 \end{aligned} \tag{6.3-26}$$

f^{ψ} denotes the boundary condition for the electrostatic potential. The first integral in (6.3-26) has to be transformed using Green's theorem in order to get rid of the second partial derivatives.

$$\int_D V \cdot \operatorname{div}(P \cdot \operatorname{grad} u) \cdot d\vec{x} = \int_{\partial D} V \cdot P \cdot \operatorname{grad} u \cdot d\vec{A} - \int_D \operatorname{grad} V \cdot (P \cdot \operatorname{grad} u) \cdot d\vec{x} \tag{6.3-27}$$

The calculation of the first integral is then straightforward for a particular choice of shape functions and weight functions. This procedure will not be outlined further here because it depends too heavily upon the choice for the above cited quantities and the actual configuration of the elements. The second integral in (6.3-26) is indeed much more crucial. To highlight this fact we have to consider first some elementary estimates for the errors associated with the finite element method.

The error induced by the polynomial approximations can be estimated with the following inequality:

$$\| u - u^a \|_{0,2} \leq C \cdot h^{k+1} \cdot \| u^{(k+1)} \|_{0,2} \tag{6.3-28}$$

with $\| \cdot \|_{0,2}$ denoting the energy norm defined as:

$$\| f(\vec{x}) \|_{0,2} = \sqrt{\int_D (f(\vec{x}))^2 \cdot d\vec{x}} \tag{6.3-29}$$

k is the degree of the approximating polynomials; h is the maximal length of the edges of the elements; $u^{(k)}$ denotes the maximum of the k-th partial derivatives. C represents a constant depending on the arcs between the edges of the elements. This constant can be bounded by:

$$C < \frac{\text{const.}}{\sin \alpha} \tag{6.3-30}$$

α denotes the lower bound for all angles in the discretization. It is therefore mandatory to avoid elements with very acute angles. For a more elaborate error analysis we refer to, e.g., [6.4], [6.51].

Remembering the results of our analytical investigations in Chapter 5 we know that the carrier concentrations exhibit a rapidly varying behavior in layer regions. We furthermore know from the results about the discretization with finite differences that the carrier concentrations can be approximated better with exponential functions than with low order polynomials. That leads to the conclusion that the finite element mesh must be extraordinarily fine in order to limit reasonably the error of the approximate solution, i.e. the right hand side of (6.3-28).

In view of these considerations the classical finite element method as described here can be seen to be significantly inferior to the exponentially fitted difference method in treating the semiconductor equations. However, it is fair to note that the classical finite difference method, where the carrier concentrations would be approximated piecewise linearly, behaves as badly as the classical finite element method. Only the exponential fitting of the carrier concentrations is responsible for the superiority of the special finite difference scheme.

For the one-dimensional semiconductor equations an exponentially fitted finite element method has been given in [6.11] and Mock [6.43] has analyzed this scheme thoroughly. An extension of this scheme to two or three space dimensions is not known at present. In [6.24], [6.45] this subject has been treated too, for different applications though.

For these reasons we shall not discuss the discretization of the continuity equations by the classical finite element method. Several workers (whom I will not name) have tried hard – and I guess there are still some activities – to circumvent the above sketched problems and to develop special finite element schemes which are competitive with the exponentially fitted finite difference method. In principle there are three possibilities to tackle the problem. One can use a different set of dependent variables, different elemental shape functions or different weight functions for the residual integrals.

Considering the dependent variables (cf. Section 5.2) it may look attractive to use the set $(\psi, \varphi_n, \varphi_p)$ instead of (ψ, n, p). However, two drawbacks are associated with these dependent variables. First, the various residual integrals arising in the discretization cannot be solved easily in case of more than one space dimension. Therefore, numerical integration usually is performed which behaves in general very poorly for exponentially varying integrands. Secondly, a polynomial approximation for φ_n and φ_p does not properly reflect pure diffusion currents unless a very fine mesh is used. The carrier concentrations tend to have a linear behavior and not an exponential variation which would be predicted by, for instance, linearly varying φ_n, φ_p.

The use of exponentially fitted weight functions for the residual integrals may be an appropriate way to design a finite element method suitable for the semiconductor equations [6.55]. However, I am not aware at present of any investigations in this direction.

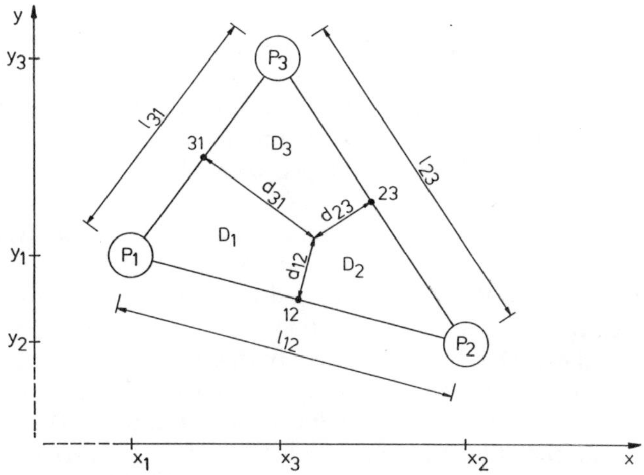

Fig. 6.3-6. Nomenclature adopted for hybrid finite element discretization

Some activities can be observed in the use of appropriate shape functions. In the development of the FIELDAY program [6.6], [6.8] the authors have used a hybrid method to link exponentially fitted carrier concentrations to the finite element method. A triangular mesh is used in this approach, and the carrier concentrations are assumed to be similar to those used in the exponentially fitted finite difference method. By adopting the nomenclature shown in Fig. 6.3-6 the scaled electron current (the treatment for holes is fully analogous) flowing along the edges is assumed to be:

$$I_n|_{12} = d_{12} \cdot D_n|_{12} \cdot \frac{B\left(\dfrac{\psi_1 - \psi_2}{Ut}\right) \cdot n_1 - B\left(\dfrac{\psi_2 - \psi_1}{Ut}\right) \cdot n_2}{l_{12}} \tag{6.3-31}$$

$$I_n|_{23} = d_{23} \cdot D_n|_{23} \cdot \frac{B\left(\dfrac{\psi_2 - \psi_3}{Ut}\right) \cdot n_2 - B\left(\dfrac{\psi_3 - \psi_2}{Ut}\right) \cdot n_3}{l_{23}} \tag{6.3-32}$$

$$I_n|_{31} = d_{31} \cdot D_n|_{31} \cdot \frac{B\left(\dfrac{\psi_3 - \psi_1}{Ut}\right) \cdot n_3 - B\left(\dfrac{\psi_1 - \psi_3}{Ut}\right) \cdot n_1}{l_{31}} \tag{6.3-33}$$

$B(x)$ denotes the Bernoulli function as defined in Section 6.1. The l_{ij} are the lengths of the edges and the d_{ij} are the lengths of the perpendicular bisectors. With the approximations (6.3-31) to (6.3-33) the elemental continuity equations are formed by associating an incremental area D_i within each element to the node P_i closest to that area.

$$I_n|_{31} - I_n|_{12} = D_1 \cdot R_1 \tag{6.3-34}$$

$$I_n|_{12} - I_n|_{23} = D_2 \cdot R_2 \tag{6.3-35}$$

$$I_n|_{23} - I_n|_{31} = D_3 \cdot R_3 \tag{6.3-36}$$

R_i denotes the scaled generation/recombination rate at node P_i. This hybrid method, which should be understood as a finite difference method on a triangular mesh, has proved to work satisfactorily for many applications. One prerequisite is that the triangulation be performed in such a manner that none of the triangles is obtuse. For an obtuse triangle one of the perpendicular bisectors falls outside the element which gives the respective current component a negative weight thus causing an inconsistent formulation of current continuity. However, triangulation of a domain without obtuse triangles can certainly be achieved. An analysis of the convergence properties of this hybrid method is not known at present.

In [6.31] an attempt to derive an exponentially fitted finite element method from a more mathematical basis has been discussed. In this approach rectangular elements and transition bands composed of three triangles have been used (cf. Fig. 6.3-5). The electrostatic potential is approximated with bilinear and linear shape functions in the rectangular and triangular elements, respectively. For the carrier concentrations in rectangular elements the following shape functions, which are bilinear in Bernoulli-like functions, have been suggested.

$$\begin{aligned}
\varphi_1^n(\xi,\eta) &= \left(1 - C\left(\xi, \frac{\partial\psi}{\partial\xi}\right)\right) \cdot \left(1 - C\left(\eta, \frac{\partial\psi}{\partial\eta}\right)\right) \\
\varphi_2^n(\xi,\eta) &= C\left(\xi, \frac{\partial\psi}{\partial\xi}\right) \cdot \left(1 - C\left(\eta, \frac{\partial\psi}{\partial\eta}\right)\right) \\
\varphi_3^n(\xi,\eta) &= C\left(\xi, \frac{\partial\psi}{\partial\xi}\right) \cdot C\left(\eta, \frac{\partial\psi}{\partial\eta}\right) \\
\varphi_4^n(\xi,\eta) &= \left(1 - C\left(\xi, \frac{\partial\psi}{\partial\xi}\right)\right) \cdot C\left(\eta, \frac{\partial\psi}{\partial\eta}\right)
\end{aligned} \tag{6.3-37}$$

$$\begin{aligned}
\varphi_1^p(\xi,\eta) &= \left(1 - C\left(\xi, -\frac{\partial\psi}{\partial\xi}\right)\right) \cdot \left(1 - C\left(\eta, -\frac{\partial\psi}{\partial\eta}\right)\right) \\
\varphi_2^p(\xi,\eta) &= C\left(\xi, -\frac{\partial\psi}{\partial\xi}\right) \cdot \left(1 - C\left(\eta, -\frac{\partial\psi}{\partial\eta}\right)\right) \\
\varphi_3^p(\xi,\eta) &= C\left(\xi, -\frac{\partial\psi}{\partial\xi}\right) \cdot C\left(\eta, -\frac{\partial\psi}{\partial\eta}\right) \\
\varphi_4^p(\xi,\eta) &= \left(1 - C\left(\xi, -\frac{\partial\psi}{\partial\xi}\right)\right) \cdot C\left(\eta, -\frac{\partial\psi}{\partial\eta}\right)
\end{aligned} \tag{6.3-38}$$

with:

$$C(x,y) = \frac{1 - \exp(x \cdot y)}{1 - \exp(y)} \tag{6.3-39}$$

For the triangular elements in the transition bands the hybrid formulation (6.3-31) to (6.3-36) has been used. This method seems to be quite convincing at first glance; however, several drawbacks become visible by closer inspection. First, the electric field components have been assumed to be constant within the elements which is inconsistent with the bilinear shape functions for the electrostatic potential. Furthermore, the shape functions (6.3-37) and (6.3-38) depend, even nonlinearly, on the electrostatic potential which itself is a dependent variable. This is indeed uncommon in the concept of the finite element method (cf. [6.51]) as has been pointed out explicitly in [6.12]. This method has been used with limited success only.

6.4 The Transient Problem

In this section we consider the treatment of the basic semiconductor equations for the case when the boundary condition for the electrostatic potential becomes time dependent. The partial derivatives with respect to time of the carrier concentrations then not vanish and the basic equations in scaled form will then read:

$$0 = \lambda^2 \cdot \operatorname{div} \operatorname{grad} \psi - (n - p - C) \tag{6.4-1}$$

$$\frac{\partial n}{\partial t} = \operatorname{div}(D_n \cdot \operatorname{grad} n - \mu_n \cdot n \cdot \operatorname{grad} \psi) - R(\psi, n, p) \tag{6.4-2}$$

$$\frac{\partial p}{\partial t} = \operatorname{div}(D_p \cdot \operatorname{grad} p + \mu_p \cdot p \cdot \operatorname{grad} \psi) - R(\psi, n, p) \tag{6.4-3}$$

These equations are posed in a "cylindrical" domain $D \times [0, T]$ subject to initial conditions at $t=0$:

$$\begin{aligned} \psi(\vec{x}, 0) &= \psi_0 \\ n(\vec{x}, 0) &= n_0 \\ p(\vec{x}, 0) &= p_0 \end{aligned} \tag{6.4-4}$$

and boundary conditions on ∂D, $0 < t < T$:

$$\begin{aligned} f_1(\psi, n, p, \vec{x}, t)|_{\partial D} &= 0 \\ f_2(\psi, n, p, \vec{x})|_{\partial D} &= 0 \\ f_3(\psi, n, p, \vec{x})|_{\partial D} &= 0 \end{aligned} \tag{6.4-5}$$

A discussion of the various possible boundary conditions has been given in Section 5.1. Therefore, we use the implicit notation (6.4-5) here.

The spatial operators, i.e., the right hand sides of (6.4-1), (6.4-2) and (6.4-3), have been dealt with in the previous sections. We, therefore, shall frequently use the abbreviated notation:

$$0 = F_1(\psi, n, p) \tag{6.4-6}$$

$$\frac{\partial n}{\partial t} = F_2(\psi, n, p) \tag{6.4-7}$$

$$\frac{\partial p}{\partial t} = F_3(\psi, n, p) \tag{6.4-8}$$

One difficulty considering the solution of (6.4-6) to (6.4-8) is that the Poisson equation (6.4-6) does not contain time derivatives of the dependent variables. Therefore, the direct application of simple "black box" methods for a solution is not feasible [6.25]. Several authors, e.g. [6.26], have circumvented this problem via brute force by introducing an artificial time derivative into (6.4-6).

$$\gamma \cdot \frac{\partial \psi}{\partial t} = F_1(\psi, n, p) \tag{6.4-9}$$

γ is a "sufficiently" small parameter. However, such an approach can not really be recommended since the error introduced thereby is independent of the time discretization to be used and, therefore, cannot be controlled by, for instance, properly choosing the size of the time steps.

Mock [6.39] has suggested a rigorous approach to introduce time derivatives into the Poisson equation. By differentiating (6.4-1) with respect to time we obtain:

$$0 = \lambda^2 \cdot \operatorname{div} \operatorname{grad} \frac{\partial \psi}{\partial t} - \frac{\partial n}{\partial t} + \frac{\partial p}{\partial t} \tag{6.4-10}$$

Then one substitutes (6.4-7) and (6.4-8) into (6.4-10) and rearranges terms:

$$\lambda^2 \cdot \operatorname{div} \operatorname{grad} \frac{\partial \psi}{\partial t} = F_2(\psi, n, p) - F_3(\psi, n, p) \tag{6.4-11}$$

The system comprised of (6.4-11), (6.4-7), (6.4-8) is equivalent to (6.4-6), (6.4-7), (6.4-8) provided that the initial data (6.4-4) satisfy the Poisson equation and the boundary conditions (6.4-5) at $t=0$. A mathematical analysis of this method can be found in [6.39], [6.42], [6.43].

In the following we shall discuss the time discretization of the basic equations. However, all algorithms will be presented in semidiscrete form, i.e. the dependent variables are discretized with respect to time but they are left continuous in the space variables. To simplify the notation we shall use the following abbreviations:

$$d_m = t_{m+1} - t_m \tag{6.4-12}$$

$$u_m = u(\vec{x}, t_m) \tag{6.4-13}$$

The simplest time discretization scheme is the (fully explicit) forward Euler method (cf. [6.36]). However, this method requires a severe restriction on the time step $d_m = 0(h^2 + k^2)$ (cf. [6.50]) which is not feasible in practice. Therefore, one should refrain from using the fully explicit method.

The probably simplest semi-implicit time discretization method is to solve repeatedly each of the basic equations using the "best available" values for the dependent variables (except in the recombination term cf. [6.43] as follows:

$$\frac{n_{m+1} - n_m}{d_m} - \operatorname{div}(D_n \cdot \operatorname{grad} n_{m+1} - \mu_n \cdot n_{m+1} \cdot \operatorname{grad} \psi_m) + R(\psi_m, n_m, p_m) = 0 \tag{6.4-14}$$

$$\frac{p_{m+1} - p_m}{d_m} - \operatorname{div}(D_p \cdot \operatorname{grad} p_{m+1} + \mu_p \cdot p_{m+1} \cdot \operatorname{grad} \psi_m) + R(\psi_m, n_m, p_m) = 0 \tag{6.4-15}$$

$$\lambda^2 \cdot \operatorname{div} \operatorname{grad} \psi_{m+1} - (n_{m+1} - p_{m+1} - C) = 0 \tag{6.4-16}$$

For this method one has to solve three linear equations at each time step. However, a strong condition on the time steps d_m is also required to guarantee stability [6.43], namely:

$$d_m < \min\left(\frac{2 \cdot \lambda^2}{\mu_n \cdot n + \mu_p \cdot p}\right) \tag{6.4-17}$$

This restriction is in general so prohibitive that the method is inapplicable for practical purpose, too.

A stable, uncoupled scheme can be obtained from the equivalent system (6.4-11), (6.4-7) and (6.4-8). The discretized equations read:

$$\lambda^2 \cdot \operatorname{div} \operatorname{grad} \frac{\psi_{m+1} - \psi_m}{d_m} - F_2(\psi_{m+1}, n_m, p_m) + F_3(\psi_{m+1}, n_m, p_m) = 0 \tag{6.4-18}$$

$$\frac{n_{m+1} - n_m}{d_m} - F_2(\psi_{m+1}, n_{m+1}, p_m) = 0 \tag{6.4-19}$$

$$\frac{p_{m+1}-p_m}{d_m}-F_3(\psi_{m+1},n_{m+1},p_{m+1})=0 \tag{6.4-20}$$

Mock [6.43] has proved that this method is stable independently of the size of the time steps d_m. However, computational experiments indicate that this method tends to damp transients too rapidly [6.41]. One has to require that the residual of the Poisson equation remains small during the iterations, i.e.:

$$E_{m+1}=\lambda^2\cdot\operatorname{div}\operatorname{grad}\psi_{m+1}-(n_{m+1}-p_{m+1}-C) \tag{6.4-21}$$

The time step d_m has to be chosen in such a way that (6.4-21), which is initially zero since the initial data are required to satisfy the Poisson equation, always remains below a prescribed accuracy bound. Note that (6.4-18) is nonlinear in ψ_{m+1} and the continuity equations (6.4-19), (6.4-20) are linear in n_{m+1}, p_{m+1}, respectively.
Another uncoupled scheme makes use of a "stabilizing" term in the Poisson equation. The continuity equations are discretized like (6.4-14), (6.4-15).

$$\frac{n_{m+1}-n_m}{d_m}-F_2(\psi_m,n_{m+1},p_m)=0 \tag{6.4-22}$$

$$\frac{p_{m+1}-p_m}{d_m}-F_3(\psi_m,n_{m+1},p_{m+1})=0 \tag{6.4-23}$$

$$\lambda^2\cdot\operatorname{div}\operatorname{grad}\psi_{m+1}-r\cdot(\psi_{m+1}-\psi_m)-(n_{m+1}-p_{m+1}-C)=0 \tag{6.4-24}$$

r is an appropriate positive and bounded damping function [6.46], [6.54]. During the iterations one has to keep, in analogy to the previous method, the residual of the Poisson equation, i.e. the stabilizing term, small by using sufficiently small timesteps.

$$E_{m+1}=r\cdot(\psi_{m+1}-\psi_m) \tag{6.4-25}$$

One particular choice for r which is motivated by the well known iterative scheme by Gummel [6.16] has been given in [6.39] (see also Section 7.2).

$$r=\frac{n_{m+1}+p_{m+1}}{U_t} \tag{6.4-26}$$

The method (6.4-22) to (6.4-24) with r given by (6.4-26) is unconditionally stable. However, an unphysical oscillatory behavior of the solution can be observed [6.3], [6.40] which makes the method almost unusable for practical applications.
A slightly different stabilizing function r has been tested in [6.3].

$$r=\frac{1}{U_t}\cdot\left(\frac{n_{m+1}}{B\left(\dfrac{\psi_{m+1}-\psi_m}{U_t}\right)}+\frac{p_{m+1}}{B\left(\dfrac{\psi_m-\psi_{m+1}}{U_t}\right)}\right) \tag{6.4-27}$$

$B(x)$ denotes the Bernoulli function (cf. Section 6.1). The Poisson equation (6.4-24) is, most unattractively, changed to a nonlinear equation in ψ_{m+1} which can be regarded as a substantial drawback. However, the oscillatory behavior of the solution is significantly reduced although it can still be observed.

If one is willing to spend considerable computer resources, full backward time differencing (backward Euler method) gives most satisfactory results. The discretized equations read in this case:

$$F_1(\psi_{m+1}, n_{m+1}, p_{m+1}) = 0 \tag{6.4-28}$$

$$\frac{n_{m+1} - n_m}{d_m} - F_2(\psi_{m+1}, n_{m+1}, p_{m+1}) = 0 \tag{6.4-29}$$

$$\frac{p_{m+1} - p_m}{d_m} - F_3(\psi_{m+1}, n_{m+1}, p_{m+1}) = 0 \tag{6.4-30}$$

This classical approach is well known to be unconditionally stable for arbitrarily large time steps d_m. The accuracy in time is readily monitored [6.38] by the local truncation error of the discrete approximations for the partial derivatives of the carrier concentrations in (6.4-29), (6.4-30). The main drawback of this fully implicit method can be seen in the large system of nonlinear algebraic equations which has to be solved at each time step. For the uncoupled approaches it is instead only necessary to solve three systems each of which has a rank of only one third of the rank of the full system, and, additionally, at least two systems (continuity equations) are linear. However, from my personal experience I have the opinion that the implicit method is worth the extra computational burden since the results obtained using the fully implicit scheme achieve the desired "numerical" reliability required for practical applications which are above purely academic examples.
If we substitute for the spatial operators in (6.4-28), (6.4-29), (6.4-30) the discretization schemes obtained by the exponentially fitted finite differences, we end up with the completely discretized equations (6.4-31), (6.4-32), (6.4-33).

$$\begin{aligned}
&\psi_{i,j-1,m+1} \cdot \lambda^2 \cdot \frac{h_{i-1}+h_i}{2 \cdot k_{j-1}} + \\
&+\psi_{i-1,j,m+1} \cdot \lambda^2 \cdot \frac{k_{j-1}+k_j}{2 \cdot h_{i-1}} - \\
&-\psi_{i,j,m+1} \cdot \lambda^2 \cdot \left(\frac{h_{i-1}+h_i}{2 \cdot k_{j-1}} + \frac{k_{j-1}+k_j}{2 \cdot h_{i-1}} + \frac{k_{j-1}+k_j}{2 \cdot h_i} + \frac{h_{i-1}+h_i}{2 \cdot k_j}\right) + \\
&+\psi_{i+1,j,m+1} \cdot \lambda^2 \cdot \frac{k_{j-1}+k_j}{2 \cdot h_i} + \\
&+\psi_{i,j,m+1} \cdot \lambda^2 \cdot \frac{h_{i-1}+h_i}{2 \cdot k_j} - \\
&-(n_{i,j,m+1} - p_{i,j,m+1}) \cdot \frac{h_{i-1}+h_i}{2} \cdot \frac{k_{j-1}+k_j}{2} = -C_{i,j} \cdot \frac{h_{i-1}+h_i}{2} \cdot \frac{k_{j-1}+k_j}{2}
\end{aligned} \tag{6.4-31}$$

$$n_{i,j-1,m+1} \cdot D_n|_{i,j-1/2,m} \cdot B\left(\frac{\psi_{i,j-1,m+1}-\psi_{i,j,m+1}}{Ut}\right) \cdot \frac{h_{i-1}+h_i}{2 \cdot k_{j-1}} +$$

$$+n_{i-1,j,m+1} \cdot D_n|_{i-1/2,j,m} \cdot B\left(\frac{\psi_{i-1,j,m+1}-\psi_{i,j,m+1}}{Ut}\right) \cdot \frac{k_{j-1}+k_j}{2 \cdot h_{i-1}} -$$

$$-n_{i,j,m+1} \cdot \left(D_n|_{i,j-1/2,m} \cdot B\left(\frac{\psi_{i,j,m+1}-\psi_{i,j-1,m+1}}{Ut}\right) \cdot \frac{h_{i-1}+h_i}{2 \cdot k_{j-1}} +\right.$$

$$+D_n|_{i-1/2,j,m} \cdot B\left(\frac{\psi_{i,j,m+1}-\psi_{i-1,j,m+1}}{Ut}\right) \cdot \frac{k_{j-1}+k_j}{2 \cdot h_{i-1}} +$$

$$+D_n|_{i+1/2,j,m} \cdot B\left(\frac{\psi_{i,j,m+1}-\psi_{i+1,j,m+1}}{Ut}\right) \cdot \frac{k_{j-1}+k_j}{2 \cdot h_i} +$$

$$+D_n|_{i,j+1/2,m} \cdot B\left(\frac{\psi_{i,j,m+1}-\psi_{i,j+1,m+1}}{Ut}\right) \cdot \frac{h_{i-1}+h_i}{2 \cdot k_j} +$$

$$\left.+\frac{1}{d_m} \cdot \frac{h_{i-1}+h_i}{2} \cdot \frac{k_{j-1}+k_j}{2}\right)+$$

$$+n_{i+1,j,m+1} \cdot D_n|_{i+1/2,j,m} \cdot B\left(\frac{\psi_{i+1,j,m+1}-\psi_{i,j,m+1}}{Ut}\right) \cdot \frac{k_{j-1}+k_j}{2 \cdot h_i} +$$

$$+n_{i,j+1,m+1} \cdot D_n|_{i,j+1/2,m} \cdot B\left(\frac{\psi_{i,j+1,m+1}-\psi_{i,j,m+1}}{Ut}\right) \cdot \frac{h_{i-1}+h_i}{2 \cdot k_j} =$$

$$=\left(R_{i,j,m}-\frac{n_{i,j,m}}{d_m}\right) \cdot \frac{h_{i-1}+h_i}{2} \cdot \frac{k_{j-1}+k_j}{2} \tag{6.4-32}$$

$$p_{i,j-1,m+1} \cdot D_p|_{i,j-1/2,m} \cdot B\left(\frac{\psi_{i,j,m+1}-\psi_{i,j-1,m+1}}{Ut}\right) \cdot \frac{h_{i-1}+h_i}{2 \cdot k_{j-1}} +$$

$$+p_{i-1,j,m+1} \cdot D_p|_{i-1/2,j,m} \cdot B\left(\frac{\psi_{i,j,m+1}-\psi_{i-1,j,m+1}}{Ut}\right) \cdot \frac{k_{j-1}+k_j}{2 \cdot h_{i-1}} -$$

$$-p_{i,j,m+1} \cdot \left(D_p|_{i,j-1/2,m} \cdot B\left(\frac{\psi_{i,j-1,m+1}-\psi_{i,j,m+1}}{Ut}\right) \cdot \frac{h_{i-1}+h_i}{2 \cdot k_{j-1}} +\right.$$

$$+D_p|_{i-1/2,j,m} \cdot B\left(\frac{\psi_{i-1,j,m+1}-\psi_{i,j,m+1}}{Ut}\right) \cdot \frac{k_{j-1}+k_j}{2 \cdot h_{i-1}} +$$

$$+D_p|_{i+1/2,j,m} \cdot B\left(\frac{\psi_{i+1,j,m+1}-\psi_{i,j,m+1}}{Ut}\right) \cdot \frac{k_{j-1}+k_j}{2 \cdot h_i} +$$

$$+D_p|_{i,j+1/2,m} \cdot B\left(\frac{\psi_{i,j+1,m+1}-\psi_{i,j,m+1}}{Ut}\right) \cdot \frac{h_{i-1}+h_i}{2 \cdot k_j} +$$

$$\left.+\frac{1}{d_m} \cdot \frac{h_{i-1}+h_i}{2} \cdot \frac{k_{j-1}+k_j}{2}\right)+$$

$$+p_{i+1,j,m+1}\cdot D_p|_{i+1/2,j,m}\cdot B\left(\frac{\psi_{i,j,m+1}-\psi_{i+1,j,m+1}}{U t}\right)\cdot\frac{k_{j-1}+k_j}{2\cdot h_i}+$$

$$+p_{i,j+1,m+1}\cdot D_p|_{i,j+1/2,m}\cdot B\left(\frac{\psi_{i,j,m+1}-\psi_{i,j+1,m+1}}{U t}\right)\cdot\frac{h_{i-1}+h_i}{2\cdot k_j}=$$

$$=\left(R_{i,j,m}-\frac{n_{i,j,m}}{d_m}\right)\cdot\frac{h_{i-1}+h_i}{2}\cdot\frac{k_{j-1}+k_j}{2} \tag{6.4-33}$$

The expressions for the carrier mobilities and the generation/recombination rate can in general be discretized at the m-th time level in any of the schemes when the solution at the $m+1$-st time level is sought. The time scales associated with carrier mobilities and generation/recombination are usually significantly larger than the time steps required to obtain acceptable truncation errors.

It is to note that the system of algebraic equations arising from the discretization of the transient problem is significantly easier to solve. One reason is that the partial time derivatives help to stabilize the center point coefficient of the spatial discretization. Furthermore, the solution at the previous time step is usually a very good initial guess for the solution at the next time step. Note also that a solution module for the static problem can be straightforwardly used for the transient problem because the discretized equations have exactly the same structure; only minor modifications in the assembly of the coefficients have to be performed.

All algorithms which have been presented in this section are accurate to first order in time, i.e. the local truncation error is $O(d)$. It is certainly possible to construct schemes which are of higher order of accuracy in time, e.g., the Crank-Nicholson method, [6.34], [6.36], [6.47], [6.50]. However, it is fairly difficult to keep the discretization of the spatial operators consistent with higher order time discretizations. I shall therefore refrain from discussing such algorithms.

6.5 Designing a Mesh

Since subregions of strong variation of the dependent variables (ψ, n, p) alternate with regions where these quantities behave smoothly (i.e. their gradients are of moderate size) different orders of magnitude of mesh sizes are mandatory for these regions. Any discretization scheme for the spatial operators must therefore be capable of locally switching from a coarse mesh to a fine mesh. For the finite difference method outlined in Section 6.1 this is accomplished by the generalization to finite boxes (cf. Section 6.2). For the finite element method it is implicit from the concept to have a non-uniform mesh.

It is impossible for most applications of the basic semiconductor equations to specify an efficient and appropriate mesh a-priori. The solution has to be known in order to design a numerically suitable mesh with as few as possible mesh points. Furthermore, the user of a device simulation program, usually an engineer, is not interested (and certainly not paid for) to quarrel with the design of a mesh. Therefore an adaptive mesh selection is desired and mathematically formulated criteria are, obviously, required. Such criteria have to satisfy two conditions in general. First, they should not cause more meshpoints than are really necessary to achieve a certain accuracy. Secondly, they should guarantee that a prescribed accuracy is really obtained once they are fulfilled.

The classical way to design adaptive mesh refinement procedures is to equidistribute the local truncation error of the discretization scheme. This statement, however, is about all what can be said in a rigorous manner. In practice, additional or more expedient criteria are required which can be straightforwardly implemented in a computer program; these criteria might be inexact mathematically, though. It has been shown in [6.32] that it is not possible in practice to equidistribute the local truncation error in layer regions because too many meshpoints are required. However, the singular perturbation analysis (cf. Section 5.6) shows that the exponentially fitted finite difference scheme which we have discussed in Section 6.1 approximates the reduced problem ($\lambda=0$ i.e. zero space charge approximation) implicitly even when the local truncation error is not equidistributed inside the layer regions. Therefore, the solutions obtained are qualitatively correct in the whole simulation domain and quantitatively acceptable if the local truncation error is only equidistributed outside the layer regions. The currents flowing through the contacts are only affected to a minor degree by the local truncation error in the layer regions [6.37], [6.44].

Note that the concept of equidistributing the local truncation error is certainly applicable for any discretization scheme. However, for the classical finite difference scheme (i.e., linear behavior assumed for the carrier concentrations between adjacent mesh points) or the classical finite element method the number of required mesh points is overwhelmingly large.

From the above given considerations we can deduce the following practical guidelines:

- Introduce a minimum mesh spacing on the order of the perturbation parameter λ.
- Calculate for each mesh point the maximum modulus of the partial derivatives of the scaled space charge and the scaled generation/recombination rate.
- Multiply these numbers with the scaled area associated with the respective mesh point. One obtains thereby a pessimistic estimate for the local truncation error of the exponentially fitted scheme.
- Insert mesh points wherever the estimates of the local truncation error are above a prescribed desired final accuracy and the minimum mesh spacing has not been reached. Practical values for the final accuracy are $O(10^{-2})$.
- Recompute the solution and adapt the mesh again if necessary.

These guidelines only qualitatively take into account the local truncation errors from the continuity equations (by adjusting the mesh to the magnitude of the generation/recombination rate). More accurately (but rather cumbersome) the magnitude of the first and second derivatives of the current densities should be used for designing the mesh (cf. (6.1-47), (6.1-48)).

From practical experience I can recommend introducing not too many new mesh points at every adaption step. A good value should be on the order of ten percent new points. The initial mesh can indeed be very coarse in the case of adaptive mesh refinement. It has been proven in [6.32] that even for a very coarse mesh the reduced problem is approximated by the exponentially fitted scheme. Therefore, it suffices that the initial mesh be fine enough to resolve the geometry and the doping profile. It is certainly possible to minimize the number of adaption processes by specifying the initial mesh with a-priori knowledge about the solution, which is available for almost all classical devices, e.g., MOS transistors, bipolar transistors. However, for structures like the parasitic thyristor in CMOS which is responsible for latch-up I cannot imagine where an appropriate and efficient mesh is designed a-priori and does not require an adaption step.

A comment should be given on the edges of a boundary (e.g., points A to H in Fig. 6.1-1). Such points represent singular points at which the local truncation error has to be expected, in general, to be much larger than at all other points. A discussion of this problem together with a splendid review of the literature has been given in [6.14]. However, the basic results indicate that the best pragmatic and feasible approach should be to use a very fine mesh in the vicinity of such points. This, non-rigorous, strategy is supported by the results obtained by careful investigations by Laux and Lomax [6.30].

One problem which arises when inserting new mesh points is that the solution has to be interpolated for the new mesh points in order to have an initial guess for the subsequent computations. The only interpolation scheme I can recommend is to use directly the difference scheme for the newly introduced mesh points with the solution at the "old" mesh points interpreted as Dirichlet data. This interpolation requires in general the solution of a nonlinear system of algebraic equations with a rank equal to the number of newly introduced mesh points times the number of variables at each node. However, the equations in this scheme are only weakly coupled, if at all, which simplifies the solution significantly. A naive linear interpolation of the dependent variables to obtain an initial guess at the new mesh points will greatly increase the effort to be spent for the subsequent solution of the nonlinear equations on the entire mesh; it should therefore be avoided.

A final remark should be given on the first initial guess for the iterative solution of the discretized semiconductor equations. The best recommendable procedure is to solve first the reduced problem, i.e., to assume zero space charge for the initial solution. Many computer experiments have proved that the solution of the full problem will then be determined with a minimum of computational effort. This approach can be elaborated further by successively increasing the complexity of the full problem caused by the fairly expensive evaluation of the various models for the physical parameters (carrier mobilities, generation/recombination). A very sophisticated and constructive approach to tackle this problem for the simulation of bipolar transistors is called STEPSOLVING [6.11]. However, a concept applicable for arbitrary devices cannot be given straightforwardly and has to be developed guided by physical knowledge about the behavior of the solution for the specific structure under consideration.

6.6 References

[6.1] Adler, M. S.: A Method for Achieving and Choosing Variable Density Grids in Finite Difference Formulations and the Importance of Degeneracy and Band Gap Narrowing in Device Modeling. Proc. NASECODE I Conf., pp. 3 – 30. Dublin: Boole Press 1979.

[6.2] Adler, M. S.: A Method for Terminating Mesh Lines in Finite Difference Formulations of the Semiconductor Device Equations. Solid-State Electron. *23,* 845–853 (1980).

[6.3] Agler, W.: Die numerische Lösung der transienten Halbleitergleichungen. Diplomarbeit, Technische Universität Wien, 1983.

[6.4] Babuska, I., Rheinboldt, W. C.: A Posteriori Error Analysis of Finite Element Solutions for One-Dimensional Problems. SIAM J. Numer. Anal. *18,* No. 3, 565–589 (1981).

[6.5] Barnes, J. J.: A Two-Dimensional Simulation of MESFET's. Dissertation, University of Michigan, 1976.

[6.6] Buturla, E. M., Cottrell, P. E., Grossman, B. M., Salsburg, K. A.: Finite-Element Analysis of Semiconductor Devices: The FIELDAY Program. IBM J. Res. Dev. *25,* 218–231 (1981).

[6.7] Clough, R. W.: The Finite Element in Plane Stress Analysis. Proc. Conf. on Electronic Computation, pp. 345–378 (1960).

[6.8] Cottrell, P. E., Buturla, E. M.: Two-Dimensional Static and Transient Simulation of Mobile Carrier Transport in a Semiconductor. Proc. NASECODE I Conf., pp. 31–64. Dublin: Boole Press 1979.

[6.9] Davies, A. J.: The Finite Element Method: A First Approach. Oxford: Clarendon Press 1980.

[6.10] Doolan, E. P., Miller, J. J. H., Schilders, W. H. A.: Uniform Numerical Methods for Problems with Initial and Boundary Layers. Dublin: Boole Press 1980.

[6.11] Engl, W. L., Dirks, H.: Numerical Device Simulation Guided by Physical Approaches. Proc. NASECODE I Conf., pp. 65–93. Dublin: Boole Press 1979.

[6.12] Engl, W. L., Dirks, H. K., Meinerzhagen, B.: Device Modeling. Proc. IEEE *71,* No. 1, 10–33 (1983).

[6.13] Forsythe, G. E., Wasow, W. R.: Finite Difference Methods for Partial Differential Equations. New York: Wiley 1960.

[6.14] Fox, L.: Finite-Difference Methods in Elliptic Boundary-Value Problems. In: The State of the Art in Numerical Analysis, pp. 799–881. London: Academic Press 1977.

[6.15] Franz, A. F., Franz, G. A., Selberherr, S., Ringhofer, C., Markovich, P.: Finite Boxes – A Generalization of the Finite Difference Method Suitable for Semiconductor Device Simulation IEEE Trans. Electron Devices *ED-30,* No. 9, 1070–1082 (1983).

[6.16] Gummel, H. K.: A Self-Consistent Iterative Scheme for One-Dimensional Steady State Transistor Calculations. IEEE Trans. Electron Devices *ED-11,* 455–465 (1964).

[6.17] Hachtel, G. D., Mack, M. H., O'Brien, R. R., Speelpennig, B.: Semiconductor Analysis Using Finite Elements – Part 1: Computational Aspects. IBM J. Res. Dev. *25,* 232–245 (1981).

[6.18] Hachtel, G. D., Mack, M. H., O'Brien, R. R.: Semiconductor Analysis Using Finite Elements – Part 2: IGFET and BJT Case Studies. IBM J. Res. Dev. *25,* 246–260 (1981).

[6.19] Hart, J. F., Cheney, E. W., Lawson, C. L., Maehly, H. J.: Computer Approximations. New York: Wiley 1968.

[6.20] Hockney, R. W., Eastwood, J. W.: Computer Simulation Using Particles. New York: McGraw-Hill 1981.

[6.21] Hrenikoff, A.: Solution of Problems in Elasticity by the Framework Method. J. Appl. Mech. *A8,* 169–175 (1941).

[6.22] Kellogg, R. B.: Analysis of a Difference Approximation for a Singular Perturbation Problem in Two Dimensions. Proc. BAIL I Conf., pp. 113–117. Dublin: Boole Press 1980.

[6.23] Kellogg, R. B., Shubin, G. R., Stephens, A. B.: Uniqueness and the Cell Reynolds Number. SIAM J. Numer. Anal. *17,* No. 6, 733–739 (1980).

[6.24] Kellogg, R. B., Han, H.: The Finite Element Method for a Singular Perturbation Problem Using Enriched Subspaces. Report BN-978, University of Maryland, 1981.

[6.25] Kraut, E. A., Murphy, W. D.: Application of Parabolic Partial Differential Equations to Semiconductor Device Modeling. Proc. NASECODE III Conf., pp. 150–154. Dublin: Boole Press 1983.

[6.26] Kreskowsky, J. P., Grubin, H. L.: Numerical Solution of the Transient, Multidimensional Semiconductor Equations Using the LBI Techniques. Proc. NASECODE III Conf., pp. 155–160. Dublin: Boole Press 1983.

[6.27] Kumar, R., Chamberlain, S. G., Roulston, D. J.: An Algorithm for Two-Dimensional Simulation of Reverse-Biased Beveled p-n Junctions. Solid-State Electron. *24,* 309–311 (1981).

[6.28] Kumar, R., Roulston, D. J., Chamberlain, S. G.: Accurate Two-Dimensional Simulation of Double-Beveled p-n Junctions. Solid-State Electron. *24,* 377–379 (1981).

[6.29] Laux, S. E.: Two-Dimensional Simulation of Gallium-Arsenide MESFET's Using the Finite-Element Method. Dissertation, University of Michigan, 1981.

[6.30] Laux, S. E., Lomax, R. J.: Numerical Investigation of Mesh Size Convergence Rate of the Finite Element Method in MESFET Simulation. Solid-State Electron. *24,* 485 – 493 (1981).

[6.31] Machek, J., Selberherr, S.: A Novel Finite Element Approach to Device Modelling. IEEE Trans. Electron Devices *ED-30,* No. 9, 1083 – 1092 (1983).

[6.32] Markowich, P. A., Ringhofer, C. A., Selberherr, S., Lentini, M.: A Singular Perturbation Approach for the Analysis of the Fundamental Semiconductor Equations. IEEE Trans. Electron Devices *ED-30,* No. 9, 1165 – 1180 (1983).

[6.33] Markowich, P. A., Ringhofer, C. A., Selberherr, S.: A Singular Perturbation Approach for the Analysis of the Fundamental Semiconductor Equations. Report 2482, MRC, University of Wisconsin, 1983.

[6.34] Marsal, D.: Die Numerische Lösung partieller Differentialgleichungen. Mannheim: Bibliographisches Institut 1976.

[6.35] McHenry, D.: A Lattice Analogy for the Solution of Plane Stress Problems. J. Inst. Civ. Eng. *21,* 59 – 82 (1943).

[6.36] Meis, T., Marcowitz, U.: Numerische Behandlung partieller Differentialgleichungen. Berlin-Heidelberg-New York: Springer 1978.

[6.37] Mock, M. S.: On the Computation of Semiconductor Device Current Characteristics by Finite Difference Methods. J. Eng. Math. *7,* No. 3, 193 – 205 (1973).

[6.38] Mock, M. S.: An Initial Value Problem from Semiconductor Device Theory. SIAM J. Math. Anal. *5,* No. 4, 597 – 612 (1974).

[6.39] Mock, M. S.: Time Discretization of a Nonlinear Initial Value Problem. J. Comp. Phys. *21,* 20 – 37 (1976).

[6.40] Mock, M. S.: The Charge-Neutral Approximation and Time Dependent Simulation. Proc. NASECODE I Conf., pp. 120 – 135. Dublin: Boole Press 1979.

[6.41] Mock, M. S.: A Time-Dependent Numerical Model of the Insulated-Gate Field-Effect Transistor. Solid-State Electron. *24,* 959 – 966 (1981).

[6.42] Mock, M. S.: The Stability Problem for Time-Dependent Models. In: An Introduction to the Numerical Analysis of Semiconductor Devices and Integrated Circuits, pp. 63 – 67. Dublin: Boole Press 1981.

[6.43] Mock, M. S.: Analysis of Mathematical Models of Semiconductor Devices. Dublin: Boole Press 1983.

[6.44] Mock, M. S.: Convergence and Accuracy in Stationary Numerical Models. In: An Introduction to the Numerical Analysis of Semiconductor Devices and Integrated Circuits, pp. 58 – 62. Dublin: Boole Press 1981.

[6.45] O'Riordan, E.: Finite Element Methods for Singularly Perturbed Problems. Proc. BAIL II Conf., pp. 52 – 57. Dublin: Boole Press 1982.

[6.46] Ortega, J. M., Rheinboldt, W. C.: Iterative Solution of Nonlinear Equations in Several Variables. New York: Academic Press 1970.

[6.47] Parter, S. V.: Numerical Methods for Partial Differential Equations. New York: Academic Press 1979.

[6.48] Scharfetter, D. L., Gummel, H. K.: Large-Signal Analysis of a Silicon Read Diode Oscillator. IEEE Trans. Electron Devices *ED-16,* 64 – 77 (1969).

[6.49] Schwarz, H. R.: Methode der finiten Elemente. Stuttgart: Teubner 1980.

[6.50] Smith, G. D.: Numerical Solution of Partial Differential Equations: Finite Difference Methods. Oxford: Clarendon Press 1978.

[6.51] Strang, G., Fix, G. J.: An Analysis of the Finite Element Method. Englewood Cliffs, N.J.: Prentice-Hall 1973.

[6.52] Sutherland, A. D.: An Algorithm for Treating Interface Surface Charge in the Two-Dimensional Discretization of Poisson's Equation for the Numerical Analysis of Semiconductor Devices such as MOSFET's. Solid-State Electron. *23,* 1085 – 1087 (1980).

[6.53] Szuhar, M.: Accurate Interface Handling for Mathematical Simulation of MOS Devices. Solid-State Electron. *25,* No. 9, 963 – 965 (1982).

[6.54] Zarantonello, E. H.: Solving Functional Equations by Contractive Averaging. Report 160, MRC, University of Wisconsin, 1960.

[6.55] Zienkiewicz, O. C.: The Finite Element Method. London: McGraw-Hill 1977.

7 The Solution of Systems of Nonlinear Algebraic Equations

The main result obtained in the preceding chapter is that discretization of the basic semiconductor equations yields a large system of nonlinear algebraic equations with the values of the dependent variables of the differential equations at discrete points as unknowns. For the considerations in this chapter we adopt the following nomenclature for the system of discretized equations:

$$\vec{F}(\vec{w})=0 \tag{7-1}$$

with

$$\vec{F}=\begin{pmatrix}\vec{f}_\psi(\vec{w})\\ \vec{f}_n(\vec{w})\\ \vec{f}_p(\vec{w})\end{pmatrix} \tag{7-2}$$

and

$$\vec{w}=\begin{pmatrix}\vec{\psi}\\ \vec{n}\\ \vec{p}\end{pmatrix} \tag{7-3}$$

$\vec{F}$ is a vector function of rank three which itself consists of the vector functions $\vec{f}_\psi$, $\vec{f}_n$ and $\vec{f}_p$. These vector functions correspond to the discrete approximations for the Poisson equation and the continuity equations, respectively. The vector of unknowns $\vec{w}$ is also comprised by three vectors which are formed by the values of the electrostatic potential $\vec{\psi}$, electron concentration $\vec{n}$ and hole concentration $\vec{p}$ at discrete points of the simulation geometry. We shall assume that the rank of all three vector functions $\vec{f}_\psi, \vec{f}_n, \vec{f}_p$ and the three vectors $\vec{\psi}, \vec{n}, \vec{p}$ equals n. This is not a necessary assumption but it will simplify the notation. It may well happen for practical applications that the rank of $\vec{\psi}$ differs from the rank of $\vec{n}$ and $\vec{p}$ (e.g., when the Laplace equation is solved in an insulator). For our purpose, the scalar rank of $\vec{F}$ and $\vec{w}$ is $3 \cdot n$.

The unknowns represent, of course, different quantities if other dependent variables for the differential equations are chosen (cf. Section 5.2). However, we restrict ourselves to (ψ, n, p) as set of dependent variables, which indeed is not essential for the following discussion.

In general, only iterative methods are applicable for the solution of systems of nonlinear algebraic equations. The most important method, without any doubt, is

Newton's method together with some modifications. In the following section I shall review the mathematics required for the understanding of nonlinear iteration with particular emphasis on Newton's method and Newton-like methods. In the second section of this chapter we will discuss some iterative methods which have proven to be valuable for the semiconductor equations, which, however, exhibit some heuristic nature such that a rigorous mathematical characterization is not possible. Continuation methods, e.g. [7.18], which might be applicable for the semiconductor equations [7.5] are not considered here.
For the sake of simplicity in notation the vector arrows will be omitted in the following. Ambiguities in the notation have not to be feared.

7.1 Newton's Method and Extensions

First I will give a qualitative introduction to the theory of nonlinear iterative methods, which naturally leads to Newton's method. A rigorous treatment of this subject is beyond the scope of this text; the interested reader is referred to the elaborate book by Ortega and Rheinboldt [7.17].
We assume for the sake of simplicity that a solution w^* exists for the system (7-1). Furthermore, we assume that there exists a neighborhood of w^* within which no other solution exists, i.e. within which the solution is unique.
All iterative methods are based on a fixpoint equation.

$$w = M(w) \tag{7.1-1}$$

$M(w)$ must be constructed in such a way that the fixpoint w^* of (7.1-1) is a solution to (7-1). The fixpoint equation is then used directly for iteration.

$$w^{k+1} = M(w^k), \quad k = 0, 1, 2, \ldots \tag{7.1-2}$$

with:

$$\lim_{k \to \infty} \| w^k - w^* \| = 0 \tag{7.1-3}$$

(7.1-3) certainly requires specific properties of $M(w)$ and, in general, of w^0, the initial "guess" too. These prerequisites can be formulated more mathematically as follows. Suppose, there is a neighbourhood $S(w^*)$, $M(w) \in S$ for $w \in S$ and a constant $\alpha \in [0, 1[$ such that for some norm:

$$\| M(w) - w^* \| \leq \alpha \cdot \| w - w^* \| \quad \forall\, w \in S \tag{7.1-4}$$

Then the iteration (7.1-2) will converge for any $w^0 \in S(w^*)$ to w^*. The iteration is locally convergent and $M(w)$ is called a "contractive" mapping. One way to ensure that condition (7.1-4) is satisfied is to assume that the Frechet derivative $M'(w)$ of $M(w)$ exists at the fixpoint w^* and that its eigenvalues are less than one in modulus. To put it more precise, the Ostrowski theorem [7.26] says if:

$$\rho(M'(w^*)) < 1 \tag{7.1-5}$$

then the mapping $M(w)$ is contractive and has a locally uniquely defined fixpoint $w^* \in S$. $\rho(A)$ denotes the spectral radius of the linear operator A; it is defined as the maximal modulus of all eigenvalues of M. $M(w)$ is Frechet differentiable at w^* if there exists a linear operator A such that:

$$\lim_{h\to 0} \frac{\| M(w^*+h) - M(w^*) - A\cdot h \|}{\| h \|} = 0 \tag{7.1-6}$$

The linear operator A is then called the Frechet derivative and is usually denoted by:

$$M'(w^*) = A \tag{7.1-7}$$

The Frechet derivative is unique if it exists and its concrete representation is the Jacobian matrix of M. The condition (7.1-6) can be viewed as a uniformity condition which ensures that $M(w)$ has a tangent space at w^*. It is to note that the usual properties and formalisms of derivatives in one space dimension can be carried over to the n-dimensional case, although their exact definition is more complex.
The preceding mathematical statements will allow us to elegantly characterize iterative methods (although they seem to be rather formal). Guided by practical aspects the iterative methods (7.1-8) are usually given preference.

$$w^{k+1} = M(w^k) = w^k - B(w^k)^{-1} \cdot F(w^k) \tag{7.1-8}$$

We shall now characterize the requirements on $B(w)$ such that the iterative scheme (7.1-8) converges. The Frechet derivative of the right hand side of (7.1-8) evaluates to:

$$M'(w) = I - (B(w)^{-1})' \cdot F(w) - B(w)^{-1} \cdot F'(w) \tag{7.1-9}$$

In order to apply the Ostrowski theorem (7.1-5) we have to evaluate (7.1-9) at the solution w^* of $F(w)$. This yields:

$$M'(w^*) = I - B(w^*)^{-1} \cdot F'(w^*) \tag{7.1-10}$$

From (7.1-10) we can deduce that a variety of operators $B(w)$ exists such that (7.1-5) is fulfilled. The classical Newton method is defined by:

$$B(w) = F'(w) \tag{7.1-11}$$

We can directly see that $M'(w^*)$ for Newton's method has a spectral radius equal to zero which gives convergence for initial guesses sufficiently close to the solution. However, we furthermore can see that $M(w)$ does not need to be the exact Frechet derivative of $F(w)$ in order to have a convergent scheme; it suffices in general to have an approximate Jacobian. To give an example we assume that $B(w)$ is proportional by a constant γ to the exact Jacobian.

$$B(w) = \gamma \cdot F'(w) \tag{7.1-12}$$

The Frechet derivative (7.1-10) evaluates in this case to:

$$M'(w^*) = \left(1 - \frac{1}{\gamma}\right) \cdot I \tag{7.1-13}$$

We can deduce trivially from (7.1-13) that the eigenvalues of $M'(w^*)$ are $(1-1/\gamma)$. Therefore, (7.1-12) gives a locally convergent iteration scheme with $\gamma \in]0.5, \infty[$ since then the Ostrowski theorem (7.1-5) is satisfied. It is obvious, however, that the

convergence properties are significantly influenced by the quality of the approximation $B(w)$ to $F'(w)$.
It should be noted, though trivial, that the iterative scheme (7.1-8) will certainly not be implemented in that form. In order to avoid the expensive inversion of the linear operator $B(w)$ we write instead:

$$B(w^k) \cdot (w^{k+1} - w^k) = -F(w^k) \tag{7.1-14}$$

For the solution of (7.1-14) it is necessary to solve a system of linear algebraic equations. This problem will be the topic of chapter 8.
In the following we consider modifications of the Jacobian matrix which give a linear operator $B(w)$ in such a manner that the corresponding modified Newton method (7.1-14) exhibits improved convergence properties for an initial guess which is not sufficiently close to the solution w^* that the classical Newton method (7.1-11) can be applied without difficulties. The main problem associated with the classical Newton method is the tendency to overestimate the length of the actual correction step for the iterate. This phenomenon is frequently termed overshoot. In the case of the semiconductor equations this overshoot problem has often been treated by simply limiting the size of the elements of the correction vector determined by Newton's method, e. g. [7.24], or by applying some nonlinear damping function on the correction vector, e.g. [7.3]. However, these approaches have in common that not only the length of the correction vector is adapted but also the direction is altered. This leads to unpredictable convergence properties which are in general poor compared to the more mathematically founded procedures to be described below.
It may also happen that the initial guess with which the classical Newton method is started does not lie in the region of attraction. Newton's method will then not converge at all. The modifications to the classical Newton method are therefore also intended to enlarge the region of convergence for the initial guess.
Another reason for allowing a deviation of $B(w)$ from $F'(w)$ is that it can be quite difficult and expensive in terms of computer resources to evaluate the Jacobian matrix accurately. An approximate Jacobian can be appropriate for a given problem as well.
The best established modifications to avoid overshoot of the classical Newton method are given by (7.1-15) and (7.1-16), respectively.

$$B(w^k) = \frac{1}{t_k} \cdot F'(w^k) \tag{7.1-15}$$

$$B(w^k) = s_k \cdot I + F'(w^k) \tag{7.1-16}$$

t_k and s_k are properly chosen positive parameters. For $t_k = 1$, $s_k = 0$ these modified Newton methods reduce to the classical Newton method. We have now to deal with the question how to choose t_k or s_k that the modified Newton methods with (7.1-15) or (7.1-16) exhibit superior convergence properties compared to the classical Newton method.
(7.1-15) has been investigated by Deuflhard [7.6] who suggested to use t_k from the interval $]0,1]$ in such a manner that for any norm:

$$\| F'(w^k)^{-1} \cdot F(w^k - t_k \cdot F'(w^k)^{-1} \cdot F(w^k)) \| < \| F'(w^k)^{-1} \cdot F(w^k) \| \tag{7.1-17}$$

Condition (7.1-17) guarantees that the correction of the k-th iterate is an improved approximation to the final solution. This condition can be easily evaluated only if the Jacobian matrix is factored into triangular matrices because the evaluation of the argument of the norm on the left hand side of (7.1-17) is then reduced to a forward and backward substitution and the evaluation of $F(w)$. However, the value to use for t_k is a question of trial and error. Frequently one chooses the following sequence:

$$t_k = \frac{1}{2^i} \quad i = 0, 1, 2, \ldots \tag{7.1-18}$$

or the more rapidly decreasing sequence:

$$t_k = \frac{1}{2^{\frac{i \cdot (i+1)}{2}}} \quad i = 0, 1, 2, \ldots \tag{7.1-19}$$

It is obvious that the largest value of t_k should be taken with which (7.1-17) is fulfilled. Sufficiently close to the solution (7.1-17) will be satisfied with $t_k = 1$ so that the convergence properties of the classical Newton method are retained in its limit. If the triangular factors of the Jacobian matrix are not available because an iterative method is used for the solution of the linear system, (7.1-17) cannot be readily applied. One may then use (7.1-20) where $D(w^k)$ denotes the main diagonal of $F'(w^k)$ [7.5].

$$\| D(w^k)^{-1} \cdot F(w^k - t_k \cdot F'(w^k)^{-1} \cdot F(w^k)) \| < \| D(w^k)^{-1} \cdot F(w^k) \| \tag{7.1-20}$$

Bank and Rose [7.1], [7.2] have presented an elaborate investigation about the proper choice of t_k and the associated convergence properties of the modified Newton method. They suggest to use:

$$t_k = \frac{1}{1 + \kappa_k \cdot \| F(w^k) \|} \tag{7.1-21}$$

with κ_k such that:

$$1 - \frac{\| F(w^{k+1}) \|}{\| F(w^k) \|} < \delta \cdot t_k, \quad \delta \in]0, 1[\tag{7.1-22}$$

Note that t_k approaches unity in (7.1-21) when w^k approaches the solution and κ_k is bounded. Actually, only if t_k approaches unity sufficiently fast as $w \to w^*$, the modified Newton method can anticipate superlinear convergence [7.2]. For further details of this method and the prerequisite assumptions on $F(w)$ and its Jacobian matrix, refer to the papers [7.1], [7.2]. Practical experience with this method applied to the semiconductor equations with indeed convincing results has been reported by Fichtner and Rose [7.7], [7.8].

The modified Newton method (7.1-16) has been investigated by, e.g., Meyer [7.14] and with application to the semiconductor equations also by Bank and Rose [7.1]. The required assumptions on $F(w)$ and its Jacobian are usually less stringent for (7.1-16) than for (7.1-15), which gives a larger field of applications for (7.1-16). Meyer has shown that (7.1-16) will converge monotonically to the solution w^* of $F(w) = 0$ if s_k is chosen in such a manner that (7.1-16) is diagonally dominant. Bank and Rose suggest to use:

$$s_k = \sigma_k \cdot \| F(w^k) \| , \quad \sigma_k > 0 \tag{7.1-23}$$

The σ_k are chosen such that:

$$\| F(w^{k+1}) \| < \| F(w^k) \| \tag{7.1-24}$$

However, considering the computer experimentes given in [7.1] the modified Newton method with (7.1-15) appears to be superior to (7.1-16) in the special application of the semiconductor equations. This statement coincides also with my personal experience.

Experiments [7.10] indicate that it can be quite advantageous to combine (7.1-15) and (7.1-16) as has also been anticipated in [7.2]. This leads to the iteration operator:

$$B(w^k) = \frac{1}{t_k} \cdot \left(s_k \cdot I + F'(w^k)\right) \tag{7.1-25}$$

(7.1-25) is in particular superior to (7.1-15) if $F'(w^k)$ represents an ill conditioned matrix.

One question which ultimately arises in the programming of Newton's method or any of its extensions is the adequacy of an approximation or the correctness of the Jacobian matrix. Particularly if parameters are involved in $F(w)$ which are complex functions of w, one most likely intends to ignore some of the partial derivatives as these can be very awkward to evaluate. One example in the case of the semiconductor equations represent the carrier mobilities which in general are functions of the dependent variables. For the simulation of silicon devices it is usually possible to neglect the partial derivatives of the carrier mobilities with an insignificant loss in the rate of convergence. However, for simulations with negative differential carrier mobilities (e. g. GaAs devices) the convergence may be significantly retarded [7.9]. Condition (7.1-26) can be very useful in checking the quality of the iteration operator $B(w)$.

$$\lim_{\alpha \to 0} \frac{1}{\alpha} \cdot \| F\left(w^k - \alpha \cdot B(w^k)^{-1} \cdot F(w^k)\right) - F(w^k) \| = \text{const.} > 0 \tag{7.1-26}$$

Some further remarks on checking the calculation of iteration operators and gradients can be found in [7.23].

In order to enable a maximum of flexibility in the specification of the models for (physical) parameters, for instance user supplied external functions, it may be necessary to calculate derivatives by numerical differentiation. From practical experience I can recommend the algorithm by Curtis and Reid [7.4] for the automatic choice of the step lengths when approximating first derivatives by first differences. This algorithm seeks a moderate value for the ratio of the truncation error of the difference approximation divided by the round-off error due to finite computer arithmetic. For details one is referred to the original work [7.4].

A final remark should be given on termination criteria for modified Newton methods. Usually one can find:

$$\| w^{k+1} - w^k \| < \varepsilon \cdot \| w^{k+1} \| \tag{7.1-27}$$

ε is a properly chosen relative accuracy parameter. However, (7.1-27) alone may be inadequate since it can be satisfied far too early in the case of a strongly damped iteration scheme. In [7.1] the following additional criterion has been used:

$$\| w^{k+1} - w^k \| < \frac{1}{2} \cdot \| w^k - w^{k-1} \| \tag{7.1-28}$$

This criterion guarantees that the actual correction step is sufficiently (half) small compared to the last correction step. Another possibility which I recommend is to check directly the residual of the nonlinear system.

$$\| F(w^{k+1}) \| < \delta \tag{7.1-29}$$

δ is a properly chosen absolute accuracy. It has to be defined in consistency with the scaling of $F(w)=0$. For the semiconductor equations scaled by the factors given in Section 5.5 δ is $O(10^{-10})$.

7.2 Iterative Methods

For Newton's method or its variants it is required to solve a system of linear algebraic equations at each iteration step. The question arises naturally how accurately these linear systems have to be solved since the result obtained thereby is just an incremental correction to the intermediate approximation of the solution. The accuracy of the solution of the linear systems required to preserve the convergence properties of Newton's method has been investigated in, e.g., [7.20]. However, with some modifications the solution of the linear system for each Newton step can be significantly simplified with an often acceptable decrease of the convergence rate. The overall solution of the nonlinear system can be much cheaper because of the simplifications for the linear systems although the number of iterations required to solve the nonlinear system is usually increased.
We shall first discuss the derivation of an SOR-Newton method [7.17] for the semiconductor equations. To simplify the notation we define the nonlinear system:

$$F(\psi, n, p) = \begin{pmatrix} F_1(\psi, n, p) \\ F_2(\psi, n, p) \\ F_3(\psi, n, p) \end{pmatrix} = 0 \tag{7.2-1}$$

F_1 denotes the Poisson equation; F_2 and F_3 are the continuity equations for electrons and holes, respectively. The correction vector for the k-th iterate is given by:

$$\begin{aligned} \delta\psi^k &= \psi^{k+1} - \psi^k \\ \delta n^k &= n^{k+1} - n^k \\ \delta p^k &= p^{k+1} - p^k \end{aligned} \tag{7.2-2}$$

Newton's method at the k-th step can then be formally written as:

$$\begin{pmatrix} \frac{\partial F_1}{\partial \psi} & \frac{\partial F_1}{\partial n} & \frac{\partial F_1}{\partial p} \\ \frac{\partial F_2}{\partial \psi} & \frac{\partial F_2}{\partial n} & \frac{\partial F_2}{\partial p} \\ \frac{\partial F_3}{\partial \psi} & \frac{\partial F_3}{\partial n} & \frac{\partial F_3}{\partial p} \end{pmatrix}^k \cdot \begin{pmatrix} \delta \psi^k \\ \delta n^k \\ \delta p^k \end{pmatrix} = - \begin{pmatrix} F_1(\psi^k, n^k, p^k) \\ F_2(\psi^k, n^k, p^k) \\ F_3(\psi^k, n^k, p^k) \end{pmatrix} \tag{7.2-3}$$

Under the assumption that the Jacobian matrix is definite and that all blocks in the main diagonal of (7.2-3) are non-singular one can use a classical block iteration scheme (iteration index m) for the solution of the k-th Newton step (cf. Section 8.3):

$$\begin{pmatrix} \frac{\partial F_1}{\partial \psi} & 0 & 0 \\ \frac{\partial F_2}{\partial \psi} & \frac{\partial F_2}{\partial n} & 0 \\ \frac{\partial F_3}{\partial \psi} & \frac{\partial F_3}{\partial n} & \frac{\partial F_3}{\partial p} \end{pmatrix}^k \cdot \begin{pmatrix} \delta \psi^k \\ \delta n^k \\ \delta p^k \end{pmatrix}^{m+1} = - \begin{pmatrix} F_1(\psi^k, n^k, p^k) \\ F_2(\psi^k, n^k, p^k) \\ F_3(\psi^k, n^k, p^k) \end{pmatrix} - \begin{pmatrix} 0 & \frac{\partial F_1}{\partial n} & \frac{\partial F_1}{\partial p} \\ 0 & 0 & \frac{\partial F_2}{\partial p} \\ 0 & 0 & 0 \end{pmatrix}^k \begin{pmatrix} \delta \psi^k \\ \delta n^k \\ \delta p^k \end{pmatrix}^m \tag{7.2-4}$$

Since the coefficient matrix of (7.2-4) is block lower triangular, one can decouple the system into three linear systems (7.2-5), (7.2-6) and (7.2-7) which have to be solved sequentially.

$$\frac{\partial F_1^k}{\partial \psi} \cdot \delta \psi^{km+1} = -F_1(\psi^k, n^k, p^k) - \frac{\partial F_1^k}{\partial n} \cdot \delta n^{km} - \frac{\partial F_1^k}{\partial p} \cdot \delta p^{km} \tag{7.2-5}$$

$$\frac{\partial F_2^k}{\partial n} \cdot \delta n^{km+1} = -F_2(\psi^k, n^k, p^k) - \frac{\partial F_2^k}{\partial \psi} \cdot \delta \psi^{km+1} - \frac{\partial F_2^k}{\partial p} \cdot \delta p^{km} \tag{7.2-6}$$

$$\frac{\partial F_3^k}{\partial p} \cdot \delta p^{km+1} = -F_3(\psi^k, n^k, p^k) - \frac{\partial F_3^k}{\partial \psi} \cdot \delta \psi^{km+1} - \frac{\partial F_3^k}{\partial n} \cdot \delta n^{km+1} \tag{7.2-7}$$

This iteration method has the advantage that the equations can be solved sequentially. To end up with the Block-SOR-Newton method one has to resubstitute the series expansions on the right hand side of (7.2-5), (7.2-6), (7.2-7) and to introduce a relaxation parameter $\omega \in]0.2[$:

$$\frac{\partial F_1^k}{\partial \psi} \cdot \delta \psi^{km+1} = -\omega \cdot F_1(\psi^k, n^k + \delta n^{km}, p^k + \delta p^{km}) \tag{7.2-8}$$

$$\frac{\partial F_2^k}{\partial n} \cdot \delta n^{km+1} = -\omega \cdot F_2(\psi^k + \delta \psi^{km+1}, n^k, p^k + \delta p^{km}) \tag{7.2-9}$$

$$\frac{\partial F_3^k}{\partial p} \cdot \delta p^{km+1} = -\omega \cdot F_3(\psi^k + \delta \psi^{km+1}, n^k + \delta n^{km+1}, p^k) \tag{7.2-10}$$

This method converges linearly [7.17]. However, thorough investigations need still be performed in order to properly judge the convergence properties of this method applied to the semiconductor equations. Experiments have proven that it can be applied to the semiconductor equations [7.10]. A similar algorithm has been proposed in [7.22]. However, the convergence properties are not too convincing if the set (ψ, n, p) is used as dependent variables. Note that the block $\partial F_1/\partial \psi$ is zero for the reduced problem which is the reason for the poor performance of this method on the full problem. It can be speculated that a suitable linearization scheme should implicitly be able to solve the reduced problem. With the set (ψ, u, v) as dependent variables the performance of this Block-SOR-Newton method is expected to be much better.

Particularly suited for the semiconductor equations is a block nonlinear iterative algorithm which has been first suggested by Gummel [7.12]. This algorithm is therefore most frequently called Gummel's method. It is motivated by the semiconductor equations in the dependent variables (ψ, u, v), which read in scaled semi-implicit form:

$$\lambda^2 \cdot \operatorname{div} \operatorname{grad} \psi - \left(\exp\left(\frac{\psi}{Ut}\right) \cdot u - \exp\left(-\frac{\psi}{Ut}\right) \cdot v - C\right) = 0 \tag{7.2-11}$$

$$\operatorname{div} \vec{J}_n - R = 0 \tag{7.2-12}$$

$$\operatorname{div} \vec{J}_p + R = 0 \tag{7.2-13}$$

Gummel's method works as follows: Given (ψ^k, u^k, v^k), ψ^{k+1} is computed by solving

$$\lambda^2 \cdot \operatorname{div} \operatorname{grad} \psi^{k+1} - \left(\exp\left(\frac{\psi^{k+1}}{Ut}\right) \cdot u^k - \exp\left(-\frac{\psi^{k+1}}{Ut}\right) \cdot v^k - C\right) = 0 \tag{7.2-14}$$

subject to the given boundary conditions for the electrostatic potential. The u^{k+1} and v^{k+1} are computed from the continuity equations (7.2-12), (7.2-13) together with the corresponding boundary conditions.

$$\operatorname{div} \vec{J}_n(\psi^{k+1}, u^{k+1}) - R(\psi^{k+1}, u^k, v^k) = 0 \tag{7.2-15}$$

$$\operatorname{div} \vec{J}_p(\psi^{k+1}, v^{k+1}) - R(\psi^{k+1}, u^k, v^k) = 0 \tag{7.2-16}$$

To avoid the numerical problems associated with the dependent variables (ψ, u, v) it is recommended to resubstitute the expressions for the carrier concentrations which gives:

$$\begin{aligned} \lambda^2 \cdot \operatorname{div} \operatorname{grad} \psi^{k+1} - \Bigg(n^k \cdot \left(\exp\left(\frac{\psi^{k+1} - \psi^k}{Ut}\right) - 1\right) - \\ - p^k \cdot \left(\exp\left(\frac{\psi^k - \psi^{k+1}}{Ut}\right) - 1\right)\Bigg) - (n^k - p^k - C) = 0 \end{aligned} \tag{7.2-17}$$

$$\operatorname{div} \vec{J}_n(\psi^{k+1}, n^{k+1}) - R(\psi^{k+1}, n^k, p^k) = 0 \tag{7.2-18}$$

$$\operatorname{div} \vec{J}_p(\psi^{k+1}, p^{k+1}) - R(\psi^{k+1}, n^k, p^k) = 0 \tag{7.2-19}$$

(7.2-17) represents a nonlinear differential equation which itself has to be solved iteratively in each step by a Newton-like method. Actually, in the original work of

Gummel [7.12] only one Newton iteration has been considered for the solution of (7.2-17) which is equivalent to solving the linearized problem:

$$\lambda^2 \cdot \operatorname{div} \operatorname{grad} \psi^{k+1} - \left(\frac{n^k + p^k}{Ut} \cdot (\psi^{k+1} - \psi^k)\right) - (n^k - p^k - C) = 0 \tag{7.2-20}$$

The continuity equations (7.2-18) and (7.2-19) are decoupled linear differential equations in n^{k+1} and p^{k+1}, respectively, if any existing nonlinearities caused by the carrier mobilities are neglected. In the original form [7.12] the generation/recombination rate is evaluated at the beginning of each step with the best available values for n and p (cf. (7.2-18), (7.2-19)). It has been frequently observed that this procedure causes difficulties when the generation/recombination rate is substantial. Therefore (7.2-18), (7.2-19) should be replaced by:

$$\operatorname{div} \vec{J}_n(\psi^{k+1}, n^{k+1}) - R(\psi^{k+1}, n^{k+1}, p^k) = 0 \tag{7.2-21}$$

$$\operatorname{div} \vec{J}_p(\psi^{k+1}, p^{k+1}) - R(\psi^{k+1}, n^{k+1}, p^{k+1}) = 0 \tag{7.2-22}$$

At present this form is commonly established (cf. [7.16]).
Gummel's method has proven to be extraordinarily valuable in practice. Convergence can be observed starting with a fairly poor initial guess with a sometimes spectacular rate in many applications. For some applications, e.g., high injection, onset of avalanche, however, it may fail to converge. An investigation of the underlying reasons for its performance has therefore been the goal of various mathematicians, e.g., [7.13], [7.15], [7.16], with limited success though. Following the treatment by Mock [7.16] Gummel's method can be classified as iterating a mapping of the following type:

$$\lambda^2 \cdot \operatorname{div} \operatorname{grad} \psi^{k+1} - r \cdot (\psi^{k+1} - \psi^k) - (n^k - p^k - C) = 0 \tag{7.2-23}$$

with the side conditions (7.2-21), (7.2-22). r is a suitably chosen positive damping function. Problems of this category have been investigated in [7.25]. The particular damping function for Gummel's method is:

$$r = \frac{1}{Ut} \cdot (n^k + p^k) \tag{7.2-24}$$

For biasing conditions close to thermal equilibrium one can prove mathematically that iteration methods of the type (7.2-23) converge for appropriate choices of damping functions r. Most unfortunately, the damping function (7.2-24) violates the required assumptions (even close to the equilibrium solution (cf. [7.16]). Thus, a general proof of the convergence of Gummel's method or constructive statements about the region of attraction for the initial guess are still missing.
An alternate choice for r is directly motivated by (7.2-17):

$$r = \frac{1}{Ut} \cdot \left(\frac{n^k}{B\left(\dfrac{\psi^{k+1} - \psi^k}{Ut}\right)} + \frac{p^k}{B\left(\dfrac{\psi^k - \psi^{k+1}}{Ut}\right)}\right) \tag{7.2-25}$$

$B(x)$ is the Bernoulli function (cf. Section 6.1). With this damping function (7.2-23) is changed to a nonlinear differential equation. However, the region of attraction and the rate of convergence have been observed to be positively affected, e.g., [7.11]. Some further remarks on that subject can be found in [7.16]; a general theory, however, does not exist at present.

Several attempts have been made to improve the rate of convergence of Gummel's method which, as mentioned, can become small for high levels of the current densities. In [7.21] an overrelaxation technique has been combined with Gummel's method for the simulation of MOSFET's. In [7.19] some modifications have been performed to obtain improved convergence properties for the simulation of MOSFET's at the onset of avalanche breakdown. In [7.7], [7.8] a block nonlinear iterative scheme has been used which is strongly related to Gummel's method. However, a general concept for improving Gummel's method cannot be given since all claimed improvements, which is at least my impression, are limited to very specific applications.

7.3 References

[7.1] Bank, R. E., Rose, D. J.: Parameter Selection for Newton-like Methods Applicable to Nonlinear Partial Differential Equations. SIAM J. Numer. Anal. *17,* No. 6, 806–822 (1980).

[7.2] Bank, R. E., Rose, D. J.: Global Approximate Newton Methods. Numer. Math. *37,* 279–295 (1981).

[7.3] Brown, G. W., Lindsay, B. W.: The Numerical Solution of Poisson's Equation for Two-Dimensional Semiconductor Devices. Solid-State Electron. *19,* 991–992 (1976).

[7.4] Curtis, A. R., Reid, J. K.: The Choice of Step Length when Using Differences to Approximate Jacobian Matrices. J. Inst. Math. Appl. *13,* 121–126 (1974).

[7.5] DenHeijer, C., Polak, S. J., Schilders, W. H. A.: A Continuation Method for the Calculation of Potentials and Currents in Semiconductors. Proc. NASECODE II Conf., pp. 182–187. Dublin: Boole Press 1981.

[7.6] Deuflhard, P.: A Modified Newton Method for the Solution of Ill-Conditioned Systems of Nonlinear Equations with Application to Multiple Shooting. Numer. Math. *22,* 289–315 (1974).

[7.7] Fichtner, W., Rose, D. J.: On the Numerical Solution of Nonlinear Elliptic PDEs Arising from Semiconductor Device Modeling. Report 80-2111-12, Bell Laboratories, 1980.

[7.8] Fichtner, W., Rose, D. J.: On the Numerical Solution of Nonlinear Elliptic PDEs Arising from Semiconductor Device Modeling. In: Elliptic Problem Solvers, pp. 277–284. New York: Academic Press 1981.

[7.9] Franz, A. F., Franz, G. A., Selberherr, S., Markowich, P.: The Influence of Various Mobility Models on the Iteration Process and Solution of the Basic Semiconductor Equations. Proc. NASECODE III Conf., pp. 117–121. Dublin: Boole Press 1983.

[7.10] Franz, A. F., Franz, G. A., Selberherr, S., Ringhofer, C., Markowich, P.: Finite Boxes – A Generalization of the Finite Difference Method Suitable for Semiconductor Device Simulation. IEEE Trans. Electron Devices *ED-30,* No. 9, 1070–1082 (1983).

[7.11] Greenfield, J. A., Hansen, S. E., Dutton, R. W.: Two-Dimensional Analysis for Device Modeling. Report G-201-7, Stanford University, 1980.

[7.12] Gummel, H. K.: A Self-Consistent Iterative Scheme for One-Dimensional Steady State Transistor Calculations. IEEE Trans. Electron Devices *ED-11,* 455–465 (1964).

[7.13] Markowich, P. A.: A Singular Perturbation Analysis of the Fundamental Semiconductor Device Equations. Habilitation, Technische Universität Wien, 1983.

[7.14] Meyer, G. H.: On Solving Nonlinear Equations with a One-Parameter Operator Imbedding. SIAM J. Numer. Anal. *5,* 739–752 (1968).

[7.15] Mock, M. S.: Asymptotic Behaviour of Solutions of Transport Equations for Semiconductor Devices. J. Math. Anal. Appl. *49,* 215–255 (1975).

[7.16] Mock, M. S.: Analysis of Mathematical Models of Semiconductor Devices. Dublin: Boole Press 1983.

[7.17] Ortega, J. M., Rheinboldt, W. C.: Iterative Solution of Nonlinear Equations in Several Variables. New York: Academic Press 1970.

[7.18] Richter, S. L., DeCarlo, R. A.: Continuation Methods: Theory and Applications. IEEE Trans. Circuits and Systems *CAS-30,* No. 6, 347–352 (1983).

[7.19] Schütz, A., Selberherr, S., Pötzl, H. W.: A Two-Dimensional Model of the Avalanche Effect in MOS Transistors. Solid-State Electron. *25,* 177–183 (1982).

[7.20] Sherman, A. H.: On Newton-Iterative Methods for the Solution of Systems of Nonlinear Equations. SIAM J. Numer. Anal. *15,* 755–771 (1978).

[7.21] Sutherland, A. D.: On the Use of Overrelaxation in Conjunction with Gummel's Algorithm to Speed the Convergence in a Two-Dimensional Computer Model for MOSFET's. IEEE Trans. Electron Devices *ED-27,* 1297–1298 (1980).

[7.22] Wang, C. T.: A Re-Extrapolation Technique in Newton-Sor Computer Simulation of Semiconductor Devices. Solid-State Electron. *25,* No. 11, 1083–1087 (1982).

[7.23] Wolfe, P.: Checking the Calculation of Gradients. ACM Trans. Mathematical Software *8,* No. 4, 337–343 (1982).

[7.24] Zaluska, E. J., Dubock, P. A., Kemhadhan, H. A.: Practical 2-Dimensional Bipolar-Transistor-Analysis Algorithm. Electron. Lett. *9,* 599–600 (1973).

[7.25] Zarantonello, E. H.: Solving Functional Equations by Contractive Averaging. Report 160, MRC, University of Wisconsin, 1960.

[7.26] Zienkiewicz, O. C.: The Finite Element Method. London: McGraw-Hill 1977.

8 The Solution of Sparse Systems of Linear Equations

For the solution of the nonlinear equations representing the discretized semiconductor equations it is required to solve repeatedly a linear system of algebraic equations. The coefficient matrices of these systems are said to be sparse because sufficiently many zero elements exist making it worthwhile to use special techniques which avoid storing and calculating with the zero elements. Actually, there are only very few non-zero elements and it is almost mandatory to account specifically for these elements. Unfortunately, this implies a significant overhead on organization for the non-zero elements, which is a well-feared source of problems in the design and coding of actual programs.

In this chapter we deal with questions about sparse matrices, which are of particular relevance for the semiconductor equations. In the first section a few comments on the direct solution (Gaussian elimination) of sparse systems are made. In Section 8.2, which is closely related to the direct methods, important symmetric permutation procedures are reviewed, i.e., ordering methods, which contribute essentially or even determine the efficiency of direct methods. Iterative methods, particularly relaxation methods, are outlined in Section 8.3 with emphasis on the underlying principles. Some comments on highly specialized iterative methods are given in Section 8.4 and Section 8.5. The final considerations are devoted to acceleration methods which can be applied to some basic iterative schemes in order to improve their convergence properties.

8.1 Direct Methods

All direct methods for the solution of sparse systems of linear equations

$$A \cdot x = b \tag{8.1-1}$$

are based on variants of Gaussian elimination. One seeks in common with all methods a factorization of the coefficient matrix A of the form:

$$P \cdot A \cdot Q = L \cdot U \tag{8.1-2}$$

P and Q are permutation matrices; and L and U is a lower and an upper triangular matrix, respectively. If A is symmetric, the so-called Cholesky decomposition can be used.

$$P \cdot A \cdot P^T = L \cdot D \cdot L^T \tag{8.1-3}$$

P is again a permutation matrix; L is a lower triangular matrix and D is a diagonal matrix. If A is indefinite, D may need to be a block diagonal matrix with blocks of order 1 and 2.

The linear systems arising in the iterative solution of the nonlinear discretized semiconductor equations can usually be decomposed without taking care for numerical stability by pivoting. However, the general criterion to omit pivoting, that the matrix A is positive definite, is not necessarily given if an exponentially fitted discretization scheme is used. Nevertheless, many years of experience by various researchers certify the above given statement. Therefore, one can fully decouple the computation of a suitable permutation matrix P from the numerical decomposition. The permutation of A can be sought only for minimizing the fill (cf. Section 8.2). However, the permutation must be symmetric, since the coefficients originally on the main diagonal of A must stay on the main diagonal in order to maintain stability, i.e., $Q = P^T$ in (8.1-2).

Prior to starting the factorization of a matrix all non-zero elements should be of comparable size in order to minimize difficulties with finite computer arithmetic. The simplest approach which is sufficient for the semiconductor problem is diagonal scaling.

$$D^{-1} \cdot A \cdot x = D^{-1} \cdot b \tag{8.1-4}$$

or:

$$D^{-1/2} \cdot A \cdot D^{-1/2} \cdot D^{1/2} \cdot x = D^{-1/2} \cdot b \tag{8.1-5}$$

The matrix D is a diagonal matrix formed by the main diagonal of A. (8.1-4) is a simple row scaling; (8.1-5) is a row and column scaling, and it is only applicable if all main diagonal elements of A are positive which certainly can be achieved easily. Note that the scaled matrix $D^{-1/2} \cdot A \cdot D^{-1/2}$ remains symmetric if A is symmetric. Both scaling procedures, (8.1-4) and (8.1-5) produce a coefficient matrix with the main diagonal elements scaled to unity. A more sophisticated scaling has been presented in, e.g., [8.8], however, this should not be required for the semiconductor problem. Scaling of the coefficient matrix is not only recommended for direct methods, but also for iterative methods as convergence properties will be improved in general. It should be noted furthermore, that it is not necessary to store the main diagonal elements of the scaled matrix since all are equal to unity.

In recent years research on sparse matrix problems has indeed reached a healthy state. It is difficult, if not arrogant, to try to summarize the results about data structures, decomposition details, etc. for the multitude of existing variants spanning wide a range. Duff has given a (compressed) survey in 1977 with more than 600 references [8.17]. I will therefore restrict myself to the citation of key publications. The present state of the art is well documented in [8.19], [8.22]. Indexing techniques and data structures for decomposition have been specifically reviewed in [8.27], [8.75]. An excellent survey about the available software for sparse matrices is presented in [8.23]. Experience and algorithmic details for implementation of sparse matrix software on vector computers are documented in, e.g., [8.20], [8.21], [8.53].

It should probably be noted that various sophisticated algorithms have been developed for linear systems arising from the discretization of the Poisson equation, e.g., [8.6], [8.7], [8.13], [8.29], [8.52], more general elliptic equations with constant coefficients, e.g., [8.5] and separable elliptic equations, e.g., [8.4], [8.85]. However, these algorithms will not be discussed here because of their fairly strong specialization.

8.2 Ordering Methods

A crucial question for the direct solution of a sparse linear system is the order of the equations for factorization or elimination. When a sparse matrix is factored, it normally suffers fill. Under the usual assumption that exact numerical cancellation does not occur, the factors of a matrix taken together are usually not as sparse as the matrix itself. In this section I shall review some of the most important strategies in computing a permutation matrix P for transforming a linear system $A \cdot X = B$ into:

$$(P \cdot A \cdot P^T) \cdot (P \cdot X) = (P \cdot B) \tag{8.2-1}$$

We use capital letters X and B for the solution vector and the right hand side. The rank of the permutation matrix P must, obviously, equal the rank of A. Note that for any permutation matrix the transformed coefficient matrix $P \cdot A \cdot P^T$ is symmetric or definite if and only if A is symmetric or definite. In general, the permuted matrix $P \cdot A \cdot P^T$ exhibits a different fill, and an appropriate choice of P can often reduce the fill enormously.

We shall demonstrate the impact of reordering the equations (i.e., permuting the matrix) with two simple examples. Fig. 8.2-1 shows a rectangular mesh with $NX = 8$ vertical and $NY = 5$ horizontal meshlines. The meshpoints have been numbered sequentially by rows. The total number of points which equals the rank of the matrix is 40.

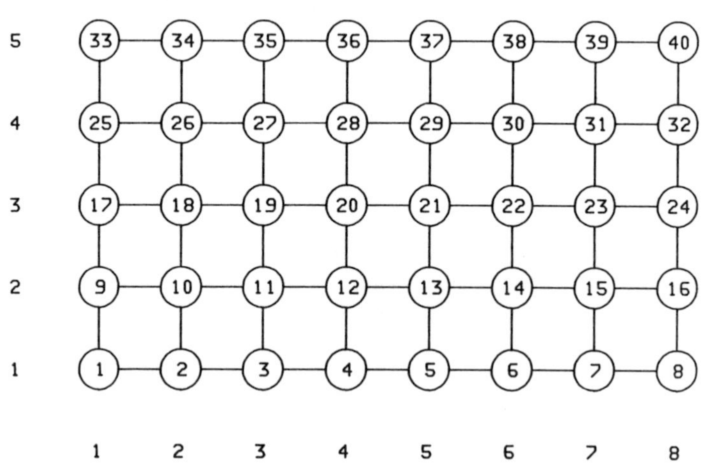

Fig. 8.2-1. Numbered graph for natural ordering by rows (five-point example)

I have to digress to review a minimum of notion from graph theory in order to compactly describe the various ordering algorithms. An excellent introduction to that subject can be found in [8.18]. Let $N=\{N_1, N_2, \ldots, N_n\}$ be the set of n nodes numbered from 1 to n. The pair (N_i, N_j) of two different nodes is called an edge between node N_i and node N_j. A graph G consists of the set N and a subset of all possible edges. A graph G is termed "undirected" if, for all edges (N_i, N_j), the edges (N_j, N_i) belong to the graph too. Otherwise, if an edge (N_i, N_j) belongs to the graph but not edge (N_j, N_i), G is called a "directed" graph. We shall, in the following, consider only undirected graphs. Any structurally symmetric matrix A, i.e., $A_{i,j} \neq 0$ if and only if $A_{j,i} \neq 0$, can be associated with an undirected graph $G(A)$. Each non-zero matrix element $A_{i,j}$ $(i \neq j)$ corresponds to the edge (N_i, N_j). The entries $A_{i,i}$ on the main diagonal of A have to be treated explicitly. We assume $A_{i,i} \neq 0$ which is always fulfilled for linear systems arising from the linearization of a discrete approximation to elliptic partial differential equations.

Fig. 8.2-1 can now be interpreted also as a graph (to come back to our examples) for a linear algebraic system. This linear system would have the following regular structure for an individual equation:

$$a_k \cdot X_{k-NX} + b_k \cdot X_{k-1} + c_k \cdot X_k + d_k \cdot X_{k+1} + e_k \cdot X_{k+NX} = B_k \qquad \forall k = 1, NX \cdot NY \tag{8.2-2}$$

The coefficients a_k, b_k, c_k, d_k, e_k do not exist for every equation. In particular we have:

$$\begin{aligned}
a_k &= A_{k,k-NX} && \text{exists for } k = NX+1, NX \cdot NY \\
b_k &= A_{k,k-1} && \text{exists for } k = 2, NX \cdot NY \wedge (k-1) \bmod NX \neq 0 \\
c_k &= A_{k,k} && \text{exists for } k = 1, NX \cdot NY \\
d_k &= A_{k,k+1} && \text{exists for } k = 1, NX \cdot NY \wedge k \bmod NX \neq 0 \\
e_k &= A_{k,k+NX} && \text{exists for } k = 1, NX \cdot (NY-1)
\end{aligned} \tag{8.2-3}$$

In a probably more transparent, two dimensional mesh oriented notation the individual equation reads:

$$a_{i,j} \cdot X_{i,j-1} + b_{i,j} \cdot X_{i-1,j} + c_{i,j} \cdot X_{i,j} + d_{i,j} \cdot X_{i+1,j} + e_{i,j} \cdot X_{i,j+1} = B_{i,j} \qquad \forall i = 1, NX \wedge j = 1, NY \tag{8.2-4}$$

The conditions for existence of the coefficients are then:

$$\begin{aligned}
a_{i,j} &= A_{(j-1) \cdot NX+i, (j-2) \cdot NX+i} && \text{exists for } i = 1, NX && \wedge\ j = 2, NY \\
b_{i,j} &= A_{(j-1) \cdot NX+i, (j-1) \cdot NX+i-1} && \text{exists for } i = 2, NX && \wedge\ j = 1, NY \\
c_{i,j} &= A_{(j-1) \cdot NX+i, (j-1) \cdot NX+i} && \text{exists for } i = 1, NX && \wedge\ j = 1, NY \\
d_{i,j} &= A_{(j-1) \cdot NX+i, (j-1) \cdot NX+i+1} && \text{exists for } i = 1, NX-1 && \wedge\ j = 1, NY \\
e_{i,j} &= A_{(j-1) \cdot NX+i, j \cdot NX+i} && \text{exists for } i = 1, NX && \wedge\ j = 1, NY-1
\end{aligned} \tag{8.2-5}$$

The main diagonal elements c_k or $c_{i,j}$ exist, of course, for all equations. This structure of the equations is always obtained by discretization of elliptic partial differential equations in two space dimensions with classical five-point finite differences (cf. Section 6.1). The existing coefficients $a_{i,j}$, $b_{i,j}$, $d_{i,j}$ and $e_{i,j}$ can be described by edges $(N_{i,j}, N_{i,j-1})$, $(N_{i,j}, N_{i-1,j})$, $(N_{i,j+1,j})$ and $(N_{i,j}, N_{i,j+1})$, respectively. In Fig. 8.2-1 it is impossible to distinguish between, for instance, an edge $(N_{i,j}, N_{i,j+1})$ and edge $(N_{i,j+1}, N_{i,j})$ as this figure is a representation of an undirected graph. However, both edges correspond to a non-zero element in a structurally symmetric matrix A. In addition, for a (numerically) symmetric matrix, $A = A^T$, the two non-zero elements corresponding to these two edges are equal, too. The coefficient matrix corresponding to the graph in Fig. 8.2-1 is shown in Fig. 8.2-2. The first column in front of the matrix denotes the numbering of the equations. The pair in the second column are the indices of the equations from a two-dimensional mesh interpretation. The asterisks above the main diagonal denote elements which become non-zero during a factorization, if the equations are treated in exactly the order given in the first column without row or column interchange during the factorization. As the matrix is structurally symmetric I have decided to show in the upper triangular part of the matrix the coefficients and the fill whereas in the lower triangular part only the non-zero elements prior to factorization are shown. The fill is certainly structurally symmetric too. All coefficients of the matrix in Fig. 8.2-2 should have the index pair given in the second column of each row. However, to have a more transparent representation of the overall pattern, this index pair has been omitted.

```
 1 1,1  ( c d             e                                                               )
 2 2,1  ( b c d           * e                                                             )
 3 3,1  (   b c d         * * e                                                           )
 4 4,1  (     b c d       * * * e                                                         )
 5 5,1  (       b c d     * * * * e                                                       )
 6 6,1  (         b c d   * * * * * e                                                     )
 7 7,1  (           b c d * * * * * * e                                                   )
 8 8,1  (             b c * * * * * * * e                                                 )
 9 1,2  ( a               c d * * * * * * e                                               )
10 2,2  (   a             b c d * * * * * * e                                             )
11 3,2  (     a             b c d * * * * * * e                                           )
12 4,2  (       a             b c d * * * * * * e                                         )
13 5,2  (         a             b c d * * * * * * e                                       )
14 6,2  (           a             b c d * * * * * * e                                     )
15 7,2  (             a             b c d * * * * * * e                                   )
16 8,2  (               a             b c * * * * * * * e                                 )
17 1,3  (                 a               c d * * * * * * e                               )
18 2,3  (                   a             b c d * * * * * * e                             )
19 3,3  (                     a             b c d * * * * * * e                           )
20 4,3  (                       a             b c d * * * * * * e                         )
21 5,3  (                         a             b c d * * * * * * e                       )
22 6,3  (                           a             b c d * * * * * * e                     )
23 7,3  (                             a             b c d * * * * * * e                   )
24 8,3  (                               a             b c * * * * * * * e                 )
25 1,4  (                                 a               c d * * * * * * e               )
26 2,4  (                                   a             b c d * * * * * * e             )
27 3,4  (                                     a             b c d * * * * * * e           )
28 4,4  (                                       a             b c d * * * * * * e         )
29 5,4  (                                         a             b c d * * * * * * e       )
30 6,4  (                                           a             b c d * * * * * * e     )
31 7,4  (                                             a             b c d * * * * * * e   )
32 8,4  (                                               a             b c * * * * * * * e )
33 1,5  (                                                 a               c d * * * * * * )
34 2,5  (                                                   a             b c d * * * * * )
35 3,5  (                                                     a             b c d * * * * )
36 4,5  (                                                       a             b c d * * * )
37 5,5  (                                                         a             b c d * * )
38 6,5  (                                                           a             b c d * )
39 7,5  (                                                             a             b c d )
40 8,5  (                                                               a             b c )
```

Fig. 8.2-2. Matrix for natural ordering by rows (five-point example)

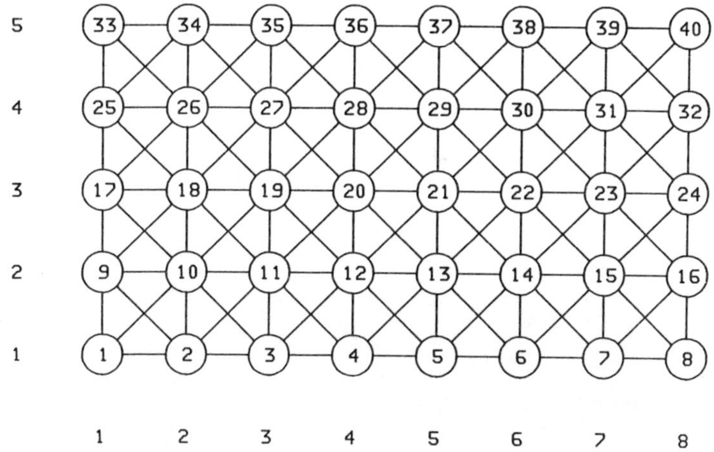

Fig. 8.2-3. Numbered graph for natural ordering by rows (nine-point example)

```
 1 1,1 ( c d             e i                                                             )
 2 2,1 ( b c d           h e i                                                           )
 3 3,1 (   b c d         * h e i                                                         )
 4 4,1 (     b c d       * * h e i                                                       )
 5 5,1 (       b c d     * * * h e i                                                     )
 6 6,1 (         b c d   * * * * h e i                                                   )
 7 7,1 (           b c d * * * * * h e i                                                 )
 8 8,1 (             b c * * * * * * h e                                                 )
 9 1,2 ( a g             c d * * * * * * e i                                             )
10 2,2 ( f a g           b c d * * * * * h e i                                           )
11 3,2 (   f a g           b c d * * * * * h e i                                         )
12 4,2 (     f a g           b c d * * * * * h e i                                       )
13 5,2 (       f a g           b c d * * * * * h e i                                     )
14 6,2 (         f a g           b c d * * * * * h e i                                   )
15 7,2 (           f a g           b c d * * * * * h e i                                 )
16 8,2 (             f a             b c * * * * * * h e                                 )
17 1,3 (                 a g             c d * * * * * * e i                             )
18 2,3 (                 f a g           b c d * * * * * h e i                           )
19 3,3 (                   f a g           b c d * * * * * h e i                         )
20 4,3 (                     f a g           b c d * * * * * h e i                       )
21 5,3 (                       f a g           b c d * * * * * h e i                     )
22 6,3 (                         f a g           b c d * * * * * h e i                   )
23 7,3 (                           f a g           b c d * * * * * h e i                 )
24 8,3 (                             f a             b c * * * * * * h e                 )
25 1,4 (                                 a g             c d * * * * * * e i             )
26 2,4 (                                 f a g           b c d * * * * * h e i           )
27 3,4 (                                   f a g           b c d * * * * * h e i         )
28 4,4 (                                     f a g           b c d * * * * * h e i       )
29 5,4 (                                       f a g           b c d * * * * * h e i     )
30 6,4 (                                         f a g           b c d * * * * * h e i   )
31 7,4 (                                           f a g           b c d * * * * * h e i )
32 8,4 (                                             f a             b c * * * * * * h e )
33 1,5 (                                                 a g             c d * * * * * * )
34 2,5 (                                                 f a g           b c d * * * * * )
35 3,5 (                                                   f a g           b c d * * * * )
36 4,5 (                                                     f a g           b c d * * * )
37 5,5 (                                                       f a g           b c d * * )
38 6,5 (                                                         f a g           b c d * )
39 7,5 (                                                           f a g           b c d )
40 8,5 (                                                             f a             b c )
```

Fig. 8.2-4. Matrix for natural ordering by rows (nine-point example)

The graph corresponding to the second example which we shall consider here and the coefficient matrix are shown in Fig. 8.2-3 and Fig. 8.2-4, respectively. The conventions for the figures are the same as outlined above for the first example. The individual equation for this linear system reads:

$$\begin{aligned} & f_{i,j}\cdot X_{i-1,j-1}+a_{i,j}\cdot X_{i,j-1}+g_{i,j}\cdot X_{i+1,j-1}+ \\ & +b_{i,j}\cdot X_{i-1,j}+c_{i,j}\cdot X_{i,j}+d_{i,j}\cdot X_{i+1,j}+ \\ & +h_{i,j}\cdot X_{i-1,j-1}+e_{i,j}\cdot X_{i,j+1}+i_{i,j}\cdot X_{i+1,j+1}=B_{i,j} \\ & \forall i=1,NX \wedge j=1,NY \end{aligned} \tag{8.2-6}$$

with:

$$\begin{aligned} f_{i,j}&=A_{(j-1)\cdot NX+i,(j-2)\cdot NX+i-1} && \text{exists for } i=2,NX && \wedge\ j=2,NY \\ a_{i,j}&=A_{(j-1)\cdot NX+i,(j-2)\cdot NX+i} && \text{exists for } i=1,NX && \wedge\ j=2,NY \\ g_{i,j}&=A_{(j-1)\cdot NX+i,(j-2)\cdot NX+i+1} && \text{exists for } i=1,NX-1 && \wedge\ j=2,NY \\ b_{i,j}&=A_{(j-1)\cdot NX+i,(j-1)\cdot NX+i-1} && \text{exists for } i=2,NX && \wedge\ j=1,NY \\ c_{i,j}&=A_{(j-1)\cdot NX+i,(j-1)\cdot NX+i} && \text{exists for } i=1,NX && \wedge\ j=1,NY \\ d_{i,j}&=A_{(j-1)\cdot NX+i,(j-1)\cdot NX+i+1} && \text{exists for } i=1,NX-1 && \wedge\ j=1,NY \\ h_{i,j}&=A_{(j-1)\cdot NX+i,j\cdot NX+i-1} && \text{exists for } i=2,NX && \wedge\ j=1,NY-1 \\ e_{i,j}&=A_{(j-1)\cdot NX+i,j\cdot NX+i} && \text{exists for } i=1,NX && \wedge\ j=1,NY-1 \\ i_{i,j}&=A_{(j-1)\cdot NX+i,j\cdot NX+i+1} && \text{exists for } i=1,NX-1 && \wedge\ j=1,NY-1 \end{aligned} \tag{8.2-7}$$

Such a structure of equations is obtained by, for instance, discretization with nine-point finite differences or rectangular finite elements using bilinear shape functions. We shall term this example the nine-point example whereas the previous example will be called the five-point example.

The number of non-zero elements prior to factorization is 174 and 286 for the matrices in Fig. 8.2-2 and Fig. 8.2-4, respectively. Both matrices are banded for the case of the natural ordering of the equations which allows the application of classical algorithms for banded linear systems. The total number of non-zero elements after factorization is 566 and 622, respectively. The bandwidth $\rho(A)$ of a matrix A is defined as:

$$\rho(\mathrm{A})=\max_{A_{i,j}\neq 0}|i-j| \tag{8.2-8}$$

In the first example the bandwidth evaluates to $NX=8$; in the second example it is $NX+1=9$. By reducing the bandwidth of a matrix through renumbering of the equations, we also reduce, in general, the fill. Therefore, it is always preferable if a natural ordering of the equations is performed such that reordering runs in that direction in which the maximum length of the columns or rows is smaller. For our examples all rows have a length of $NX=8$ and all columns have a length of $NY=5$. This is certainly not the case if the simulation geometry is non-rectangular. Fig. 8.2-5 and Fig. 8.2-6 show the numbered graph and the permuted matrix for the five-point

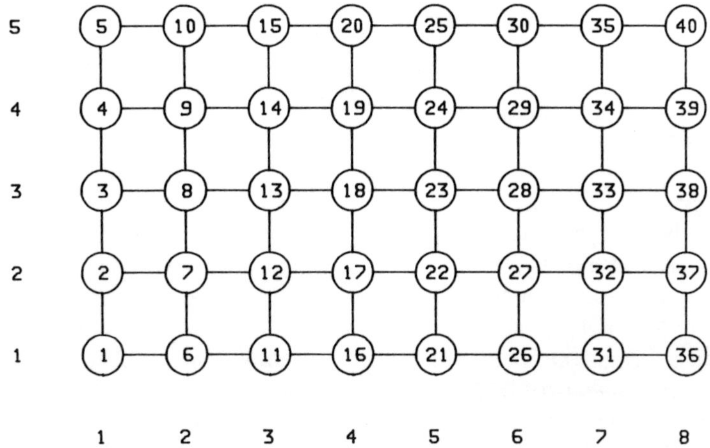

Fig. 8.2-5. Numbered graph for natural ordering by columns (five-point example)

```
 1 1,1  (c e       d                                                                     )
 2 1,2  (a c e     * d                                                                   )
 3 1,3  (  a c e   * * d                                                                 )
 4 1,4  (    a c e * * * d                                                               )
 5 1,5  (      a c * * * * d                                                             )
 6 2,1  (b         c e * * * d                                                           )
 7 2,2  (  b       a c e * * * d                                                         )
 8 2,3  (    b       a c e * * * d                                                       )
 9 2,4  (      b       a c e * * * d                                                     )
10 2,5  (        b       a c * * * * d                                                   )
11 3,1  (          b         c e * * * d                                                 )
12 3,2  (            b       a c e * * * d                                               )
13 3,3  (              b       a c e * * * d                                             )
14 3,4  (                b       a c e * * * d                                           )
15 3,5  (                  b       a c * * * * d                                         )
16 4,1  (                    b         c e * * * d                                       )
17 4,2  (                      b       a c e * * * d                                     )
18 4,3  (                        b       a c e * * * d                                   )
19 4,4  (                          b       a c e * * * d                                 )
20 4,5  (                            b       a c * * * * d                               )
21 5,1  (                              b         c e * * * d                             )
22 5,2  (                                b       a c e * * * d                           )
23 5,3  (                                  b       a c e * * * d                         )
24 5,4  (                                    b       a c e * * * d                       )
25 5,5  (                                      b       a c * * * * d                     )
26 6,1  (                                        b         c e * * * d                   )
27 6,2  (                                          b       a c e * * * d                 )
28 6,3  (                                            b       a c e * * * d               )
29 6,4  (                                              b       a c e * * * d             )
30 6,5  (                                                b       a c * * * * d           )
31 7,1  (                                                  b         c e * * * d         )
32 7,2  (                                                    b       a c e * * * d       )
33 7,3  (                                                      b       a c e * * * d     )
34 7,4  (                                                        b       a c e * * * d   )
35 7,5  (                                                          b       a c * * * * d )
36 8,1  (                                                            b         c e * * * )
37 8,2  (                                                              b       a c e * * )
38 8,3  (                                                                b       a c e * )
39 8,4  (                                                                  b       a c e )
40 8,5  (                                                                    b       a c )
```

Fig. 8.2-6. Permuted matrix for natural ordering by columns (five-point example)

example where the equations are ordered by columns. The bandwidth is reduced to 5 and the total number of non-zero elements after factorization is 410 which is significantly smaller than in case of ordering by rows. The situation for the nine-point example is fully analogous. We obtain a bandwidth of 6 and 454 non-zero elements after factorization. The numbered graph and the permuted coefficient matrix are shown in Fig. 8.2-7 and Fig. 8.2-8, respectively. The numbering of the nodes is obviously identical to the one shown in Fig. 8.2-5. This ordering is, considering all bandwidth oriented ordering procedures, an optimal ordering. The bandwidth cannot be further reduced for the nine-point example. The (maximum) bandwidth in the case of the five-point example cannot be reduced either; however, the average bandwidth which may influence the fill significantly can be reduced. To measure the average bandwidth a quantity termed the "profile" of a matrix is frequently used. The profile $P(A)$ of a structurally symmetric matrix A with $\operatorname{rank}(A)=n$ is defined as:

$$P(A)=\sum_{i=1}^{n}[i-\min(j: A_{i,j}\neq 0)] \tag{8.2-9}$$

A matrix permuted to have a small bandwidth does not necessarily have a small profile and vice versa. If one uses a classical band elimination technique which does not account for the local bandwidth of an individual equation, the optimum ordering has already been attained by the natural ordering by columns. However, the account for the profile during factorization is not at all a problem in principle but just a matter of programming effort which might be strongly rewarded by the additional reduction in the fill.

For all ordering methods used in our examples the node at mesh location (1, 1) is the starting node. Equally well qualified starting nodes are those at locations (1, 5), (8, 1) and (8, 5). However, the sparsity pattern of the factored matrix would not be altered by choosing any other starting node from the qualified set.

One well established algorithm to reduce bandwidth and profile has been proposed by Cuthill and McKee [8.9]. This algorithm works as follows. First, one takes a node with small (minimum) degree from the graph $G(A)$ and labels this node as the starting node with number one. The degree of node N_i is the number of edges which start from this node; it is thus the number of non-zero off-diagonal elements of the i-th equation. Secondly, one determines all neighboring nodes of the starting node. These are the nodes which are connected by edges to the starting node. These nodes are numbered sequentially with increasing degree. If nodes have the same degree, they are numbered in arbitrary order. The nodes which have been numbered in this step form the first "level". In the next step one determines successively for all nodes of the first level (in the order of their numbering) their neighboring nodes which have not been numbered yet. These nodes are again numbered in order of increasing degree by arbitrarily breaking ties; they form the second level. In subsequent steps one proceeds analogously until all nodes are numbered. Then the bandwidth obtained by the ordering is calculated and the whole ordering procedure is carried out again with a different starting node from a set of possible candidates with small degree until an ordering with a satisfactory small bandwith is obtained.

The major shortcoming of this algorithm is that it has to be applied repeatedly for different starting nodes in order to find an ordering with minimal bandwidth. It is a

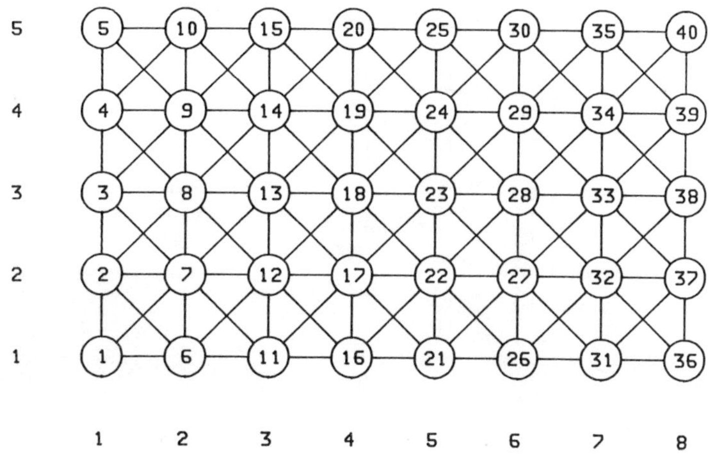

Fig. 8.2-7. Numbered graph for natural ordering by columns (nine-point example)

```
 1 1,1  ( c e       d i                                                                   )
 2 1,2  ( a c e     g d i                                                                 )
 3 1,3  (   a c e   * g d i                                                               )
 4 1,4  (     a c e * * g d i                                                             )
 5 1,5  (       a c * * * g d                                                             )
 6 2,1  ( b h       c e * * * d i                                                         )
 7 2,2  ( f b h     a c e * * g d i                                                       )
 8 2,3  (   f b h     a c e * * g d i                                                     )
 9 2,4  (     f b h     a c e * * g d i                                                   )
10 2,5  (       f b       a c * * * g d                                                   )
11 3,1  (           b h       c e * * * d i                                               )
12 3,2  (           f b h     a c e * * g d i                                             )
13 3,3  (             f b h     a c e * * g d i                                           )
14 3,4  (               f b h     a c e * * g d i                                         )
15 3,5  (                 f b       a c * * * g d                                         )
16 4,1  (                     b h       c e * * * d i                                     )
17 4,2  (                     f b h     a c e * * g d i                                   )
18 4,3  (                       f b h     a c e * * g d i                                 )
19 4,4  (                         f b h     a c e * * g d i                               )
20 4,5  (                           f b       a c * * * g d                               )
21 5,1  (                               b h       c e * * * d i                           )
22 5,2  (                               f b h     a c e * * g d i                         )
23 5,3  (                                 f b h     a c e * * g d i                       )
24 5,4  (                                   f b h     a c e * * g d i                     )
25 5,5  (                                     f b       a c * * * g d                     )
26 6,1  (                                         b h       c e * * * d i                 )
27 6,2  (                                         f b h     a c e * * g d i               )
28 6,3  (                                           f b h     a c e * * g d i             )
29 6,4  (                                             f b h     a c e * * g d i           )
30 6,5  (                                               f b       a c * * * g d           )
31 7,1  (                                                   b h       c e * * * d i       )
32 7,2  (                                                   f b h     a c e * * g d i     )
33 7,3  (                                                     f b h     a c e * * g d i   )
34 7,4  (                                                       f b h     a c e * * g d i )
35 7,5  (                                                         f b       a c * * * g d )
36 8,1  (                                                             b h       c e * * * )
37 8,2  (                                                             f b h     a c e * * )
38 8,3  (                                                               f b h     a c e * )
39 8,4  (                                                                 f b h     a c e )
40 8,5  (                                                                   f b       a c )
```

Fig. 8.2-8. Permuted matrix for natural ordering by columns (nine-point example)

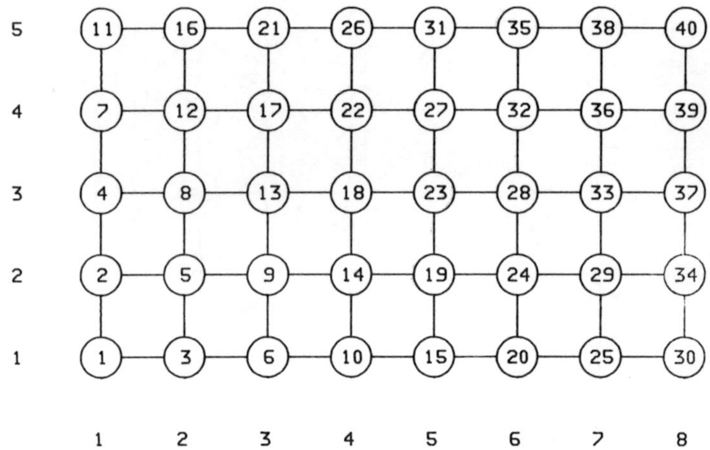

Fig. 8.2-9. Numbered graph for Cuthill-McKee ordering (five-point example)

```
 1 1,1  ( c e d                                                                           )
 2 1,2  ( a c * e d                                                                       )
 3 2,1  ( b   c * e d                                                                     )
 4 1,3  (   a   c * * e d                                                                 )
 5 2,2  (   b a   c * * e d                                                               )
 6 3,1  (     b     c * * e d                                                             )
 7 1,4  (       a     c * * * e d                                                         )
 8 2,3  (       b a     c * * * e d                                                       )
 9 3,2  (         b a     c * * * e d                                                     )
10 4,1  (           b       c * * * e d                                                   )
11 1,5  (             a       c * * * * d                                                 )
12 2,4  (             b a       c * * * e d                                               )
13 3,3  (               b a       c * * * e d                                             )
14 4,2  (                 b a       c * * * e d                                           )
15 5,1  (                   b         c * * * e d                                         )
16 2,5  (                     b a       c * * * * d                                       )
17 3,4  (                       b a       c * * * e d                                     )
18 4,3  (                         b a       c * * * e d                                   )
19 5,2  (                           b a       c * * * e d                                 )
20 6,1  (                             b         c * * * e d                               )
21 3,5  (                               b a       c * * * * d                             )
22 4,4  (                                 b a       c * * * e d                           )
23 5,3  (                                   b a       c * * * e d                         )
24 6,2  (                                     b a       c * * * e d                       )
25 7,1  (                                       b         c * * * e d                     )
26 4,5  (                                         b a       c * * * * d                   )
27 5,4  (                                           b a       c * * * e d                 )
28 6,3  (                                             b a       c * * * e d               )
29 7,2  (                                               b a       c * * * e d             )
30 8,1  (                                                 b         c * * * e             )
31 5,5  (                                                   b a       c * * * d           )
32 6,4  (                                                     b a       c * * e d         )
33 7,3  (                                                       b a       c * * e d       )
34 8,2  (                                                         b a       c * * e       )
35 6,5  (                                                             b a     c * * d     )
36 7,4  (                                                               b a     c * e d   )
37 8,3  (                                                                 b a     c * e   )
38 7,5  (                                                                     b a   c * d )
39 8,4  (                                                                       b a   c e )
40 8,5  (                                                                           b a c )
```

Fig. 8.2-10. Permuted matrix for Cuthill-McKee ordering (five-point example)

matter of experience to suggest checking all nodes with minimum degree as being appropriate starting nodes for the Cuthill-McKee algorithm. However, there exist graphs of linear systems for which the minimum bandwidth is not obtained when starting with a node with minimum degree [8.9].

Fig. 8.2-9 and Fig. 8.2-10 show the numbering of the nodes and the permuted coefficient matrix for the five-point example. The total number of non-zero elements after factorization is 370 which is significantly lower than the value obtained for the best natural ordering. One can observe that the nodes are numbered by diagonals of the graph. The numbering by diagonals has been known to be superior to the ordering by rows or columns for five-point problems on rectangular domains prior to the existence of the Cuthill-McKee algorithm [8.66]. However, the Cuthill-McKee algorithm is generally applicable whereas an extension of diagonal indexing to general problems is not possible.

Fig. 8.2-11 and Fig. 8.2-12 show the numbered graph and the permuted matrix for the nine-point example. The total number of non-zero elements after factorization is 518. Thus the natural ordering by columns is preferable; the effort spent for the Cuthill-McKee algorithm is not rewarded in this example.

George [8.35] observed that the profile of a matrix could frequently be further reduced by reversing the ordering obtained by the Cuthill-McKee algorithm ($i \rightarrow n-i+1$, $i=1,2,\ldots,n$). This algorithm is then called the "reverse" Cuthill-McKee algorithm. It has been proved in [8.63] that the reverse Cuthill-McKee algorithm can never increase the profile compared to the Cuthill-McKee algorithm; it produces an ordering which is at least as good. The bandwidth, of course, is not affected by reversing the ordering. In the case of the five-point example the sparsity pattern is not influenced by reversing the Cuthill-McKee ordering. The profile and fill are therefore, obviously, the same too.

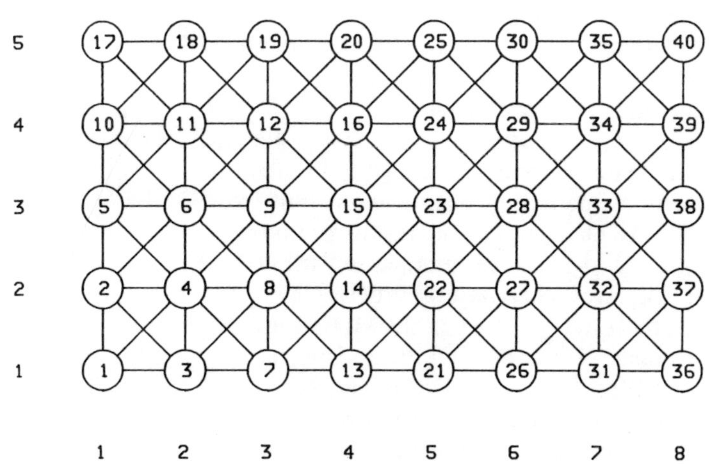

Fig. 8.2-11. Numbered graph for Cuthill-McKee ordering (nine-point example)

```
 1 1,1  (c e d i                                                                         )
 2 1,2  (a c g d e i                                                                     )
 3 2,1  (b h c e * * d i                                                                 )
 4 2,2  (f b a c h e g d i                                                               )
 5 1,3  (  a   g c d * * * e i                                                           )
 6 2,3  (  f   a b c * g d h e i                                                         )
 7 3,1  (    b h     c e * * * * d i                                                     )
 8 3,2  (    f b   h a c e * * * g d i                                                   )
 9 3,3  (      f   b   a c * h e * g d i                                                 )
10 1,4  (        a g       c d * * * * * e i                                             )
11 2,4  (        f a     g b c d * * * * h e i                                           )
12 3,4  (          f     a   b c * * g d * h e i                                         )
13 4,1  (            b h         c e * * * * * * d i                                     )
14 4,2  (            f b h       a c e * * * * * g d i                                   )
15 4,3  (              f b     h   a c e * * * * * g d i                                 )
16 4,4  (                f     b     a c * * h e * * g d i                               )
17 1,5  (                  a g           c d * * * * * * *                               )
18 2,5  (                  f a g         b c d * * * * * *                               )
19 3,5  (                    f a     g     b c d * * * * *                               )
20 4,5  (                      f     a       b c * * * g d                               )
21 5,1  (                        b h             c e * * * d i                           )
22 5,2  (                        f b h           a c e * * g d i                         )
23 5,3  (                          f b h           a c e * * g d i                       )
24 5,4  (                            f b     h       a c e * * g d i                     )
25 5,5  (                              f     b         a c * * * g d                     )
26 6,1  (                                        b h       c e * * * d i                 )
27 6,2  (                                        f b h     a c e * * g d i               )
28 6,3  (                                          f b h     a c e * * g d i             )
29 6,4  (                                            f b h     a c e * * g d i           )
30 6,5  (                                              f b       a c * * * g d           )
31 7,1  (                                                  b h       c e * * * d i       )
32 7,2  (                                                  f b h     a c e * * g d i     )
33 7,3  (                                                    f b h     a c e * * g d i   )
34 7,4  (                                                      f b h     a c e * * g d i )
35 7,5  (                                                        f b       a c * * * g d )
36 8,1  (                                                            b h       c e * * * )
37 8,2  (                                                            f b h     a c e * * )
38 8,3  (                                                              f b h     a c e * )
39 8,4  (                                                                f b h     a c e )
40 8,5  (                                                                  f b       a c )
```

Fig. 8.2-12. Permuted matrix for Cuthill-McKee ordering (nine-point example)

Fig. 8.2-13 and Fig. 8.2-14 show the numbered graph and the permuted matrix for the nine-point example. The number of non-zero elements after factorization is now 464 which is indeed a significant improvement compared to the unreversed ordering. The fill is now almost as small as for the best natural ordering.

An ordering algorithm which produces a bandwidth and profile which are comparable and often even better than these obtained with the reverse Cuthill-McKee algorithm has been proposed by Gibbs et al. [8.45]. This algorithm, in general, requires significantly less computation time. Excellent explanations of the algorithm can be found in [8.45] and [8.67]; I shall refrain from a repetition here. Hints on the implementation of this algorithm and FORTRAN code have been given in [8.59], [8.60], [8.67]. The results obtained by this algorithm for the five-point example are identical to the reverse Cuthill-McKee ordering. For the nine-point example the (reversed) natural ordering by columns will be obtained which is indeed better than the reverse Cuthill-McKee ordering.

A further decrease of the bandwidth and profile of a matrix can sometimes be obtained with the iterative algorithm by Rosen [8.77]. The basic step of this algorithm consists of interchanging the row-column pair(s) of the matrix, which determines the bandwidth, with other row-column pairs such that the profile, or bandwidth if possible, is reduced. In the case of our examples a further reduction of the profile (bandwidth) as already obtained is not possible.

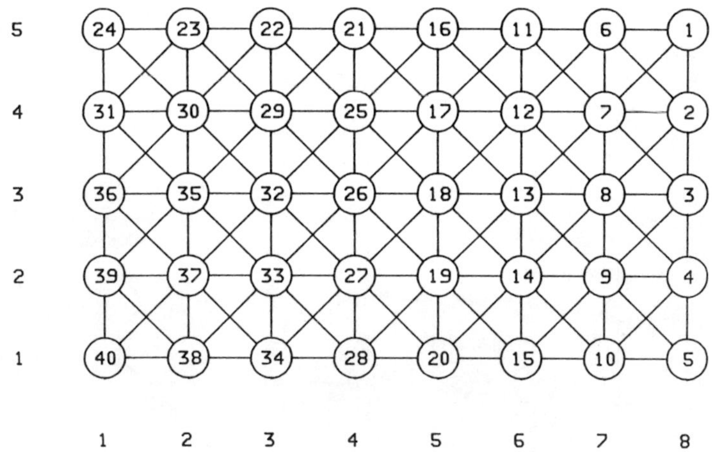

Fig. 8.2-13. Numbered graph for reverse Cuthill-McKee ordering (nine-point example)

```
 1 8,5  (c a       b f                                                                       )
 2 8,4  (e c a     h b f                                                                     )
 3 8,3  (  e c a   * h b f                                                                   )
 4 8,2  (    e c a * * h b f                                                                 )
 5 8,1  (      e c * * * h b                                                                 )
 6 7,5  (d g       c a * * * b f                                                             )
 7 7,4  (i d g     e c a * * h b f                                                           )
 8 7,3  (  i d g     e c a * * h b f                                                         )
 9 7,2  (    i d g     e c a * * h b f                                                       )
10 7,1  (      i d       e c * * * h b                                                       )
11 6,5  (          d g       c a * * * b f                                                   )
12 6,4  (          i d g     e c a * * h b f                                                 )
13 6,3  (            i d g     e c a * * h b f                                               )
14 6,2  (              i d g     e c a * * h b f                                             )
15 6,1  (                i d       e c * * * h b                                             )
16 5,5  (                    d g       c a * * * b       f                                   )
17 5,4  (                    i d g     e c a * * h       b f                                 )
18 5,3  (                      i d g     e c a * *       h b f                               )
19 5,2  (                        i d g     e c a *       * h b f                             )
20 5,1  (                          i d       e c *       * * h b                             )
21 4,5  (                              d g       c b     a * * * f                           )
22 3,5  (                                        d c b   g * * * a f                         )
23 2,5  (                                          d c b * * * * g a f                       )
24 1,5  (                                            d c * * * * * g a                       )
25 4,4  (                              i d g     e h     c a * * b * * f                     )
26 4,3  (                                i d g           e c a * h * * b f                   )
27 4,2  (                                  i d g           e c a * * * h b f                 )
28 4,1  (                                    i d             e c * * * * h b                 )
29 3,4  (                                        i e h   d g     c b * a * * f               )
30 2,4  (                                          i e h         d c b g * * a f             )
31 1,4  (                                            i e           d c * * * g a             )
32 3,3  (                                                i d g   e h   c a * b * f           )
33 3,2  (                                                  i d g       e c a h * b f         )
34 3,1  (                                                    i d         e c * * h b         )
35 2,3  (                                                        i e h d g   c b a * f       )
36 1,3  (                                                          i e       d c g * a       )
37 2,2  (                                                              i d g e h c a b f )
38 2,1  (                                                                i d     e c h b )
39 1,2  (                                                                        i e d g c a )
40 1,1  (                                                                            i d e c )
```

Fig. 8.2-14. Permuted matrix for reverse Cuthill-McKee ordering (nine-point example)

Another class of ordering algorithms are the so called dissection algorithms. These algorithms do not minimize the bandwidth or profile of a matrix but seek directly to make the fill small.

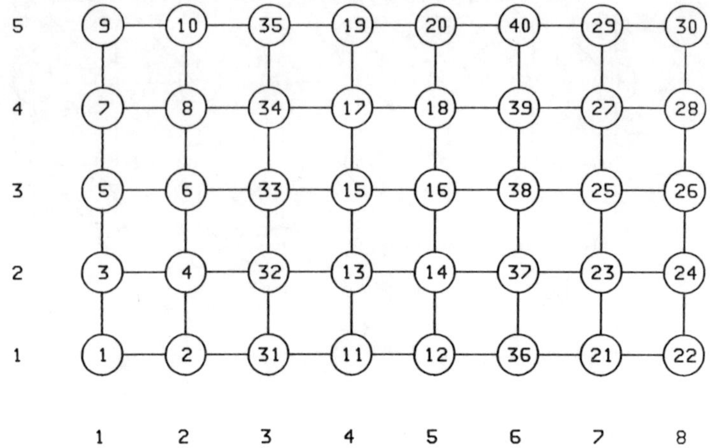

Fig. 8.2-15. Numbered graph for one-way dissection ordering (five-point example)

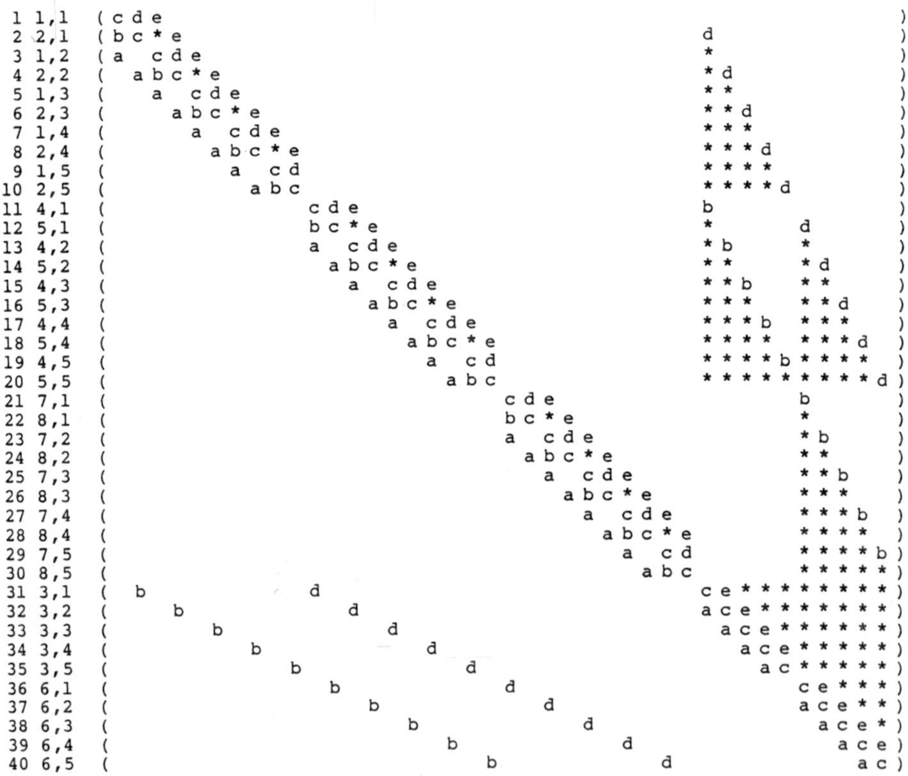

Fig. 8.2-16. Permuted matrix for one-way dissection ordering (five-point example)

The first algorithm we shall briefly discuss here is the one-way dissection algorithm [8.40]. First, one chooses m so-called separators with as few nodes as possible which partition the graph into $m+1$ independent parts which all have about the same size. The nodes in the $m+1$ independent parts are numbered first with, for instance, the reverse Cuthill-McKee algorithm. Then the nodes on the m separators are numbered sequentially. Fig. 8.2-15 shows the graph for the five-point example numbered after the one-way dissection algorithm. As separators the third and sixth column have been taken. The permuted matrix is shown in Fig. 8.2-16. The total number of non-zero elements after factorization is 452 which is indeed not small. The permuted matrix for the nine-point example is shown in Fig. 8.2-17. The numbering of the nodes is identical to the one shown in Fig. 8.2-15 for five-point example. The total number of non-zero elements after factorization is 540. However, our examples are not very well suited for the one-way dissection algorithm. They should only be considered here to help understand the underlying ideas of the algorithm. An automatic algorithm to find the one-way dissection ordering for an irregular graph is described in [8.40]. Very efficient schemes on how to store the matrix to be factored in block form and its factors are discussed in [8.37]. For very large problems the one-way dissection ordering is asymptotically inferior to the nested dissection ordering which will be described next.

```
 1 1,1  (c d e i                                                                         )
 2 2,1  (b c h e                                                         d i             )
 3 1,2  (a g c d e i                                                     * *             )
 4 2,2  (f a b c h e                                                     g d i           )
 5 1,3  (    a g c d e i                                                 * * *           )
 6 2,3  (    f a b c h e                                                 * g d i         )
 7 1,4  (        a g c d e i                                             * * * *         )
 8 2,4  (        f a b c h e                                             * * g d i       )
 9 1,5  (            a g c d                                             * * * * *       )
10 2,5  (            f a b c                                             * * * g d       )
11 4,1  (                    c d e i                                     b h             )
12 5,1  (                    b c h e                                     * *       d i   )
13 4,2  (                    a g c d e i                                 f b h     * *   )
14 5,2  (                    f a b c h e                                 * * *     g d i )
15 4,3  (                        a g c d e i                             * f b h   * * * )
16 5,3  (                        f a b c h e                             * * * *   * g d i )
17 4,4  (                            a g c d e i                         * * f b h * * * * )
18 5,4  (                            f a b c h e                         * * * * * * * g d i )
19 4,5  (                                a g c d                         * * * f b * * * * * )
20 5,5  (                                f a b c                         * * * * * * * * g d )
21 7,1  (                                        c d e i                           b h       )
22 8,1  (                                        b c h e                           * *       )
23 7,2  (                                        a g c d e i                       f b h     )
24 8,2  (                                        f a b c h e                       * * *     )
25 7,3  (                                            a g c d e i                   * f b h   )
26 8,3  (                                            f a b c h e                   * * * *   )
27 7,4  (                                                a g c d e i               * * f b h )
28 8,4  (                                                f a b c h e               * * * * * )
29 7,5  (                                                    a g c d               * * * f b )
30 8,5  (                                                    f a b c               * * * * * )
31 3,1  (  b   h             d   i                                   c e * * * * * * * * )
32 3,2  (  f   b   h         g   d   i                               a c e * * * * * * * )
33 3,3  (      f   b   h         g   d   i                             a c e * * * * * * )
34 3,4  (          f   b   h         g   d   i                             a c e * * * * * )
35 3,5  (              f   b             g   d                               a c * * * * * )
36 6,1  (                      b   h             d   i                         c e * * * )
37 6,2  (                      f   b   h         g   d   i                     a c e * * )
38 6,3  (                          f   b   h         g   d   i                   a c e * )
39 6,4  (                              f   b   h         g   d   i                 a c e )
40 6,5  (                                  f   b             g   d                   a c )
```

Fig. 8.2-17. Permuted matrix for one-way dissection ordering (nine-point example)

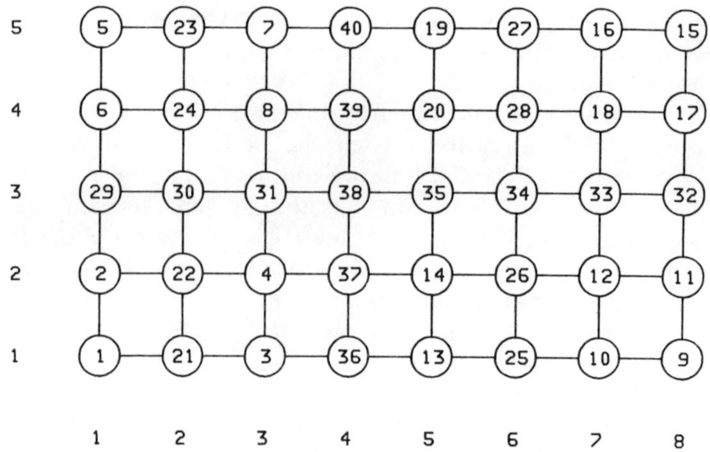

Fig. 8.2-18. Numbered graph for nested dissection ordering (five-point example)

```
 1 1,1  ( c e                                     d                                       )
 2 1,2  ( a c                                     * d             e                       )
 3 3,1  (     c e                                 b                             d         )
 4 3,2  (     a c                                 * b                 e         * d       )
 5 1,5  (         c a                                 d                                   )
 6 1,4  (         e c                                 * d         a                       )
 7 3,5  (             c a                             b                                 d )
 8 3,4  (             e c                             * b             a               d * )
 9 8,1  (                 c b e                                                           )
10 7,1  (                 d c * e                         b                               )
11 8,2  (                 a   c b                         *             e                 )
12 7,2  (                   a d c                         * b           * e               )
13 5,1  (                         c e                     d                     b         )
14 5,2  (                         a c                     * d                 e * b       )
15 8,5  (                             c b a                                               )
16 7,5  (                             d c * a                 b                           )
17 8,4  (                             e   c b                 *         a                 )
18 7,4  (                               e d c                 * b       * a               )
19 5,5  (                                     c a             d                         b )
20 5,4  (                                     e c             * d             a       b * )
21 2,1  ( b   d                                   c e             *   *         * *       )
22 2,2  (   b   d                                 a c             * e *         * *       )
23 2,5  (         b   d                               c a         *   *               * * )
24 2,4  (           b   d                             e c         * a *               * * )
25 6,1  (                   d     b                       c e           * *   * * *       )
26 6,2  (                       d   b                     a c           * * e * * *       )
27 6,5  (                               d     b                 c a         * *   *       * * )
28 6,4  (                                   d   b               e c         * * a *       * * )
29 1,3  (   a       e                                             c d *         * *   * * )
30 2,3  (                                           a   e         b c d         * *   * * )
31 3,3  (       a       e                                           b c         * * d * * )
32 8,3  (                     a           e                             c b * * * *   * * )
33 7,3  (                       a           e                           d c b * * *   * * )
34 6,3  (                                                   a   e         d c b * *   * * )
35 5,3  (                           a           e                           d c * * b * * )
36 4,1  (     b                   d                                             c e * * * )
37 4,2  (       b                   d                                           a c e * * )
38 4,3  (                                                             b       d   a c e * )
39 4,4  (               b                       d                                   a c e )
40 4,5  (             b                       d                                       a c )
```

Fig. 8.2-19. Permuted matrix for nested dissection ordering (five-point example)

```
 1 1,1  ( c e                                     d i                                     )
 2 1,2  ( a c                                     g d             e i                     )
 3 3,1  (     c e                                 b h                           d i       )
 4 3,2  (     a c                                 f b               h e         g d i     )
 5 1,5  (         c a                                 d g                                 )
 6 1,4  (         e c                                 i d         a g                     )
 7 3,5  (             c a                             b f                             g d )
 8 3,4  (             e c                             h b           f a           g d i )
 9 8,1  (                 c b e h                                                         )
10 7,1  (                 d c i e                         b h                             )
11 8,2  (                 a f c b                         * *           e h               )
12 7,2  (                 g a d c                         f b           i e h             )
13 5,1  (                         c e                     d i                   b h       )
14 5,2  (                         a c                     g d               i e f b h     )
15 8,5  (                             c b a f                                             )
16 7,5  (                             d c g a                   b f                       )
17 8,4  (                             e h c b                   * *         a f           )
18 7,4  (                             i e d c                   h b         g a f         )
19 5,5  (                                     c a               d g                   f b )
20 5,4  (                                     e c               i d           g a     f b h )
21 2,1  ( b h d i                                 c e             * * *           * * *     )
22 2,2  ( f b g d                                 a c             h e i           * * *     )
23 2,5  (         b f d g                             c a         * * *               * * * )
24 2,4  (         h b i d                             e c         f a g               * * * )
25 6,1  (                   d   i b h                       c e           * * * * * * *     )
26 6,2  (                   g   d f b                       a c           * i e h * * *     )
27 6,5  (                               d   g b f               c a       * * * *     * * * )
28 6,4  (                               i   d h b               e c       * g a f     * * * )
29 1,3  (   a       e                               g   i         c d *           * * * * * )
30 2,3  (   f   g   h   i                           a   e         b c d           * * * * * )
31 3,3  (       a       e                           f   h           b c           * g d i * )
32 8,3  (                     a f         e h                           c b * * * * * * * )
33 7,3  (                     g a         i e               f   h       d c b * * * * * * )
34 6,3  (                       g   f       i   h           a   e         d c b * * * * * )
35 5,3  (                           a           e           g   i           d c * f b h * )
36 4,1  (     b h                 d i                                           c e * * * )
37 4,2  (     f b                 g d                                 h       i a c e * * )
38 4,3  (       f       h           g           i                     b       d   a c e * )
39 4,4  (             h b                       i d                   f       g     a c e )
40 4,5  (             b f                       d g                                     a c )
```

Fig. 8.2-20. Permuted matrix for nested dissection ordering (nine-point example)

In essence, in the nested dissection algorithm, one basic step is applied repeatedly. This step consists of choosing a separator which, as nearly as possible, equally partitions a graph into two parts. The nodes in the two parts are numbered before those on the separator. This basic step is applied successively for the subgraphs until no further partitioning is possible. Fig. 8.2-18 shows the numbering of the nodes obtained by nested dissection for the five-point example. The numbering for the nine-point example is identical. The first separator is the vertical line at location four. The nodes on this separator are numbered last. The separators for the two subgraphs lie on the horizontal line at location three; the next, final for this example, separators lie on the vertical lines at location two and six. The total graph has thus been split into eight parts which cannot be further dissected. The nodes in these parts are numbered first followed successively by the separators. Fig. 8.2-19 and Fig. 8.2-20 show the permuted matrix for the five-point and nine-point example, respectively. The total number of non-zero elements after factorization is 358 and 452, respectively. These are the lowest values we have obtained so far. An algorithm for automatic nested dissection of irregular graphs has been given in [8.36] together with an efficient storage scheme for the factorization. The impact of the nested dissection ordering on Gaussian elimination has been studied in [8.33], [8.62], [8.79]. Remarks on incomplete nested dissection, i.e., stopping partitioning of the subgraphs prior to natural termination, have been given in [8.39].

Another ordering algorithm which has proven to be very valuable is called the minimum degree algorithm. For this algorithm the factorization of the matrix is first carried out symbolically. At each elimination step the part of the matrix remaining to be factored is permuted such that a column with the fewest non-zero elements is eliminated next. More in detail, one starts with the graph of the matrix, and one picks a node with minimum degree. Then one eliminates this node and properly updates the graph of the matrix by introducing new edges which represent the created fill. The rank of the matrix to be factored, which is the number of nodes, has obviously been reduced by one by this step. Then one picks again a node with minimum degree from the new (updated) graph and proceeds with the symbolic elimination as just described until all nodes have been removed, i. e., the matrix has been completely factored in symbolic form. The order of the nodes used for this symbolic factorization is then the ordering to be taken for the actual factorization. One problem occuring during symbolic elimination is that frequently several nodes with minimum degree exist as candidates for the next elimination step. These ties are usually broken arbitrarily [8.38]. For some applications, however, my own experience indicates that breaking ties by looking ahead one elimination step for each of the nodes with minimum degree can significantly decrease the fill. The computational effort to be spent, unfortunately, increases non-negligibly. The implementation of the minimum degree algorithm is indeed a non-trivial task. Some constructive investigations can be found in, e. g., [8.38], [8.42], [8.43]. Storage schemes for the factorization of a matrix obeying the minimum degree ordering can be found in [8.38]. The numbered graph and the permuted matrix for the five-point example are given in Fig. 8.2-21 and Fig. 8.2-22, respectively. The total number of non-zero elements after factorization is 326 which is by far the smallest value of all ordering methods we have dealt with here. The situation is analogous for the nine-point example. The ordered graph and the permuted matrix are shown in Fig. 8.2-23 and Fig. 8.2-24, respectively. The total number of non-zero elements after factorization is 428 which, again, is the best value we have obtained.

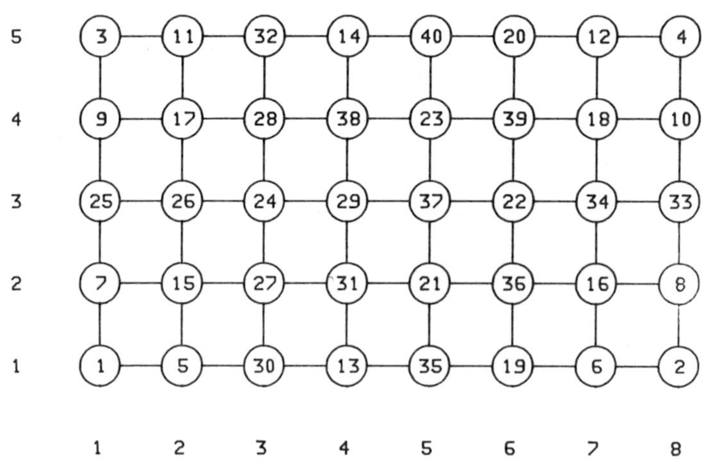

Fig. 8.2-21. Numbered graph for minimum degree ordering (five-point example)

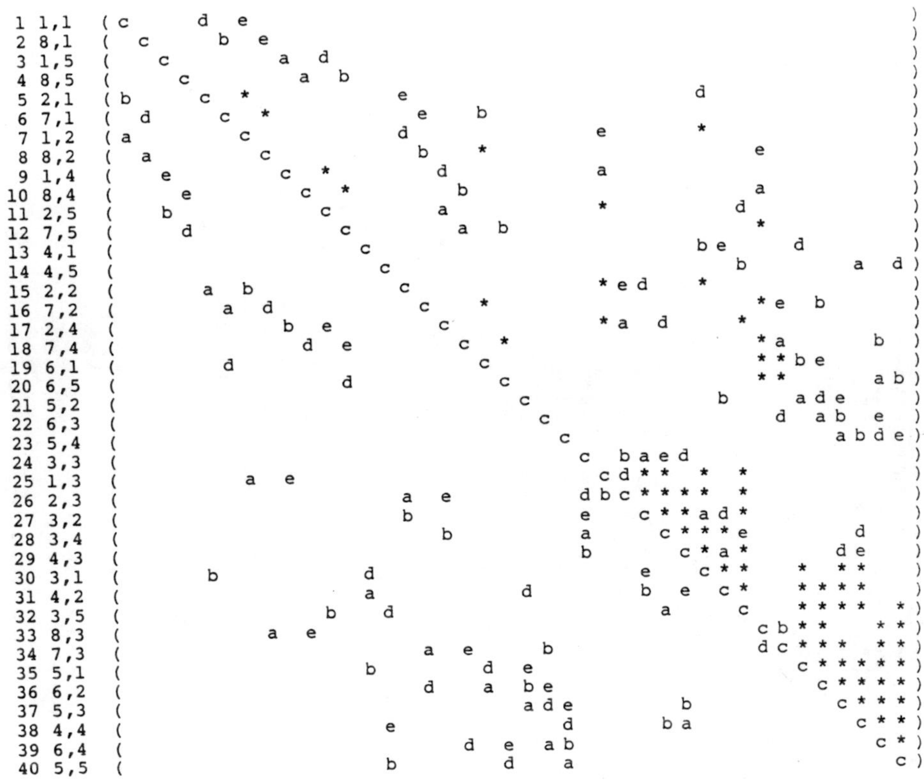

Fig. 8.2-22. Permuted matrix for minimum degree ordering (five-point example)

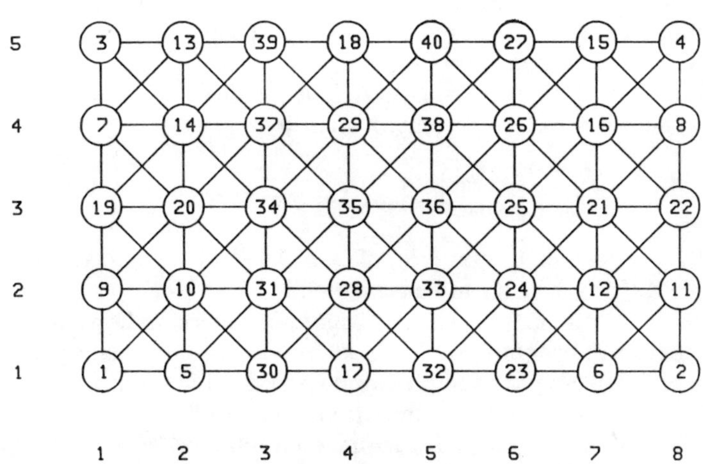

Fig. 8.2-23. Numbered graph for minimum degree ordering (nine-point example)

```
 1 1,1  ( c       d       e i                                                             )
 2 8,1  (   c       b         e h                                                         )
 3 1,5  (     c       a           d g                                                     )
 4 8,5  (       c       a           b f                                                   )
 5 2,1  ( b       c       h e                                     d i                     )
 6 7,1  (   d       c         i e         b h                                             )
 7 1,4  (     e       c           i d         a g                                         )
 8 8,4  (       e       c           h b         f a                                       )
 9 1,2  ( a       g       c d                 e i                 * *                     )
10 2,2  ( f       a       b c                 h e                 g d     i               )
11 8,2  (   a       f         c b                 h e * *                                 )
12 7,2  (   g       a         d c                 e i f b h                               )
13 2,5  (     b       f           c a         * *                               g   d     )
14 2,4  (     h       b           e c         f a                         g     d   i     )
15 7,5  (       d       g           c a         * *       f b                              )
16 7,4  (       i       d           e c         a g     f b h                              )
17 4,1  (                                 c                   e   b h d i                 )
18 4,5  (                                   c                   a               f g b d   )
19 1,3  (             e   a g       i         c d                 * *     *     *   *     )
20 2,3  (             h   f a       e         b c                 * g     d     i   *     )
21 7,3  (               i     g a       e         c d * f b h *                           )
22 8,3  (               e     a f       h         b c * * * * *                           )
23 6,1  (           d           i                   c e * * *         b h                 )
24 6,2  (           g           d               i   a c e * *         f b     h           )
25 6,3  (                       g       i       d     a c e *         * f     b   h       )
26 6,4  (                             i d       g       a c e         * *     f   b   h   )
27 6,5  (                             d g                 a c         * *     *   f   b   )
28 4,2  (                                 a                     c   f b g d h e i         )
29 4,4  (                                   e                     c         f a g b d h i )
30 3,1  (         b         h             d                     i   c e * * * * * *   *   )
31 3,2  (         f         b             g     h               d   a c * * e i * *   *   )
32 5,1  (                                 b           d i       h       c e * * * * * * * )
33 5,2  (                                 f           g d i     b       a c * h e * * * * )
34 3,3  (                   f       h           b               g i   a     c d * e * * * )
35 4,3  (                                                       a e   f   g b c d h i * * )
36 5,3  (                                             g d i   f h         a   b c * e * * )
37 3,4  (                         h b       i   f                 d           a g   c * e * )
38 5,4  (                                   h                 g d i   b           f a   c * e )
39 3,5  (                         b f       d                     g               a   c * )
40 5,5  (                                   b               g d   f                 a   c )
```

Fig. 8.2-24. Permuted matrix for minimum degree ordering (nine-point example)

One drawback of the minimum degree algorithm obtained by minimizing only the fill is that the permuted matrix appears with its non-zero elements widely spread. This can cause many "paging operations" during the actual elimination on a computer with a virtual memory operating system. However, with special programming techniques it should be possible to solve this problem.

Many linear systems arising from discretization of partial differential equations can be identified as "red/black" systems. A linear system $A \cdot X = B$ is said to be a red/black system if it can be permuted such that:

$$P \cdot A \cdot P^T = \begin{pmatrix} D_R & C_R \\ C_B & D_B \end{pmatrix} \tag{8.2-10}$$

D_R and D_B are diagonal matrices of rank n_R and n_B, $\operatorname{rank}(A) = n_R + n_B = n$, respectively. Matrices which can be permuted to a red/black system are frequently termed to possess "property A" [8.92]. Theoretical investigations of the properties of such matrices can be found in the books by Varga [8.88] and by Young [8.92]. The matrix of the five-point example can be permuted into red/black form. Such a permutation is very desirable for elimination since no fill will occur in the upper red part of the matrix as the block D_R is diagonal. One ordering which produces a red/black structure in case of the five-point example is the so-called "checkerboard" ordering. The numbering of the graph is shown in Fig. 8.2-25. The nodes have been

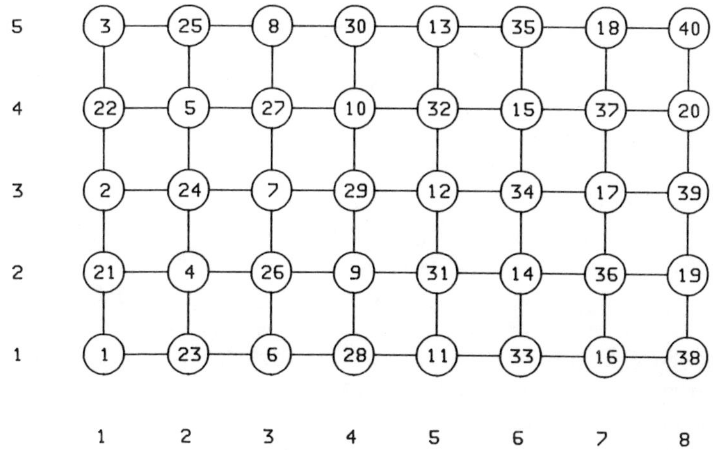

Fig. 8.2-25. Numbered graph for checkerboard ordering (five-point example)

Fig. 8.2-26. Permuted matrix for checkerboard ordering (five-point example)

marked alternatingly red and black (like a checkerboard) such that the edges starting at "red" nodes terminate only at "black" nodes. Then the red nodes are numbered before the black nodes by rows or columns, whatever consists of less nodes. The permuted matrix is shown in Fig. 8.2-26. As already mentioned, fill occurs only in the black diagonal part of the system. Furthermore the fill exhibits a banded structure which allows the application of a band elimination technique. The bandwidth of the submatrix generated by the fill is identical to the bandwidth obtained by the best natural ordering (by columns in our case), however, the rank of the submatrix is only half as large as the rank of the original matrix. The number of algebraic operations required for factorizing the banded submatrix is, therefore, also just half. The total number of non-zero elements after factorization is 342 which is larger than the value obtained with the minimum degree ordering.
Another red/black ordering applicable for the five-point example is the "alternating diagonal" ordering. Fig. 8.2-27 and Fig. 8.2-28 show the numbered graph and the permuted matrix. The total number of non-zero elements after factorization is 336 which does not seem to be a very significant improvement compared to the checkerboard ordering. However, for larger and more realistic systems the improvement in absolute numbers can be quite convincing. It should be noted that the alternating diagonal ordering can be obtained from the checkerboard ordering without moving nodes from the red to the black subsystem and vice versa. The red/black partitioning is unique considering the nodes in the red and black subsystem for a given graph if it exists. The numbering of the nodes, however, within the subsystems can be changed. Of course, one can also interchange the red and black blocks entirely; one should always use the larger block for the red subsystem. In our example both blocks have the same size, however.
For the nine-point example no red/black permutation exists. An automatic algorithm for finding the possible existance of a red/black permutation of a general sparse matrix has been given in the ITPACK subroutine package [8.47], [8.48], [8.57].

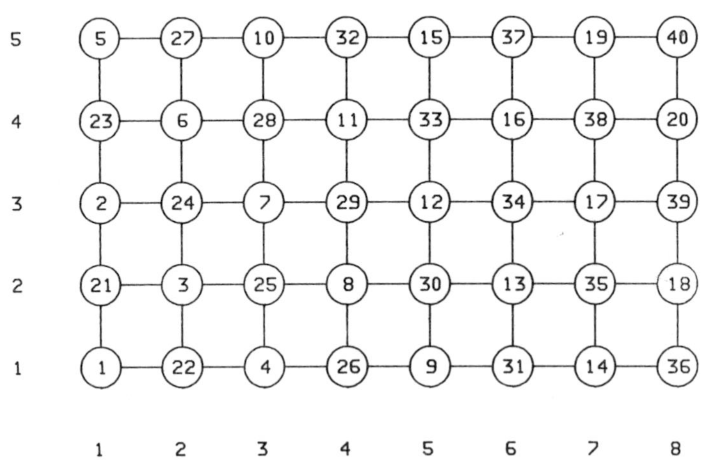

Fig. 8.2-27. Numbered graph for alternating diagonal ordering (five-point example)

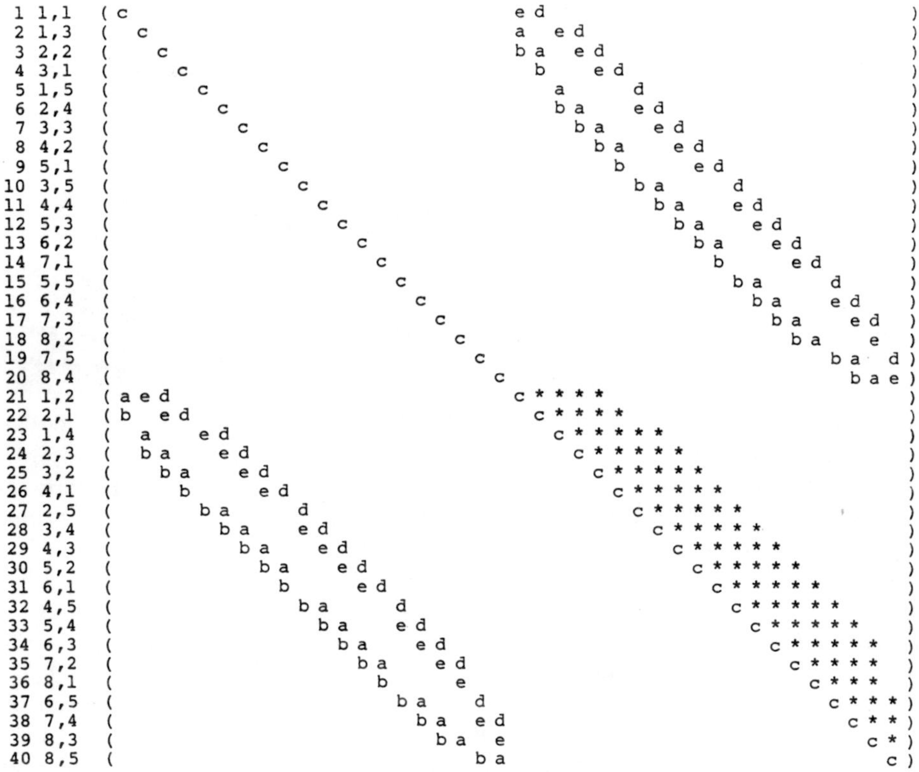

Fig. 8.2-28. Permuted matrix for alternating diagonal ordering (five-point example)

Table 8.2-1 and Table 8.2-2 summarize the results we have obtained with the ordering algorithms described for the five-point and the nine-point example, respectively.

Table 8.2-1. *Results of ordering algorithms for the five-point example*

method	non-zeros	fill [%]	bandwidth	profile
natural by rows	566	225	8	263
natural by columns	398	129	5	179
Cuthill-McKee	370	113	5	165
reverse Cuthill-McKee	same as Cuthill-McKee			
Gibbs-Poole-Stockmeyer	same as Cuthill-McKee			
one-way dissection	452	160		
nested dissection	358	106		
minimum degree	326	87		
checkerboard	342	97		
alternating diagonal	336	93		

Table 8.2-2. *Results of ordering algorithms for the nine-point example*

method	non-zeros	fill [%]	bandwidth	profile
natural by rows	622	117	9	291
natural by columns	454	59	6	207
Cuthill-McKee	506	77	9	233
reverse Cuthill-McKee	458	60	9	209
Gibbs-Poole-Stockmeyer	same as natural by columns			
one-way dissection	540	89		
nested dissection	452	58		
minimum degree	428	50		

It should be noted explicitly again that all ordering procedures presented here are designed only to make the fill which occurs during factorization small. It is assumed that column and/or row interchanges to maintain numerical stability are not necessary for the factorization. This is true only for special classes of linear systems, e. g., positive definite matrices. However, the linear systems arising from the discretization of partial differential equations by finite differences, finite boxes or finite elements usually exhibit this property. Some comments on combining the, in general contrary, constraints of minimizing the fill and maintaining numerical stability by a proper permutation matrix can be found in, e.g., [8.16], [8.93]. Algorithms for sparse Gaussian elimination with column interchanges during elimination to support numerical stability have been presented in, e.g., [8.81].

A thorough comparison of the performance of different ordering algorithms is an absolutely non-trivial task. Some comparisons using the algorithms available in the SPARSPAK subroutine package [8.34], [8.41] have been given in [8.61]. It has to be noted again that the small examples given here for didactical purpose should not mislead the reader as the absolute differences among the various methods are rather small. This issue is delightfully addressed in [8.61] and the results in this paper should contribute to raise the level of understanding concerning the complexity which real life problems can exhibit.

A final remark should be given on systems arising from the discretization by finite boxes. These systems are structurally non-symmetric. Therefore the above outlined algorithms are not directly applicable. This problem can be circumvented by adding elements which indeed are zero to the non-zero pattern of the coefficient matrix such that the resulting matrix is structurally symmetric. In other words that means to use the graph of $A+A^T$ for the ordering which is certainly symmetric instead of using just the graph of A. The number of non-zero elements artificially introduced thereby is usually small; it is two times the number of termination points. For the work presented in [8.31] the minimum degree algorithm has been generalized to be applicable for non-symmetric systems. This is indeed straightforward, theoretically, as only the elimination graphs have to be considered for this ordering. The complexity in efficiently coding the algorithm, however, rises significantly as one has to deal with two directed graphs instead of one undirected graph in each factorization step. Nevertheless, this effort seems to be rewarded as the obtained fill has been observed to remain surprisingly small.

8.3 Relaxation Methods

Before we can discuss iterative methods in detail it is necessary to review some of the fundamental properties of one-step stationary iterative schemes. The linear system

$$A \cdot x = b \tag{8.3-1}$$

may be rewritten with an arbitrary non-singular matrix B with $\operatorname{rank}(B) = \operatorname{rank}(A)$ as:

$$B \cdot x + (A - B) \cdot x = b \tag{8.3-2}$$

With (8.3-2) it is straightforward to obtain an iterative scheme by setting:

$$B \cdot x^{k+1} = (B - A) \cdot x^k + b \tag{8.3-3}$$

(8.3-3) can be solved, trivially, for x^{k+1}:

$$x^{k+1} = (I - B^{-1} \cdot A) \cdot x^k + B^{-1} \cdot b \tag{8.3-4}$$

For actual computations one always uses (8.3-3) instead of (8.3-4) to avoid the costly inversion of B. However, (8.3-4) is more convenient for the characterization of the properties of an iterative scheme. Fixpoint schemes like (8.3-4), for the nonlinear case though, have already been dealt with in Chapter 7, and we may carry over the results about the convergence properties. Thus, a necessary condition for the convergence of (8.3-4) is that the Frechet derivative, i.e. the Jacobian matrix, of the right hand side has a spectral radius smaller than unity.

$$\rho(I - B^{-1} \cdot A) < 1 \tag{8.3-5}$$

(8.3-5) is indeed a necessary and sufficient condition for the convergence of (8.3-3) or (8.3-4) with arbitrary initial guess x^0 since the Jacobian does not depend on the intermediate solutions x^k. Therefore, the assumption that a solution exists, which is necessary to take for the nonlinear case, can be dropped. If (8.3-5) is fulfilled, the iteration (8.3-3) will converge to the unique solution of $A \cdot x = b$.
To simplify the notation we define the matrix:

$$M = I - B^{-1} \cdot A \tag{8.3-6}$$

e^k shall denote the error vector of the k-th iterate; x^* is the exact solution.

$$e^k = x^k - x^* \tag{8.3-7}$$

By simple calculations we obtain:

$$e^k = M \cdot e^{k-1} = \ldots = M^k \cdot e^0 \tag{8.3-8}$$

Using norms on (8.3-8) we obtain:

$$\| e^k \|_2 \leq \| M^k \|_2 \cdot \| e^0 \|_2 \tag{8.3-9}$$

with the vector norm:

$$\| x \|_2 = \sqrt{x^T \cdot x} = \sqrt{\sum_{i=1}^{n} x_i^2} \tag{8.3-10}$$

and the matrix norm:

$$\| A \|_2 = \sqrt{\rho(A^T \cdot A)} \tag{8.3-11}$$

Note that if A is symmetric $A = A^T$ (8.3-11) simplifies to:

$$\| A \|_2 = \rho(A) \tag{8.3-12}$$

(8.3-9) suggests to us $\| M^k \|$ as a measure for the rate of convergence. Following the treatment of Varga [8.88] we define:

$$R(M,k) = \frac{-\ln \| M^k \|_2}{k} \tag{8.3-13}$$

$R(M,k)$ is the average rate of convergence for k iterations of the matrix M. A prerequisite for the applicability of (8.3-13) is that k is sufficiently large such that $\| M^k \|_2 < 1$. If $R(M_1,k) < R(M_2,k)$, then M_2 is iteratively faster for k iterations than M_1. The average rate of convergence gives a direct measure for the number of iterations k required to reduce the initial error vector by a factor of α.

$$\| e^k \|_2 = \frac{\| e^0 \|_2}{\alpha} \rightarrow k \geq \frac{-\ln \alpha}{R(M,k)} \tag{8.3-14}$$

(8.3-13) simplifies significantly if M is symmetric:

$$R(M,k) = -\ln \rho(M) \tag{8.3-15}$$

For non-symmetric matrices we have the following theorems available for the characterization of (8.3-13):

$$\lim_{k \to \infty} R(M,k) = -\ln \rho(M) \tag{8.3-16}$$

$$R(M,k) \leq -\ln \rho(M) \tag{8.3-17}$$

The spectral radius $\rho(M)$ is the asymptotic rate of convergence. Considering practical aspects it is the simplest measure for the convergence of a given scheme; however, it can give quite misleading information in some cases [8.76].

It is worth noting explicitly that two iterative schemes which exhibit the same asymptotic rate of convergence can behave quite differently considering the average rate of convergence. The asymptotic rate of convergence gives, obviously, no information about the path on which it is attained. In general the sequence $\| M^k \|_2$ is not monotonous. For many iterations it may even look as if a specific, indeed convergent, iterative scheme is not convergent. For a symmetric iterative scheme we do not have to face this problem as the average convergence rate is equal to the asymptotic convergence rate (8.3-15). However, most of the symmetric iterative schemes for non-symmetric linear systems are so slowly converging $\rho(M) = 1 - \varepsilon \cong 1$ that monotonous convergence can be indeed frustrating in actual applications.

Considering the great variety of available iterative methods for the solution of systems of linear equations, the relaxation methods are probably most popular. The basis for relaxation methods applied to a system $A \cdot x = b$ is a splitting of the coefficient matrix A:

$$A = D - L - U \tag{8.3-18}$$

with:

$$\begin{aligned} &D: \text{ a non-singular matrix} \\ &L: \text{ a strict lower triangular matrix} \\ &U: \text{ a strict upper triangular matrix} \end{aligned} \tag{8.3-19}$$

In many textbooks it can be found that D is required to be a diagonal matrix. This prerequisite is in fact too strong. It suffices that:

$$\begin{aligned} &D^{-1} \cdot L: \text{ a strict lower triangular matrix} \\ &D^{-1} \cdot U: \text{ a strict upper triangular matrix} \end{aligned} \tag{8.3-20}$$

However, the choice of D should be such that linear systems with D as the coefficient matrix can easily be solved. We also allow that the elements $A_{i,j}$ of A are interpreted as matrices instead of simple scalars. In that case it is only required that the main diagonal elements $A_{i,i}$ are non-singular square matrices. By allowing this interpretation we can refrain from dealing explicitly with block iterative methods [8.2], [8.10], [8.71], [8.72], [8.89]. All results to be given here hold for the case when the elements $A_{i,j}$ are matrices [8.73], [8.88].
The various relaxation methods are obtained by choosing the matrix B in (8.3-3) with the matrices D, L and U obtained by the splitting (8.3-18) of A. Basically four different methods of setting B are established. The simplest scheme is the Jacobi method [8.54]. Frequently this method is also termed the method of simultaneous displacements after [8.32]. The matrix D is used for B and the iteration matrix M can be straightforwardly evaluated:

$$B_J = D \tag{8.3-21}$$

$$M_J = D^{-1} \cdot (L + U) \tag{8.3-22}$$

The convergence properties of this method are rather poor as a matter of fact. However, if D is a (scalar) diagonal matrix, the solution of (8.3-3) with (8.3-21) can be vectorized easily which brings about an enormous gain in execution speed on vector computers in such a way that this method can be superior to more sophisticated and complex ones. Furthermore, if M_J is symmetric – which can easily be achieved when A is symmetric – M_J has only real eigenvalues, which allows the application of many convergence acceleration methods (cf. Section 8.6).
For the Gauß-Seidel method [8.69], which is also called the method of successive displacements [8.32], B and M are defined as:

$$B_{GS} = D - L \tag{8.3-23}$$

$$M_{GS} = (D - L)^{-1} \cdot U \tag{8.3-24}$$

As L is a strict lower triangular matrix it is still simple to solve (8.3-3) very efficiently. However, in order to vectorize the iteration much more effort has to be spent than for the Jacobi method.
The successive overrelaxation method (SOR) has been derived and characterized simultaneously by Frankel [8.30] and Young [8.91]. This method makes use of an iteration parameter ω, the overrelaxation parameter, which will be characterized later.

$$B_{SOR} = \frac{1}{\omega} \cdot D - L \tag{8.3-25}$$

$$M_{SOR} = (D - \omega \cdot L)^{-1} \cdot (\omega \cdot U + (1 - \omega) \cdot D) \tag{8.3-26}$$

For $\omega = 1$ the SOR method degenerates to the Gauß-Seidel method which indeed was the basis for the development of the SOR method. The matrix B_{SOR} is structurally identical with the matrix B_{GS}.
The symmetric successive overrelaxation method ($SSOR$), which was first considered by Aitken [8.1] and Sheldon [8.80], consists of one forward SOR sweep and one backward SOR sweep for each of its iterations. This may be written as:

$$\begin{aligned} B_{SSOR.1} \cdot x^{k+1/2} &= (B_{SSOR.1} - A) \cdot x^k + b \\ B_{SSOR.2} \cdot x^{k+1} &= (B_{SSOR.2} - A) \cdot x^{k+1/2} + b \end{aligned} \tag{8.3-27}$$

with:

$$\begin{aligned} B_{SSOR.1} &= \frac{1}{\omega} \cdot D - L \\ B_{SSOR.2} &= \frac{1}{\omega} \cdot D - U \end{aligned} \tag{8.3-28}$$

The iteration matrix evaluates to:

$$\begin{aligned} M_{SSOR} = (D - \omega \cdot U)^{-1} &\cdot (\omega \cdot L + (1 - \omega) \cdot D) \cdot \\ &\cdot (D - \omega \cdot L)^{-1} \cdot (\omega \cdot U + (1 - \omega) \cdot D) \end{aligned} \tag{8.3-29}$$

The convergence properties of $SSOR$ are very similar to those of SOR. However, M_{SSOR} is symmetric if A is symmetric which enables the successful application of acceleration methods, whereas M_{SOR} is always non-symmetric. The $SSOR$ method should only be used in conjunction with acceleration methods because the convergence properties differ just marginally, as mentioned, from those of the SOR method.
In the following we shall compare some of the results on convergence and the prerequisites on A for these four methods. However, we first have to recall some notions of basic linear algebra.
A matrix $A = D - L - U$ is termed consistently ordered if all the eigenvalues of

$$F(\alpha) = D^{-1} \cdot \left(\alpha \cdot L + \frac{1}{\alpha} \cdot U \right) \tag{8.3-30}$$

are independent of α for arbitrary complex $\alpha \neq 0$. A matrix with "property A" (cf. Section 8.2) can always be permuted in such a way that it is consistently ordered. However, consistently ordered matrices do not necessarily exhibit "property A".
A matrix is termed reducible if there exists a permutation matrix P such that

$$P \cdot A \cdot P^T = \begin{pmatrix} \alpha_{11} & \alpha_{12} \\ 0 & \alpha_{22} \end{pmatrix} \tag{8.3-31}$$

where α_{11} and α_{22} are square matrices. If a matrix is not reducible it is irreducible.

The first theorem I shall review here gives relations between the spectral radii of M_J and M_{GS} [8.82]. If all elements of M_J (8.3-22) are non-negative, one and only one of the following relations is valid:

$$\begin{aligned} &1.\ \rho(M_J)=\rho(M_{GS})=0 \\ &2.\ 0<\rho(M_{GS})<\rho(M_J)<1 \\ &3.\ \rho(M_J)=\rho(M_{GS})=1 \\ &4.\ 1<\rho(M_J)<\rho(M_{GS}) \end{aligned} \tag{8.3-32}$$

Thus, if the Jacobi method is convergent, then the Gauß-Seidel method is convergent too and its asymptotic convergence rate is superior.
If A is consistently ordered then:

$$\rho(M_{GS})=\rho(M_J)^2 \tag{8.3-33}$$

That means that the asymptotic rate of convergence for the Gauß-Seidel method is twice as large as for the Jacobi method.
If all elements of M_J are non-negative, A is irreducible and $\rho(M_J)<1$, then there exists a constant $\omega_{\text{opt}}\geq 1$ such that $\rho(M_{SOR})$ decreases strictly monotonously for $0<\omega\leq\omega_{\text{opt}}$. If additionally all eigenvalues of M_J are real, ω_{opt} can be computed to:

$$\omega_{\text{opt}}=\frac{2}{1+\sqrt{1-\rho(M_J)^2}} \tag{8.3-34}$$

Furthermore, the spectral radius $\rho(M_{SOR})<1$ for $0<\omega<2$ can be given explicitly:

$$\rho(M_{SOR})=\begin{cases} 0<\omega<\omega_{\text{opt}} & 1-\omega+\dfrac{\omega^2\cdot\rho(M_J)^2}{2}+\omega\cdot\rho(M_J)\cdot\sqrt{1-\omega+\dfrac{\omega^2\cdot\rho(M_J)^2}{4}} \\ \omega_{\text{opt}}\leq\omega<2 & \omega-1 \end{cases} \tag{8.3-35}$$

For $\omega=\omega_{\text{opt}}$ we obtain for the spectral radius:

$$\rho(M_{\text{sor}})=\frac{1-\sqrt{1-\rho(M_J)^2}}{1+\sqrt{1-\rho(M_J)^2}} \tag{8.3-36}$$

If A is symmetric, $\omega\in]0.2[$ and D positive definite and symmetric, then $\rho(M_{SOR})<1$ if and only if A is positive definite [8.70]. This theorem is probably the best established estimate as to whether the *SOR* method should converge or not. However, the prerequisite that A must be symmetric is indeed too strong as can be shown with a simple example. Assume that we have a linear system with A, D symmetric, positive definite such that the above given theorem is fulfilled. Then we pick an arbitrary non-singular diagonal matrix H and we rewrite the linear system:

$$A\cdot x=b\ =\ A\cdot H^{-1}\cdot H\cdot x=b\ =\ \tilde{A}\cdot\tilde{x}=b \tag{8.3-37}$$

$\tilde{A}$ is obviously not symmetric if A is symmetric. The iteration matrix $\tilde{M}_{SOR}$ is given by:

$$\tilde{M}_{SOR}=(\tilde{D}-\omega\cdot\tilde{L})^{-1}\cdot(\omega\cdot\tilde{U}+(1-\omega)\cdot\tilde{D}) \tag{8.3-38}$$

Remembering that $\tilde{A}$ is $A \cdot H^{-1}$ (8.3-38) can be rewritten since H is diagonal:

$$\tilde{M}_{SOR} = H \cdot (D - \omega \cdot L)^{-1} \cdot (\omega \cdot U + (1-\omega) \cdot D) \cdot H^{-1} \tag{8.3-39}$$

We can deduce directly that $\tilde{M}_{SOR}$ has the same eigenvalues as M_{SOR} because H transforms M_{SOR} "similar" into $\tilde{M}_{SOR}$. This result is extraordinarily important because the often found prerequisite on A to be positive definite and symmetric is not fulfilled for the matrices arising from the discrete approximations to the continuity equations (cf. Section 6.1). However, it should be a simple exercise to show that these linear systems can be transformed into symmetric, positive definite form with a diagonal matrix. That means that the frequently found statement that the discretized continuity equations cannot be solved by relaxation methods is simply wrong.

In general it is difficult to determine precisely the spectral radius of a given matrix. Therefore, some constructive remarks will be given in the following to estimate the spectral radius.

An upper bound for the spectral radius is obtained with the theorem by Gerschgorin [8.44]. For an arbitrary matrix A all its eigenvalues lie in the union of the disks:

$$|x - A_{i,i}| \leq \sum_{\substack{j=1 \\ j \neq i}}^{n} |A_{i,j}|, \quad 1 \leq i \leq n \tag{8.3-40}$$

(8.3-40) can be calculated easily; however, it can be made more explicit for the iteration matrices of the relaxation methods. These are convergent if (8.3-41) holds for A.

$$\max\left(\min_{i=1}^{n}\left(|A_{i,i}| - \sum_{\substack{j=1 \\ j \neq i}}^{n} |A_{i,j}|\right), \min_{i=1}^{n}\left(|A_{i,i}| - \sum_{\substack{j=1 \\ j \neq i}}^{n} |A_{j,i}|\right)\right) \geq 0 \tag{8.3-41}$$

The equal sign in (8.3-41) is allowed only if A is consistently ordered and if not all minima are exactly zero [8.88].

A final remark should be given on termination criteria. For practical purposes one can frequently find (8.3-42) with a properly chosen relative accuracy ε.

$$\| x^{k+1} - x^{k} \|_2 < \varepsilon \cdot \| x^{k+1} \|_2 \tag{8.3-42}$$

However, really intended is the convergence criterion:

$$\| x^{k+1} - x^{*} \|_2 < \varepsilon \cdot \| x^{k+1} \|_2 \tag{8.3-43}$$

The exact solution x^* is obviously not known during the iteration such that (8.3-43) can not be evaluated. However, (8.3-42) is indeed extremely misleading for many iterative schemes. It can be recommended to use instead of (8.3-42) the following slightly more sophisticated criterion:

$$\| x^{k+1} - x^{k} \|_2 < \varepsilon \cdot \| x^{k+1} \|_2 \cdot (1 - \rho(M)) \tag{8.3-44}$$

$\rho(M)$ can be roughly estimated with:

$$\rho(M) \cong \frac{\| x^{k+1} - x^{k} \|}{\| x^{k} - x^{k-1} \|} \quad \text{for } k \text{ sufficiently large} \tag{8.3-45}$$

For the vector norm in (8.3-45) any norm is suitable. It can be quite informative to evaluate this approximation with different norms to have some measure for the uncertainty. The $\|\cdot\|_2$ can be used as defined in (8.3-10), or the $\|\cdot\|_1$ and the $\|\cdot\|_\infty$ norm.

$$\| x \|_1 = \sum_{i=1}^{n} |x_i| \tag{8.3-46}$$

$$\| x \|_\infty = \max_{i=1}^{n} |x_i| \tag{8.3-47}$$

The approximation (8.3-45) of the spectral radius of the iteration matrix can also be used to estimate adaptively the optimum *SOR* parameter ω with (8.3-35) and (8.3-34). However, the approximation (8.3-45) should be evaluated very accurately before a parameter adaption is performed. A thorough discussion of this subject can be found in, e.g., [8.47], [8.67] and error bounds are elaborately discussed in [8.50].

8.4 Alternating Direction Methods

A widely used category of iterative methods which is particularly well established for the solution of systems arising from the discretization of partial differential equations are the so-called "alternating direction implicit" (*ADI*) iterative methods. These methods have been considered first in [8.15], [8.74]. Since then several variants have been developed. The basic idea is again a splitting of the coefficient matrix of the linear system $A \cdot x = b$:

$$A = H + V \tag{8.4-1}$$

This splitting suggests the following iterative scheme:

$$\begin{aligned} (H + \alpha_{k+1} \cdot I) \cdot x^{k+1/2} &= (\alpha_{k+1} \cdot I - V) \cdot x^k + b \\ (V + \alpha_{k+1} \cdot I) \cdot x^{k+1} &= (\alpha_{k+1} \cdot I - H) \cdot x^{k+1/2} + b \end{aligned} \tag{8.4-2}$$

α_{k+1} is a positive constant acting as an acceleration parameter. The iteration matrix M_{ADI} can be readily computed to:

$$M_{ADI} = (V + \alpha_{k+1} \cdot I)^{-1} \cdot (\alpha_{k+1} \cdot I - H) \cdot (H + \alpha_{k+1} \cdot I)^{-1} \cdot (\alpha_{k+1} \cdot I - V) \tag{8.4-3}$$

The *ADI* method is convergent for arbitrary positive α_{k+1} when the matrices H and V are positive definite or at least one of the matrices is positive definite and the other one is not negative definite.

The α_{k+1} could be chosen in such a way that the spectral radius of (8.4-3) is minimal which would minimize the asymptotic rate of convergence. However, an "optimum" sequence of the α_{k+1} may greatly improve the average rate of convergence [8.88], [8.89]. An algorithm for calculating an (in some sense) optimal sequence α_{k+1} for a given problem has been presented in, e.g., [8.14], [8.78].

The name "alternating direction implicit" is based on the way the splitting (8.4-1) has been defined in [8.15], [8.74]. The method has been developed for the solution of elliptic differential equations in two dimensions discretized with five-point

differences on a rectangular mesh. The coefficients of A which describe the influence of the horizontal neighbors in the mesh have been put into H whereas the coefficients describing the influence of the vertical neighbors have been put into V. The main diagonal of A, i.e., the coefficients of the center points, has been equally divided and put into H and V. The coefficient matrices $(H+\alpha_{k+1}\cdot I)$ and $(V+\alpha_{k+1}\cdot I)$ are, after suitable permutations, tridiagonal matrices such that the solution of the systems (8.4-2) can be carried out with direct (implicit) methods.
Sharp bounds on the convergence properties of the *ADI* method are only available if the matrices V and H commute [8.88].

$$V\cdot H=H\cdot V \tag{8.4-4}$$

One variant of the *ADI* method is to use a positive definite tridiagonal matrix [8.89] instead of the identity matrix in (8.4-2). In some cases this may greatly improve the convergence properties; however, a rigorous analysis mandating general application of this technique does not exist at present.

8.5 Strongly Implicit Methods

The basic idea of strongly implicit iterative methods is fully analogous to the concept used for any iterative method, however, with probably a slightly different interpretation. By choosing a matrix B which is a "good" approximation for A we can expect the following iterative scheme to be rapidly convergent:

$$B\cdot x^{k+1}=(B-A)\cdot x^k+b \tag{8.5-1}$$

It is obvious that by choosing $B=A$ this iterative scheme degenerates to the original system and will "converge" in one step (ignoring round-off errors). However, the prerequisite on B is that it can be factored significantly more easily than A because otherwise there would not be any advantage of the iterative scheme over the direct solution. The convergence properties and prerequisites for (8.5-1) have already been discussed in Section 8.3; any further discussion will be omitted here.
The strongly implicit iterative method by Stone [8.83] has proven to be extraordinarily useful for the semiconductor equations. Stone's algorithm is designed for linear systems arising from a five-point finite difference discretization of an elliptic equation in two space dimensions on a rectangular mesh $(NX\times NY)$. These problems lead to a linear system $A\cdot x=b$ where the individual equation in the case of natural ordering is given by:

$$\begin{aligned}A_{k,k-NX}\cdot x_{k-NX}+A_{k,k-1}\cdot x_{k-1}+A_{k,k}\cdot x_k+\\+A_{k,k+1}\cdot x_{k+1}+A_{k,k+NX}\cdot x_{k+NX}=b_k\end{aligned} \tag{8.5-2}$$

A discussion of this type of equation has already been given in Section 8.2. Stone's idea was to modify the matrix A in such a manner that the modified matrix $\hat{A}$ can be factored easily.

$$\hat{A}=A+N=L\cdot U \tag{8.5-3}$$

If $\| N \| \ll \| A \|$ then the choice $B = L \cdot U$ in (8.5-1) can be intuitively expected to give a rapidly converging iterative scheme.

$$L \cdot U \cdot x^{k+1} = N \cdot x^k + b \tag{8.5-4}$$

Stone decided that L and U should each have only three diagonals with non-zero elements.

$$L = (L_{k,k-NX}, L_{k,k-1}, L_{k,k}) \tag{8.5-5}$$

$$U = (U_{k,k} = 1, U_{k,k+1}, U_{k,k+NX}) \tag{8.5-6}$$

The product $L \cdot U = \hat{A}$ has seven non-zero diagonals.

$$\hat{A} = (\hat{A}_{k,k-NX}, \hat{A}_{k,k-NX+1}, \hat{A}_{k,k-1}, \hat{A}_{k,k}, \hat{A}_{k,k+1}, \hat{A}_{k,k+NX-1}, \hat{A}_{k,k+NX}) \tag{8.5-7}$$

The coefficients of $\hat{A}$ are fully determined by the following relations.

$$\hat{A}_{k,k-NX} = L_{k,k-NX} \tag{8.5-8}$$

$$\hat{A}_{k,k-NX+1} = L_{k,k-NX} \cdot U_{k-NX,k-NX+1} \tag{8.5-9}$$

$$\hat{A}_{k,k-1} = L_{k,k-1} \tag{8.5-10}$$

$$\hat{A}_{k,k} = L_{k,k} + L_{k,k-1} \cdot U_{k-1,k} + L_{k,k-NX} \cdot U_{k-NX,k} \tag{8.5-11}$$

$$\hat{A}_{k,k+1} = L_{k,k} \cdot U_{k,k+1} \tag{8.5-12}$$

$$\hat{A}_{k,k+NX-1} = L_{k,k-1} \cdot U_{k-1,k+NX-1} \tag{8.5-13}$$

$$\hat{A}_{k,k+NX} = L_{k,k} \cdot U_{k,k+NX} \tag{8.5-14}$$

Five diagonals of $\hat{A}$ coincide in position with diagonals of A. One would expect to have a good approximation to the matrix A if the diagonals of $\hat{A}$ and A, which coincide in position, have also the same values. In this case the matrix N would consist only of two diagonals.

$$N = (N_{k,k-NX+1}, N_{k,k+NX-1}) = (\hat{A}_{k,k-NX+1}, \hat{A}_{k,k+NX-1}) \tag{8.5-15}$$

The coefficients of L, U and N are then easily evaluated from the relations (8.5-8) to (8.5-14).

However, Stone has found that the iteration obtained by this choice does not converge as rapidly as one would expect. Therefore, he decided to decrease the magnitude of N by subtracting an approximation to N. By series expansion one can obtain (8.5-16) and (8.5-17) since the components of the solution vector are nodal values on a rectangular mesh.

$$x_{k-NX+1} = -x_k + x_{k+1} + x_{k-NX} + O(h \cdot k) \tag{8.5-16}$$

$$x_{k+NX-1} = -x_k + x_{k-1} + x_{k+NX} + O(h \cdot k) \tag{8.5-17}$$

Then a matrix H is introduced which exhibits the following structure of non-zero elements:

$$H = (H_{k,k-NX}, H_{k,k-1}, H_{k,k}, H_{k,k+1}, H_{k,k+NX}) \tag{8.5-18}$$

This is identical to the structure of A. The coefficients of H are determined in view of (8.5-16) and (8.5-17) in such a manner that $(N-H)\cdot x$ can be expected to be small. We obtain the relations:

$$H_{k,k-NX}=\alpha\cdot N_{k,k-NX+1} \tag{8.5-19}$$

$$H_{k,k-1}=\alpha\cdot N_{k,k+NX-1} \tag{8.5-20}$$

$$H_{k,k}=-\alpha\cdot(N_{k,k-NX+1}+N_{k,k+NX-1}) \tag{8.5-21}$$

$$H_{k,k+1}=\alpha\cdot N_{k,k-NX+1} \tag{8.5-22}$$

$$H_{k,k+NX}=\alpha\cdot N_{k,k+NX-1} \tag{8.5-23}$$

$\alpha\in[0.1[$ is a suitably chosen iteration parameter (cf. [8.83]). Then one uses $B=A+N+H$ in (8.5-1) which gives the following scheme after some algebraic manipulation to enable an efficient implementation:

$$\tilde{L}\cdot\tilde{U}\cdot\delta^k=b-A\cdot x^k,\ x^{k+1}=x^k+\delta^k \tag{8.5-24}$$

Since $\| N-H \|$ is expected to be smaller than $\| N \|$, $A+N-H$ can be considered to represent an improved approximation to A compared to $A+N$. The coefficients of $\tilde{L}$ and $\tilde{U}$ which have the same structure as L (8.5-5) and U (8.5-6), respectively, relate to the coefficients of A by:

$$\tilde{L}_{k,k-NX}=\frac{A_{k,k-NX}}{1+\alpha\cdot U_{k-NX,k-NX+1}} \tag{8.5-25}$$

$$\tilde{L}_{k,k-1}=\frac{A_{k,k-1}}{1+\alpha\cdot U_{k-1,k+NX-1}} \tag{8.5-26}$$

$$\begin{aligned}\tilde{L}_{k,k}&=A_{k,k}+\tilde{L}_{k,k-NX}\cdot(\alpha\cdot\tilde{U}_{k-NX,k-NX+1}-\tilde{U}_{k-NX,k})+\\&\quad+\tilde{L}_{k,k-1}\cdot(\alpha\cdot\tilde{U}_{k-1,k+NX-1}-\tilde{U}_{k-1,k})\end{aligned} \tag{8.5-27}$$

$$\tilde{U}_{k,k+1}=\frac{A_{k,k+1}-\alpha\cdot\tilde{L}_{k,k-NX}\cdot\tilde{U}_{k-NX,k-NX+1}}{\tilde{L}_{k,k}} \tag{8.5-28}$$

$$\tilde{U}_{k,k+NX}=\frac{A_{k,k+NX}-\alpha\cdot\tilde{L}_{k,k-1}\cdot\tilde{U}_{k-1,k+NX-1}}{\tilde{L}_{k,k}} \tag{8.5-29}$$

Theoretical investigations about the optimal value of the iteration parameter α are, most unfortunately, not known at present. Stone has given some suggestions in [8.83]. However, as far as I know, most authors using Stone's method have their own recipe for choosing α. Jesshope [8.55] has recommended by heuristically reasoning to use different values of α for every row of $\tilde{L}$ and $\tilde{U}$ in (8.5-24) to (8.5-27), which, as stated, increases the average rate of convergence significantly. However, the success of Stone's method has not been fully understood yet.

There exist numerous strongly implicit iterative methods which are similar in concept to Stone's method but better founded mathematically, e.g., [8.11], [8.12], [8.24], [8.25], [8.28], [8.94]. However, none of these algorithms is competitive in practice.

8.6 Convergence Acceleration of Iterative Methods

The acceleration of iterative methods has always been a desire of engineers when a specific program making use of an iterative method becomes exceedingly time consuming and therefore expensive. Several methods have been developed for the purpose of accelerating iterative methods, all of which require that the eigenvalue with maximum modulus of the iteration matrix be purely real. This prerequisite reduces the number of iterative methods which can be accelerated quite significantly. From the methods we have considered so far the Jacobi method, the *SSOR* method and the *ADI* method can possibly be accelerated since only these methods submit to the given prerequisite in theory.
The most simple acceleration method has been suggested by Lyusternik [8.64]. With the notions given in Section 8.3 we may write for the exact solution:

$$x^* = x^k - e^k \tag{8.6-1}$$

or:

$$x^* = x^{k+1} - e^{k+1} \tag{8.6-2}$$

The error vector e^{k+1} fulfills:

$$e^{k+1} = M \cdot e^k \tag{8.6-3}$$

(8.6-3) may be written as:

$$e^{k+1} = \lambda \cdot e^k + \delta^k \tag{8.6-4}$$

λ is the maximum eigenvalue of M. By eliminating the error vector in (8.6-2) with (8.6-1) and (8.6-4) we obtain:

$$x^* = x^k + \frac{x^{k+1} - x^k}{1-\lambda} + \frac{\delta^k}{1-\lambda} \tag{8.6-5}$$

If $\| \delta^k \|$ is small compared to $1-\lambda$ (that means k is sufficiently large), (8.6-5) suggests the following extrapolation:

$$x^* \cong x^k + \frac{x^{k+1} - x^k}{1-\lambda} \tag{8.6-6}$$

As all eigenvalues of the iteration matrix are required to be real and positive for this acceleration method, λ is the spectral radius of the iteration matrix, which can be estimated as sketched in Section 8.3.
A very similar approach has been suggested by Aitken [8.1]. By writing the error vector of two successive iterations we obtain:

$$e^k = x^* - x^k \cong \lambda \cdot (x^* - x^{k-1}) \tag{8.6-7}$$

$$e^{k+1} = x^* - x^{k+1} \cong \lambda \cdot (x^* - x^k) \tag{8.6-8}$$

Dividing componentwise (8.6-7) by (8.6-8) yields the following extrapolated results for the i-th component.

$$x_i^* = x_i^k + \frac{x_i^{k+1} - x_i^k}{1 - \dfrac{x_i^{k+1} - x_i^k}{x_i^k - x_i^{k-1}}} \tag{8.6-9}$$

If we compare (8.6-9) to (8.6-6), we observe that Aitken's method makes use of a componentwise estimation of the maximum eigenvalue. Practical comparisons of these two methods do not yield any results such that preference can be given to one or the other method. Aitken's method is more likely to produce oscillatory results if repeatedly applied.

All iterative methods considered so far are one-step stationary methods which consist essentially in iterating the mapping:

$$x^{k+1} = M \cdot x^k + b \tag{8.6-10}$$

One may speculate that optimal methods cannot have the structure (8.6-10) because information obtained from previous iterations is not used as feedback to improve the iteration. This idea leads to the study of more general schemes which are frequently termed semi-iterative methods [8.87].

$$y^k = N_k(x^k, x^{k-1}, \ldots, x^0) \tag{8.6-11}$$

with:

$$\lim_{k \to \infty} (y^k - x^*) = 0 \tag{8.6-12}$$

Specifically, methods which use a properly weighted sum (with real coefficients) of all preceding iterates of (8.6-10) have been established.

$$y^k = \sum_{i=0}^{k} c_i^k \cdot x^i \tag{8.6-13}$$

An almost obvious restriction on the coefficients c_i^k arises because (8.6-13) must be, in general, solution preserving. If the iteration is started with the exact solution, the iterates must not deviate from the exact solution. We obtain therefore:

$$\sum_{i=0}^{k} c_i^k = 1 \tag{8.6-14}$$

The error vector of (8.6-13) is given by:

$$y^k - x^* = \sum_{i=0}^{k} c_i^k \cdot x^i - x^* \tag{8.6-15}$$

With (8.6-14) we may rewrite (8.6-15) to:

$$y^k - x^* = \sum_{i=0}^{k} c_i^k \cdot (x^i - x^*) \tag{8.6-16}$$

By introducing the polynomial notation

$$p^k(z) = \sum_{i=0}^{k} c_i^k (z)^i \tag{8.6-17}$$

(8.6-16) can be expressed with a matrix polynomial.

$$y^k - x^* = P^k(M) \cdot (x^0 - x^*) \tag{8.6-18}$$

We now have to determine polynomials in such a manner that the spectral radius of $P^k(M)$ is minimal and $P^k(1) = 1$ (cf. (8.6-14). This choice optimizes the average rate of convergence. It has been shown (probably first) in [8.58] that the polynomials (8.6-19) are optimal for (8.6-13) if all eigenvalues of M are real.

$$P^k(z) = \frac{C^k\left(\dfrac{2 \cdot z - \lambda^{\max} - \lambda^{\min}}{\lambda^{\max} - \lambda^{\min}}\right)}{C^k\left(\dfrac{2 - \lambda^{\max} - \lambda^{\min}}{\lambda^{\max} - \lambda^{\min}}\right)} \tag{8.6-19}$$

$\lambda^{\min}$ and $\lambda^{\max}$ are the minimal and maximal eigenvalues of M; $C^k(z)$ are the Chebyshev polynomials which are defined by:

$$C^k(z) = \cos(k \cdot \operatorname{acos}(z)),\ z \in [-1, 1],\ k > 0 \tag{8.6-20}$$

Assuming a symmetric spectrum of M

$$\lambda^{\max} = -\lambda^{\min} = \rho(M) = \rho \tag{8.6-21}$$

we obtain for (8.6-19):

$$P^k(z) = \frac{C^k\left(\dfrac{z}{\rho}\right)}{C^k\left(\dfrac{1}{\rho}\right)} \tag{8.6-22}$$

With the trigonometric identity

$$\cos((k-1) \cdot \alpha) + \cos((k+1) \cdot \alpha) = 2 \cdot \cos(\alpha) \cdot \cos(k \cdot \alpha) \tag{8.6-23}$$

the well known three term recurrence relation (8.6-24) is immediately obtained for the Chebyshev polynomials (8.6-20).

$$C^{k+1}(z) = 2 \cdot z \cdot C^k(z) - C^{k-1}(z) \tag{8.6-24}$$

where:

$$C^0(z) = 1,\ C^1(z) = z \tag{8.6-25}$$

Substituting (8.6-22) into (8.6-18), remembering that $x^* = M \cdot x^* + b$ and using (8.6-24) we obtain after laborious calculations the following iterative scheme:

$$y^{k+1} = \omega_{k+1} \cdot (M \cdot y^k + b) + (1 - \omega_{k+1}) \cdot y^{k-1} \tag{8.6-26}$$

with:

$$\omega_{k+1} = 1 + \frac{C^{k-1}\left(\dfrac{1}{\rho}\right)}{C^{k+1}\left(\dfrac{1}{\rho}\right)},\quad \omega_1 = 1 \tag{8.6-27}$$

(8.6-26) is a non-stationary two-step iterative method because the iteration parameter ω_{k+1} changes from iteration to iteration and the last two iterates determine the next iterate. The recursive scheme (8.6-26) shows that it is not necessary to form the iterates x^k explicitly to determine the accelerated solutions y^k.

Compared to the *SOR* method, for instance, the Chebyshev semi-iterative method applied to the Jacobi method requires one additional vector of storage for y^{k-1}. However, the average rate of convergence is significantly improved in many applications. The asymptotic rate of convergence, in contrast, is not better than for the *SOR* method because (cf. [8.46], [8.88]):

$$\lim_{k\to\infty} \omega_k = \frac{2}{1+\sqrt{1-\rho^2}} \tag{8.6-28}$$

The adaptive calculation of ρ or $\lambda^{\max}$, $\lambda^{\min}$ for the optimal sequence of iteration parameters ω_k is dealt with in some detail in [8.47], [8.65]. We have required for the applicability of the Chebyshev semi-iterative method that all eigenvalues of M are real. Actually this is too severe a restriction. With minor modifications to the iteration parameters the Chebyshev method may also be applied to many matrices with complex eigenvalues [8.3].

The last acceleration method we are going to consider here is the conjugate gradient method. This method is frequently understood as a stand alone iterative method after its inventors Hestenes and Stiefel [8.51]. The basic idea of the conjugate gradient method is the minimization of the following functional:

$$F(z) = \frac{1}{2} \cdot (A \cdot z - b)^T \cdot A^{-1} \cdot (A \cdot z - b) \tag{8.6-29}$$

If A is positive definite, (8.6-29) is zero (and minimal) only for $z = x^*$, the solution of $A \cdot x = b$. We define the residual vector:

$$r^k = A \cdot x^k - b \tag{8.6-30}$$

(8.6-29) can now be minimized with a sequence (8.6-31) where d^k is a chosen search direction and λ_k a parameter describing the optimal length for the given search direction.

$$x^{k+1} = x^k + \lambda_k \cdot d^k \tag{8.6-31}$$

We immediately obtain with (8.6-30):

$$r^{k+1} = r^k + \lambda_k \cdot A \cdot d^k \tag{8.6-32}$$

Since the optimal value of λ_k makes $(r^{k+1})^T$ orthogonal to d^k we have:

$$0 = (r^{k+1})^T \cdot d^k \tag{8.6-33}$$

and hence we can calculate λ_k from (8.6-32), (8.6-33).

$$\lambda_k = -\frac{(d^k)^T \cdot r^k}{(d^k)^T \cdot A \cdot d^k} \tag{8.6-34}$$

The search direction is determined by the residual of the last iterate and the previously used search direction.

$$d^{k+1} = -r^{k+1} + \beta_k \cdot d^k, \quad d^0 = -r^0 \tag{8.6-35}$$

Furthermore, all search directions are required to be conjugately orthogonal with respect to the matrix A.

$$(d^{k+1})^T \cdot A \cdot d^j = 0, \; j = 0, k \tag{8.6-36}$$

From (8.6-36) we obtain in particular the sequence β_k for (8.6-35) since

$$(d^{k+1})^T \cdot A \cdot d^k = 0 \tag{8.6-37}$$

Substituting (8.6-35) into (8.6-37) gives:

$$\beta_k = \frac{(r^{k+1})^T \cdot A \cdot d^k}{(d^k)^T \cdot A \cdot d^k} \tag{8.6-38}$$

The basic conjugate gradient algorithm is thus completely defined by equations (8.6-30), (8.6-31), (8.6-34), (8.6-35) and (8.6-38). It is to note that the conjugate gradient algorithm would terminate theoretically after rank (A) iterations since not more orthogonal search directions exist (cf. (8.6-36)). However, in practice round-off errors due to the finite computer arithmetic may necessitate further iterations until the residual (8.6-30) is sufficiently small.
In the following we consider the conjugate gradient method for the acceleration of iterative methods:

$$B \cdot x^{k+1} = (B - A) \cdot x^k + b \tag{8.6-39}$$

with:

$$\rho(I - B^{-1} \cdot A) < 1 \tag{8.6-40}$$

B is furthermore assumed to be symmetric and positive definite. Under these assumptions B can be understood also as an approximation for A. Since B is symmetric and positive definite there exists a symmetric factorization of B.

$$B = C \cdot C^T \tag{8.6-41}$$

C is a lower triangular matrix. With the factors of B we may write the equivalent linear system:

$$C^{-1} \cdot A \cdot C^{-T} \cdot C^T \cdot x = C^{-1} \cdot b \tag{8.6-42}$$

We speak now of an iterative method accelerated by the conjugate gradient method if we minimize the functional of the equivalent (frequently termed preconditioned) linear system (8.6-42). After some calculations we obtain the following equations for the acceleration algorithm.

$$B \cdot s^k = r^k \tag{8.6-43}$$

$$x^{k+1} = x^k + \lambda_k \cdot e^k \tag{8.6-44}$$

$$e^{k+1} = -s^{k+1} + \beta_k \cdot e^k, \; e^0 = -s^0 \tag{8.6-45}$$

$$r^{k+1} = r^k + \lambda_k \cdot A \cdot e^k, \; r^0 = A \cdot x^0 - b \tag{8.6-46}$$

with the sequence of parameters:

$$\lambda_k = \frac{(e^k)^T \cdot r^k}{(e^k)^T \cdot A \cdot e^k} \tag{8.6-47}$$

$$\beta_k = \frac{(s^{k+1})^T \cdot A \cdot e^k}{(e^k)^T \cdot A \cdot e^k} \tag{8.6-48}$$

Note that the factors of B are not explicitly required in this algorithm. One can show that (8.6-43) to (8.6-46) simplify to the unaccelerated iterative scheme (8.6-39) for $\lambda_k = 1$ and $\beta_k = 0$. The effort per iteration for this algorithm lies in the solution of one linear system with B as coefficient matrix, a multiplication of a vector with the original coefficient matrix A and a few operations on vectors. This acceleration algorithm degenerates into the basic conjugate gradient algorithm if one chooses the identity matrix for B. Thus, the basic conjugate gradient algorithm can be interpreted as the accelerated Jacobi method for the preconditioned system:

$$D^{-1/2} \cdot A \cdot D^{-1/2} \cdot D^{1/2} \cdot x = D^{-1/2} \cdot b \tag{8.6-49}$$

D is the main diagonal of A.
I personally am absolutely convinced of the success of the conjugate gradient method for accelerating iterative methods. It is a very attractive feature that no parameters have to be estimated a priori for this method. This method is probably most efficient together with an incomplete Cholesky factorization of A used for B (cf. [8.56], [8.68], [8.86]). This algorithm has been used for the linear systems arising from the semiconductor equations in [8.90] with, as stated, extraordinary success. A generalization of the conjugate gradient method for non-symmetric systems is possible; some discussion on that subject can be found in, e.g, [8.84]. Details about efficient implementation have been presented in [8.26]. A comparison between acceleration by Chebyshev semi-iteration and the conjugate gradient method has been carried out in [8.49] with the conclusion that no general theoretical preference can be given. Moreover, conditions are presented in [8.49] for which these acceleration methods are equivalent considering convergence properties.

8.7 References

[8.1] Aitken, A. C.: Studies in Practical Mathematics V. On the Iterative Solution of a System of Linear Equations. Proc. Roy. Soc. Edinburgh Sec. *A63*, 52–60 (1950).

[8.2] Arms, R. J., Gates, L. D., Zondek, B.: A Method of Block Iteration. SIAM *4*, No. 4, 220–229 (1956).

[8.3] Axellsson, O.: Lecture Notes on Iterative Methods. Report 72.04, Chalmers University, Göteborg, 1972.

[8.4] Bank, R. E.: Marching Algorithms for Elliptic Boundary Value Problems. II: The Variable Coefficient Case. SIAM J. Numer. Anal. *14*, No. 5, 950–970 (1977).

[8.5] Bank, R. E., Rose, D. J.: Marching Algorithms for Elliptic Boundary Value Problems. I: The Constant Coefficient Case. SIAM J. Numer. Anal. *14*, No. 5, 792–829 (1977).

[8.6] Buzbee, B. L., Golub, G. H., Nielson, C. W.: On Direct Methods for Solving Poisson's Equations. SIAM J. Numer. Anal. *7*, No. 4, 627–656 (1970).

[8.7] Buzbee, B. L., Dorr, F. W., George, J. A., Golub, G. H.: The Direct Solution of the Discrete Poisson Equation on Irregular Regions. SIAM J. Numer. Anal. *8*, No. 4, 722–736 (1971).

[8.8] Curtis, A. R., Reid, J. K.: On the Automatic Scaling of Matrices for Gaussian Elimination. J. Inst. Math. Appl. *10*, 118–124 (1972).

[8.9] Cuthill, E., McKee, J.: Reducing the Bandwidth of Sparse Symmetric Matrices. Proc. ACM Conf., pp. 157–172 (1969).

[8.10] Cuthill, E. H.: A Method of Normalized Block Iteration. J. ACM *6*, 236–244 (1959).

[8.11] DeLaVallee-Poussin, F.: An Accelerated Relaxation Algorithm for Iterative Solution of Elliptic Equations. SIAM J. Numer. Anal. *5*, No. 2, 340–351 (1968).

[8.12] Diamond, M.: An Economical Algorithm for the Solution of Finite Difference Equations. Dissertation, University of Illinois, Urbana, 1971.

[8.13] Dorr, F. W.: The Direct Solution of the Discrete Poisson Equation on a Rectangle. SIAM Review *12*, No. 2, 248 – 263 (1970).

[8.14] Doss, S., Miller, K.: Dynamic ADI methods for Elliptic Equations. SIAM J. Numer. Anal. *16*, No. 5, 837 – 856 (1979).

[8.15] Douglas, J., Rachford, H. H.: On the Numerical Solution of Heat Conduction Problems in Two or Three Space Dimensions. Trans. Amer. Math. Soc. *82*, 421 – 439 (1956).

[8.16] Duff, I. S., Reid, J. K.: A Comparison of Sparsity Orderings for Obtaining a Pivotal Sequence in Gaussian Elimination. J. Inst. Math. Appl. *14*, 281 – 291 (1974).

[8.17] Duff, I. S.: A Survey of Sparse Matrix Research. Proc. IEEE *65*, No. 4, 500 – 535 (1977).

[8.18] Duff, I. S., Stewart, G. W.: Sparse Matrix Proceedings 1978. Philadelphia: SIAM 1979.

[8.19] Duff, I. S.: MA28 – A Set of Fortran Subroutines for Sparse Unsymmetric Linear Equations. Report R-8730, AERE Harwell, Oxfordshire, 1980.

[8.20] Duff, I. S., Reid, J. K.: Experience of Sparse Matrix Codes on the CRAY-1. Report CSS-116, AERE Harwell, Oxfordshire, 1981.

[8.21] Duff, I. S.: The Solution of Sparse Linear Equations on the CRAY-1. Report CSS-125, AERE Harwell, Oxfordshire, 1982.

[8.22] Duff, I. S.: Research Directions in Sparse Matrix Computations. Report R-10547, AERE Harwell, Oxfordshire, 1982.

[8.23] Duff, I. S.: A Survey of Sparse Matrix Software. Report R-10512, AERE Harwell, Oxfordshire, 1982.

[8.24] Dupont, T.: A Factorization Procedure for the Solution of Elliptic Difference Equations. SIAM J. Numer. Anal. *5*, No. 4, 753 – 782 (1968).

[8.25] Dupont, T., Kendall, R. D., Rachford, H. H.: An Approximate Factorization Procedure for Solving Self-Adjoint Elliptic Difference Equations. SIAM J. Numer. Anal. *5*, 559 – 573 (1968).

[8.26] Eisenstat, S. C.: Efficient Implementation of a Class of Preconditioned Conjugate Gradient Methods. SIAM J. Sci. Stat. Comput. *2*, No. 1, 1 – 4 (1981).

[8.27] Eisenstat, S. C., Schultz, M. H., Sherman, A. H.: Algorithms and Data Structures for Sparse Symmetric Gaussian Elimination. SIAM J. Sci. Stat. Comput. *2*, No. 2, 225 – 237 (1981).

[8.28] Evans, D. J.: Iterative Sparse Matrix Algorithms. In: Software for Numerical Mathematics, pp. 49 – 83. London: Academic Press 1974.

[8.29] Even, R. K., Wallach, Y.: On the Direct Solution of Dirichlet's Problem in Two Dimensions. Comput. *5*, 45 – 56 (1970).

[8.30] Frankel, S. P.: Convergence Rates of Iterative Treatments of Partial Differential Equations. Mat. Tables Aids Comput. *4*, 65 – 75 (1950).

[8.31] Franz, A. F., Franz, G. A., Selberherr, S., Ringhofer, C., Markowich, P.: Finite Boxes – A Generalization of the Finite Difference Method Suitable for Semiconductor Device Simulation. IEEE Trans. Electron Devices *ED-30*, No. 9, 1070 – 1082 (1983).

[8.32] Geiringer, H.: On the Solution of Systems of Linear Equations by Certain Iterative Methods. Reissner Anniversary Volume, pp. 365 – 393. Ann Arbor: J. W. Edwards 1949.

[8.33] Gentleman, W. M.: Implementing Nested Dissection. Report CS-82-03, University of Waterloo, 1982.

[8.34] George, A., Liu, J., Ng, E.: User Guide for SPARSPAK: Waterloo Sparse Linear Equations Package. Report CS-78-30, University of Waterloo, 1980.

[8.35] George, A.: Computer Implementation of the Finite Element Method. Report CS-71-208, Stanford University, 1971.

[8.36] George, A., Liu, J. W. H.: An Automatic Nested Dissection Algorithm for Irregular Finite Element Problems. SIAM J. Numer. Anal. *15*, No. 5, 1053 – 1069 (1978).

[8.37] George, A., Liu, J. W. H.: Algorithms for Matrix Partitioning and the Numerical Solution of Finite Element Systems. SIAM J. Numer. Anal. *15*, No. 2, 297 – 327 (1978).

[8.38] George, A., McIntyre, D. R.: On the Application of the Minimum Degree Algorithm to Finite Element Systems. SIAM J. Numer. Anal. *15*, No. 1, 90 – 112 (1978).

[8.39] George, A., Poole, W. G., Voigt, R. G.: Incomplete Nested Dissection for Solving n by n Grid Problems. SIAM J. Numer. Anal. *15*, No. 4, 662 – 673 (1978).

[8.40] George, A.: An Automatic One-Way Dissection Algorithm for Irregular Finite Element Problems. SIAM J. Numer. Anal. *17*, No. 6, 740 – 751 (1980).

[8.41] George, A., Liu, J. W. H.: Computer Solutions of Large Sparse Positive Definite Systems. Englewood Cliffs, N.J.: Prentice-Hall 1981.

[8.42] George, J. A., Liu, J. W. H.: A Minimal Storage Implementation of the Minimum Degree Algorithm. SIAM J. Numer. Anal. *17,* No. 2, 282 – 299 (1980).

[8.43] George, J. A., Liu, J. W. H.: A Fast Implementation of the Minimum Degree Algorithm Using Quotient Graphs. ACM Trans. Mathematical Software *6,* No. 3, 337 – 358 (1980).

[8.44] Gerschgorin, S.: Über die Abgrenzung der Eigenwerte einer Matrix. Izv. Akad. Nauk SSSR Ser. Mat. *7,* 749 – 754 (1931).

[8.45] Gibbs, N. E., Poole, W. G., Stockmeyer, P. K.: An Algorithm for Reducing the Bandwidth and Profile of a Sparse Matrix. SIAM J. Numer. Anal. *13,* No. 2, 236 – 250 (1976).

[8.46] Golub, G. H., Varga, R. S.: Chebyshev Semi-Iterative Methods, Successive Overrelaxation Methods, and Second Order Richardson Iterative Methods. Num. Math. *3,* 147 – 168 (1961).

[8.47] Grimes, R. G., Kincaid, D. R., Young, D. M.: ITPACK 2.0 User's Guide. Report CNA-139, Center for Numerical Analysis, University of Texas at Austin, 1979.

[8.48] Grimes, R. G., Kincaid, D. R., Young, D. M.: ITPACK 2A – A Fortran Implementation of Adaptive Accelerated Iterative Methods for Solving Large Sparse Linear Systems. Report CNA-164, Center for Numerical Analysis, University of Texas at Austin, 1980.

[8.49] Hageman, L. A., Franklin, T. L., Young, D. M.: On the Equivalence of Certain Iterative Acceleration Methods. SIAM J. Numer. Anal. *17,* No. 6, 852 – 873 (1980).

[8.50] Hatcher, T. R.: An Error Bound for Certain Successive Overrelaxation Schemes. SIAM J. Numer. Anal. *19,* No. 5, 930 – 941 (1982).

[8.51] Hestenes, M. R., Stiefel, E.: Method of Conjugate Gradients for Solving Linear Systems. J. Res. Nat. Bur. Stand. *49,* 409 – 436 (1952).

[8.52] Hockney, R. W.: A Fast Direct Solution of Poisson's Equation Using Fourier Analysis. J. ACM *12,* No. 3, 95 – 113 (1965).

[8.53] Huang, J. W., Wing, O.: Optimal Parallel Triangulation of a Sparse Matrix. IEEE Trans. Circuits and Systems *CAS-26,* No. 9, 726 – 732 (1979).

[8.54] Jacobi, C. G. J.: Über eine neue Auflösungsart der bei der Methode der kleinsten Quadrate vorkommenden linearen Gleichungen. Astron. Nachr. *22,* No. 523, 297 – 306 (1845).

[8.55] Jesshope, C. R.: Bipolar Transistor Modelling with Numerical Solutions to the 2-Dimensional DC and Transient Problems. Dissertation, University of Southampton, 1976.

[8.56] Kershav, D. S.: The Incomplete Cholesky Conjugate Gradient Method for the Iterative Solution of Systems of Linear Equations. J. Comp. Phys. *26,* 43 – 65 (1978).

[8.57] Kincaid, D. R., Respess, J. R., Young, D. M., Grimes, R. G.: ITPACK 2C: A FORTRAN Package for Solving Large Sparse Linear Systems by Adaptive Accelerated Iterative Methods. ACM Trans. Mathematical Software *8,* No. 3, 302 – 322 (1982).

[8.58] Lanczos, C.: Solution of Systems of Linear Equations by Minimized Iterations. J. Res. Nat. Bur. Stand. *49,* 33 – 53 (1952).

[8.59] Lewis, J. G.: Implementation of the Gibbs-Poole-Stockmeyer and Gibbs-King Algorithms. ACM Trans. Mathematical Software *8,* No. 2, 180 – 189 (1982).

[8.60] Lewis, J. G.: The Gibbs-Poole-Stockmeyer and Gibbs-King Algorithms for Reordering Sparse Matrices. ACM Trans. Mathematical Software *8,* No. 2, 190 – 194 (1982).

[8.61] Lewis, J. G.: Numerical Experiments with SPARSPAK. ACM Signum Newsletter *18,* No. 3, 12 – 22 (1983).

[8.62] Lipton, R. J., Rose, D. J., Tarjan, R. E.: Generalized Nested Dissection. SIAM J. Numer. Anal. *16,* No. 2, 346 – 358 (1979).

[8.63] Liu, W.-H., Sherman, A. H.: Comparative Analysis of the Cuthill-McKee and the Reverse Cuthill-McKee Ordering Algorithms for Sparse Matrices. SIAM J. Numer. Anal. *13,* No. 2, 198 – 213 (1976).

[8.64] Lyusternik, L. A.: A Note for the Numerical Solution of Boundary Value Problems for the Laplace Equation and for the Calculation of Eigenvalues by the Method of Nets. Trudy Inst. Math. Acad. Sci. SSSR *20,* 49 – 64 (1947).

[8.65] Manteuffel, T. A.: Optimal Parameters for Linear Second Degree Stationary Iterative Methods. SIAM J. Numer. Anal. *19,* No. 4, 833 – 839 (1982).

[8.66] Marsal, D.: Die Numerische Lösung partieller Differentialgleichungen. Mannheim: Bibliographisches Institut 1976.

[8.67] Meis, T., Marcowitz, U.: Numerische Behandlung partieller Differentialgleichungen. Berlin-Heidelberg-New York: Springer 1978.

[8.68] Munksgaard, N.: Solving Sparse Symmetric Sets of Linear Equations by Preconditioned Conjugate Gradients. ACM Trans. Mathematical Software *6*, No. 2, 206 – 219 (1980).

[8.69] Nekrasov, P. A.: Die Bestimmung der Unbekannten nach der Methode der kleinsten Quadrate bei einer sehr großen Anzahl der Unbekannten. Mat. Sob. *12*, 189 – 204 (1885).

[8.70] Ostrowski, A. M.: On the Linear Iteration Procedures for Symmetric Matrices. Rend. Mat. Appl. *14*, 140 – 163 (1954).

[8.71] Parter, S. V.: On "Two-Line" Iterative Methods for Laplace and Biharmonic Equations. Numer. Math. *1*, 240 – 252 (1959).

[8.72] Parter, S. V.: "Multi-Line" Iterative Methods for Elliptic Difference Equations and Fundamental Frequencies. Numer. Math. *3*, 305 – 319 (1961).

[8.73] Parter, S. V., Steuerwalt, M.: Block Iterative Methods for Elliptic and Parabolic Difference Equations. SIAM J. Numer. Anal. *19*, No. 6, 1173 – 1195 (1982).

[8.74] Peaceman, D. W., Rachford, H. H.: The Numerical Solution of Parabolic and Elliptic Differential Equations. J. SIAM *3*, 28 – 41 (1955).

[8.75] Pooch, U. W., Nieder, A.: A Survey of Indexing Techniques for Sparse Matrices. ACM Computing Surveys *5*, No. 2, 109 – 132 (1973).

[8.76] Rheinboldt, W. C.: Methods for Solving Systems of Nonlinear Equations. Philadelphia: SIAM 1974.

[8.77] Rosen, R.: Matrix Bandwidth Minimization. Proc. ACM Conf., pp. 585 – 595 (1968).

[8.78] Saylor, P. E., Sebastian, J. D.: Calculation of Optimum Parameters for Alternating Direction Implicit Procedures. ACM Coll. Alg., No. 460 (1972).

[8.79] Schreiber, R.: A New Implementation of Sparse Gaussian Elimination. ACM Trans. Mathematical Software *8*, No. 3, 256 – 276 (1982).

[8.80] Sheldon, J. W.: On the Numerical Solution of Elliptic Difference Equations. Math. Tables Aids Comput. *9*, 101 – 112 (1955).

[8.81] Sherman, A. H.: Algorithms for Sparse Gaussian Elimination with Partial Pivoting. ACM Trans. Mathematical Software *4*, No. 4, 330 – 338 (1978).

[8.82] Stein, P., Rosenberg, R. L.: On the Solution of Linear Simultaneous Equations by Iteration. J. London Math. Soc. *23*, 111 – 118 (1948).

[8.83] Stone, H. L.: Iterative Solution of Implicit Approximations of Multidimensional Partial Differential Equations. SIAM J. Numer. Anal. *5*, 530 – 558 (1968).

[8.84] Strickwerda, J. C.: A Generalized Conjugate Gradient Method for Non-Symmetric Systems of Linear Equations. Report 2290, MRC, University of Wisconsin, 1981.

[8.85] Swarztrauber, P. N., Sweet, R. A.: Efficient Fortran Subprograms for the Solution of Separable Elliptic Partial Differential Equations. ACM Coll. Alg., No. 541 (1979).

[8.86] VanDerVorst, H. A.: A Vectorizable Variant of Some ICCG Methods. SIAM J. Sci. Stat. Comput. *3*, No. 3, 350 – 356 (1982).

[8.87] Varga, R. S.: A Comparison of the Successive Overrelaxation Method and Semi-Iterative Methods Using Chebyshev Polynomials. J. SIAM *5*, 39 – 46 (1957).

[8.88] Varga, R. S.: Matrix Iterative Analysis. Englewood Cliffs, N.J.: Prentice-Hall 1962.

[8.89] Wachspress, E. L.: Iterative Solution of Elliptic Systems. Englewood Cliffs, N.J.: Prentice-Hall 1966.

[8.90] Wada, T., Dang, R. L. M.: Modification of ICCG Method for Application to Semiconductor Device Simulators. Electron. Lett. *18*, 265 – 266 (1982).

[8.91] Young, D. M.: Iterative Methods for Solving Partial Difference Equations of Elliptic Type. Dissertation, Harvard University, 1950.

[8.92] Young, D. M.: Iterative Solution of Large Linear Systems. New York: Academic Press 1971.

[8.93] Zlatev, Z.: On Some Pivotal Strategies in Gaussian Elimination by Sparse Technique. SIAM J. Numer. Anal. *17*, No. 1, 18 – 30 (1980).

[8.94] Zlatev, Z., Wasniewski, J., Schaumburg, K.: Comparison of Two Algorithms for Solving Large Linear Systems. SIAM J. Sci. Stat. Comput. *3*, No. 4, 486 – 501 (1982).

9 A Glimpse on Results

Actually, it is rather difficult to present examples of simulations which are both interesting for readers with experience in numerical modeling and informative for readers with just general interest in modeling but without specialized knowledge of device physics. I have chosen two examples which are intended as a fair trade-off between these objectives.

The first example (Section 9.1) will highlight problems associated with breakdown phenomena in miniaturized MOS transistors. Results of two-dimensional simulations are presented for this purpose. The influence of channel implantation on the punch-through effect is shown, which actually is a classical example for two-dimensional MOS-transistor simulation. Furthermore, the importance of adequate models for the physical parameters of the basic semiconductor equations is stressed.

In the second example (Section 9.2) a thyristor is investigated. Particularly the impact of a short between cathode and gate on the rate effect is demonstrated. Snapshots of the electrostatic potential, the electron concentration and the hole concentration obtained by a quasi-three-dimensional (cylindrical symmetry) transient simulation are presented in order to give evidence of parasitic triggering.

Thus, the ability of simulating both miniaturized devices and power devices is demonstrated.

9.1 Breakdown Phenomena in MOSFET's

Computer-aided simulation has become an urgent requirement for the design of modern MOS transistors since the upcoming of VLSI. Pure experimental investigations are very expensive, time consuming and for some problems not at all feasible. For the simulations presented in this section two state-of-the-art simulation programs have been coupled. The SUPREM program [9.2] has been used to calculate the doping profile and the MINIMOS program [9.13], [9.15] to simulate the electric behavior.

We investigate two MOS transistors with a slightly different doping profile in the channel region. Fig. 9.1-1 and Fig. 9.1-2 show the doping profile for these devices, respectively. The geometrical channel length (i.e. the width of the mask) is 1.4 μm. The junction depth is about 0.3 μm and the lateral subdiffusion of the source and the drain region is about 0.2 μm. Thus, the metallurgical channel length is about 1 μm.

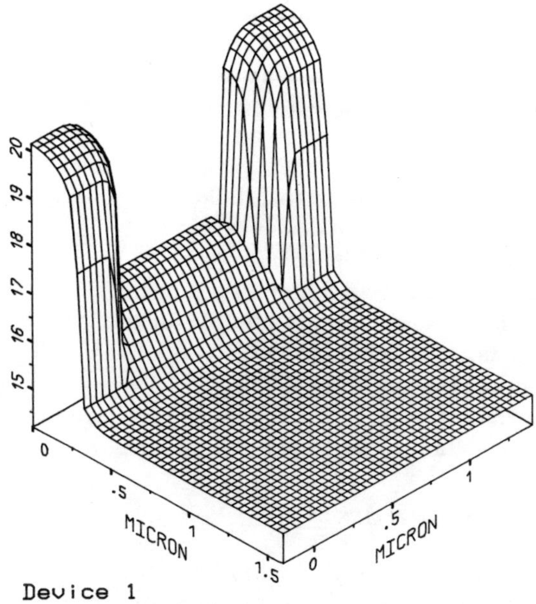

Fig. 9.1-1. Doping concentration [log., cm^{-3}] for device 1

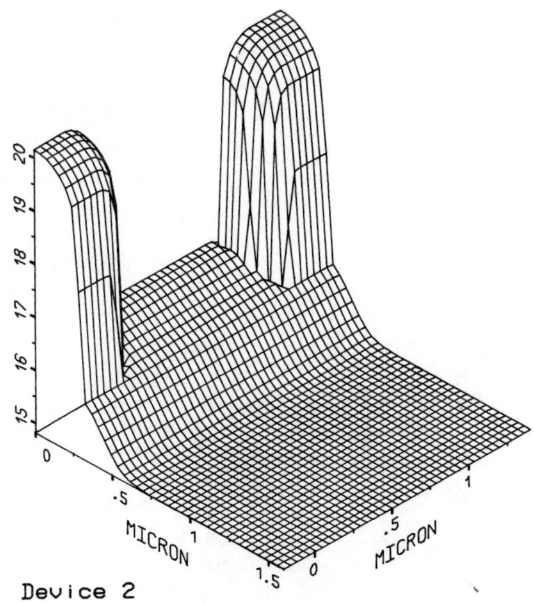

Fig. 9.1-2. Doping concentration [log., cm^{-3}] for device 2

The doping profile in the highly (arsenic) doped source/drain regions has been fitted empirically in the lateral direction (cf. Section 3.2) since the SUPREM program is at present only capable of computing one-dimensional profiles. The difference between

these two devices can be found, as mentioned, in the doping profile in the channel region. For device 1 (Fig. 9.1-1) a single channel implantation with boron has been performed, followed by a low temperature anneal. An additional second channel implantation (also boron) with a higher energy and lower dose has been simulated for device 2 (Fig. 9.1-2). This implantation has been designed to smoothly extend the first implant into the depth.
In the following we shall discuss the threshold voltage of these devices, which is obviously one of the most important design parameters. First, the threshold voltage has to be adequately defined in order to investigate its behavior. The most common definitions are based on an extrapolation of a tangent to the drain current. These methods are relatively inexact and not suitable for numerical simulation because several points of a characteristic have to be computed to allow an extrapolation. Therefore, I have chosen the following definition [9.16] which at present is also quite established. The threshold voltage U_{th} is that gate voltage at which the transistor sinks 0.1 μA times the channel width per channel length.

$$I_D(U_{GS}=U_{th}, U_{DS}, U_{SB})=0.1\ \mu\text{A}\cdot\frac{W}{L} \tag{9.1-1}$$

The threshold voltage obtained by this definition is therefore a function of the drain bias and the bulk bias. It is ensured that no threshold voltage shift versus channel length occurs for long channel tansistors, which enables a quantitative characterization of the influence of short-channel effects [9.16]. Furthermore, it is easy to determine the threshold voltage using this definition by experimental investigations as well as by simulation. The appropriateness of the constant (0.1 μA) in definition (9.1-1) can certainly be argued. However, this value has been chosen arbitrarily in view of practical experience; any other value derived by proper reasoning is just as good.

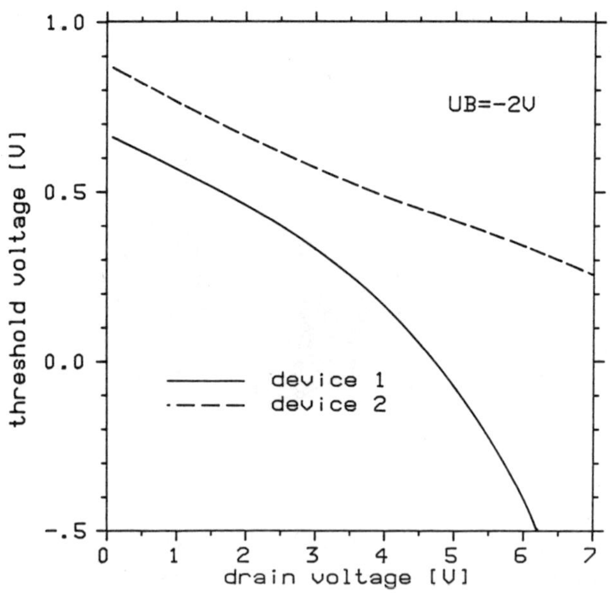

Fig. 9.1-3. Threshold voltage characteristics

Fig. 9.1-3 shows the threshold voltage versus drain voltage for device 1 (solid line) and device 2 (dashed line) at a bulk bias $U_B = -U_{SB} = 2\,V$. For low drain bias both characteristics are parallel. The characteristic for device 2 is shifted by about 200 mV upwards due to the additional channel implantation. Both characteristics are decreasing for increasing drain bias. However, this decrease is only dramatic for device 1. For a drain bias higher than about 4.75 V device 1 is "normally on" (i.e. $U_{th} < 0$), which makes this device unfit for applications in circuits with a 5 V supply. The threshold voltage for device 2 is slightly below 500 mV for a drain bias of 5 V, which is about the limit for applicability. A higher value is desirable in order to have a sufficiently large margin for technological tolerances. The decrease of the threshold voltage for device 2 is caused by drain induced barrier lowering which can be understood as a weak (acceptable) form of the punch-through effect [9.19]. Device 1 significantly exhibits the punch-through effect for high drain bias as will be illustrated in the following.
I have now chosen the operating conditions $U_{DS} = 6\,V$, $U_{GS} = 0\,V$, $U_{SB} = 2\,V$ and we shall discuss the distribution of various physical quantities in the interior of the devices.

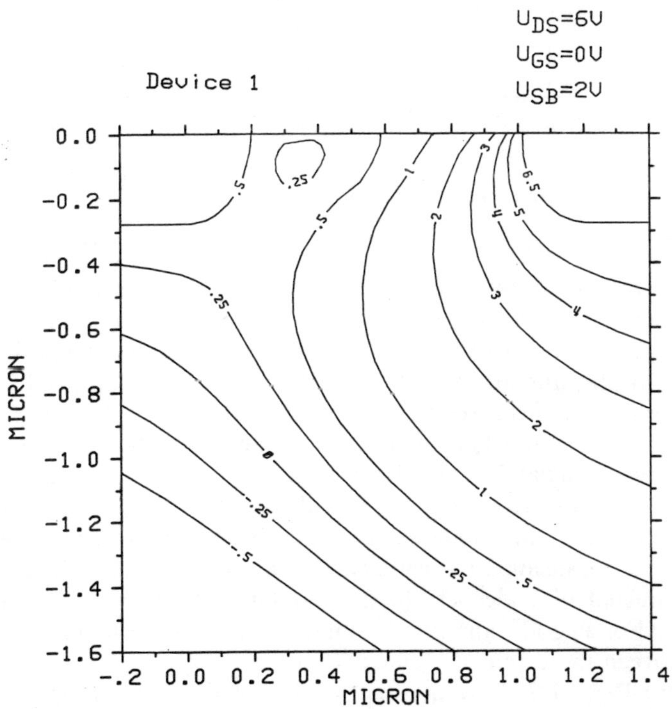

Fig. 9.1-4. Electrostatic potential [V] for device 1 ($U_{DS} = 6\,V$, $U_{GS} = 0\,V$, $U_{SB} = 2\,V$)

Fig. 9.1-4 and Fig. 9.1-5 show contourlines of the electrostatic potential at the above specified operating conditions. The source contact is at the left hand side of the figures and the drain contact at the right hand side. The 0.5 V and the 6.5 V contourline give an impression of the shape of the source and the drain region,

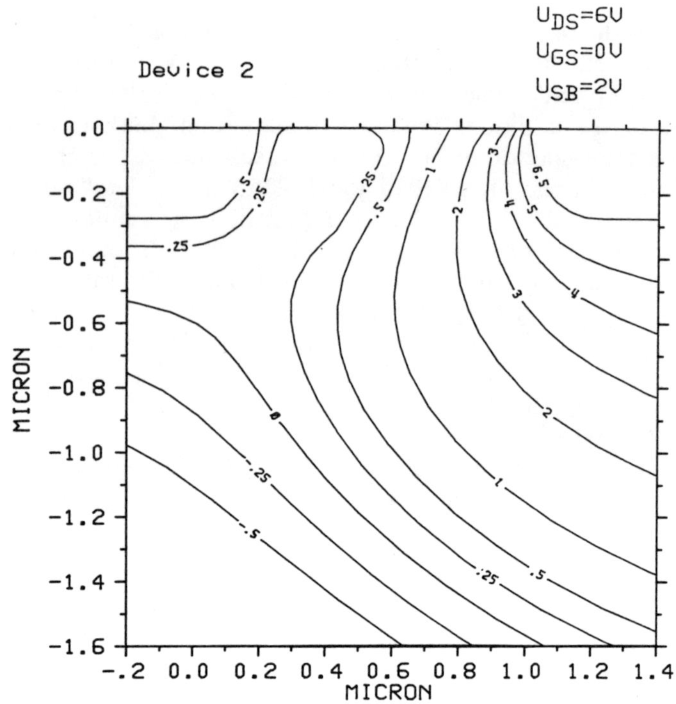

Fig. 9.1-5. Electrostatic potential [V] for device 2 ($U_{DS}=6$ V, $U_{GS}=0$ V, $U_{SB}=2$ V)

respectively. One can nicely observe a saddle point and a potential isle in Fig. 9.1-4, which is a typical indication of punch-through in weak inversion. This phenomenon has been reported for many years by authors working on two- and three-dimensional MOS transistor models, e.g., [9.3], [9.4], [9.6], [9.9], [9.10], [9.17]. This saddle point is a field free point in which current can flow only as diffusion current. From the contourlines in Fig. 9.1-5 one can deduce that a significantly pronounced potential barrier exists between source and drain to guarantee a proper subthreshold behavior. However, one can also see that the depletion region of the reverse biased drain-bulk diode extends closely to the source region below the *pn*-junctions, thus causing a parasitic current by weak punch-through. This effect can be suppressed by a deeper channel implant which, however, increases the capacitances. For a good transistor design it is one objective to find an appropriate trade-off between these contrary effects.

Fig. 9.1-6 and Fig. 9.1-7 show the distribution of the electron concentration for device 1 and device 2, respectively. The magnitude of the electron concentration in the channel region is several orders lower for device 2 than for device 1. In both figures one can imagine the pinch-off region close to the drain. One can also observe the low level of the electron concentration in the bulk corresponding to the depletion regions of the reverse biased source-bulk and drain-bulk diodes.

Fig. 9.1-8 and Fig. 9.1-9 show the lateral component of the electron current density for device 1 and device 2, respectively. These figures are qualitatively very similar.

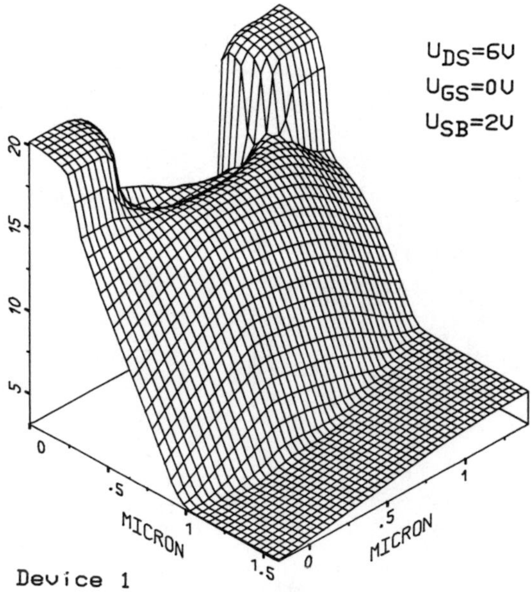

Fig. 9.1-6. Concentration of electrons [log., cm^{-3}] for device 1 ($U_{DS}=6$ V, $U_{GS}=0$ V, $U_{SB}=2$ V)

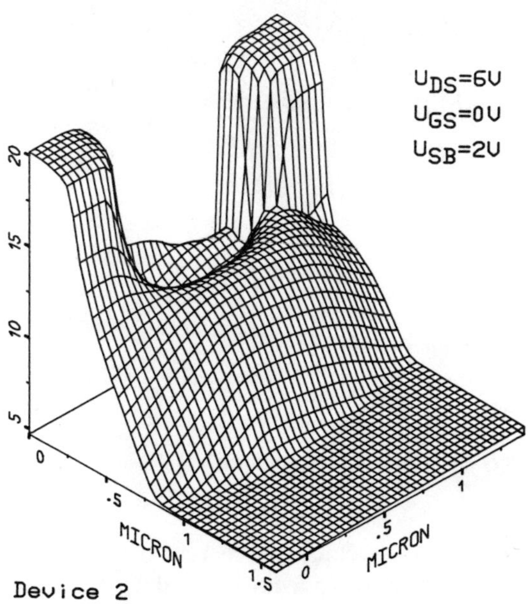

Fig. 9.1-7. Concentration of electrons [log., cm^{-3}] for device 2 ($U_{DS}=6$ V, $U_{GS}=0$ V, $U_{SB}=2$ V)

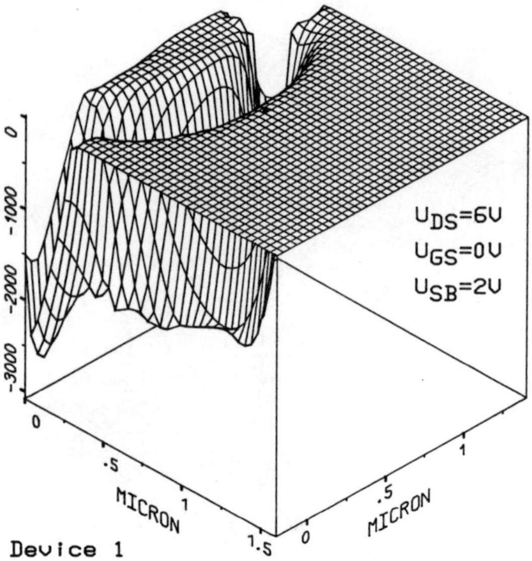

Fig. 9.1-8. Lateral component of electron current density [$A\,cm^{-2}$] for device 1 ($U_{DS}=6\,V$, $U_{GS}=0\,V$, $U_{SB}=2\,V$)

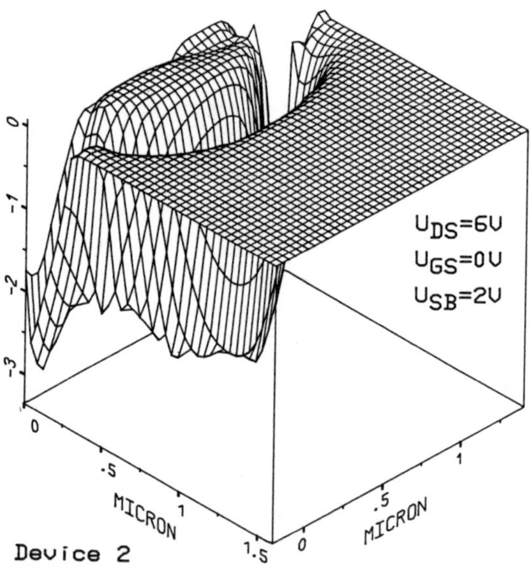

Fig. 9.1-9. Lateral component of electron current density [$A\,cm^{-2}$] for device 2 ($U_{DS}=6\,V$, $U_{GS}=0\,V$, $U_{SB}=2\,V$)

However, the current density level for device 1 is larger by about a factor of 1000. The maximal value of less than $3.5\,A\,cm^{-2}$ for device 2 is small enough for an acceptable operation. The parasitic channel almost vanishes by reducing the drain bias to 5 V.

At the bias point considered so far no significant impact ionization takes place in both devices. The reason for this fact can be found in the absolute current level which is simply too low [9.14], [9.16]. One has to remember that the impact ionization rate is proportional to the (inner) product of electric field and current density (cf. Section 4.2). To demonstrate the influence of impact ionization I have chosen the bias point $U_{DS}=6$ V, $U_{GS}=2$ V and $U_{SB}=2$ V for device 2, which lies in the regime of strong inversion. All figures shown in the following correspond to this operating condition. Actually, the simulation has been carried out twice: first with the standard model for impact ionization by Chynoweth [9.5] with the parameters by VanOverstraeten et al. [9.20] and, secondly, with impact ionization fully neglected.

Fig. 9.1-10 and Fig. 9.1-11 show the contourlines of the electrostatic potential. Almost no barrier exists between source and drain in Fig. 9.1-10, whereas an acceptable barrier is still simulated when neglecting avalanche generation. The device can practically not be operated since the amount of impact ionization is too large as will become visible from the figures discussed in the following. Note that approaches for the estimation of avalanche generation based on the evaluation of ionization integrals can be quite misleading and, therefore, have to be performed very carefully, e.g., [9.11], [9.18] since the feedback of avalanche generation on the electrostatic potential, and hence the electric field, cannot be accounted for by those methods.

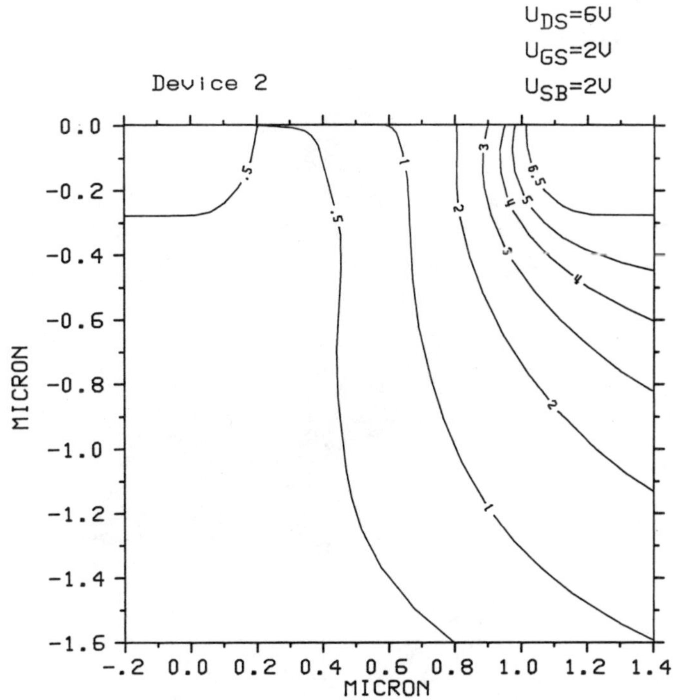

Fig. 9.1-10. Electrostatic potential [V] ($U_{DS}=6$ V, $U_{GS}=2$ V, $U_{SB}=2$ V)

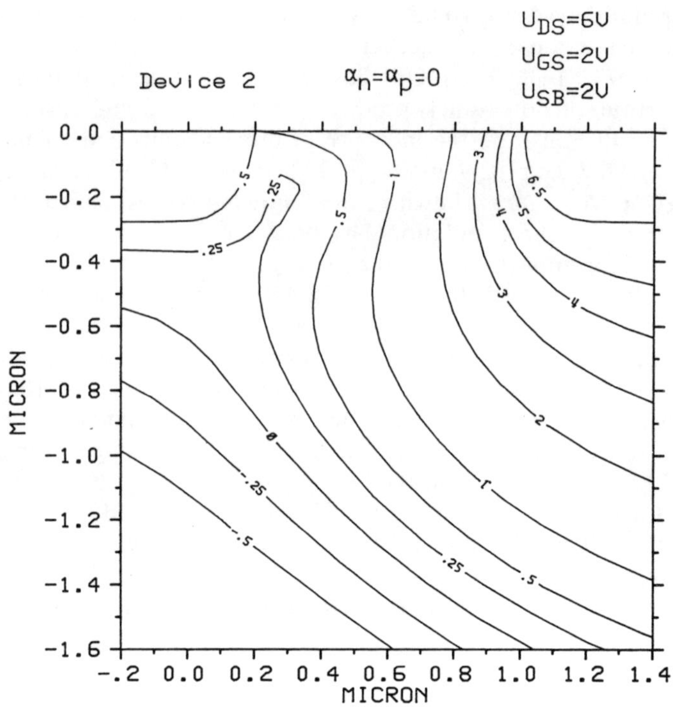

Fig. 9.1-11. Electrostatic potential [V] ($U_{DS}=6$ V, $U_{GS}=2$ V, $U_{SB}=2$ V), avalanche generation neglected

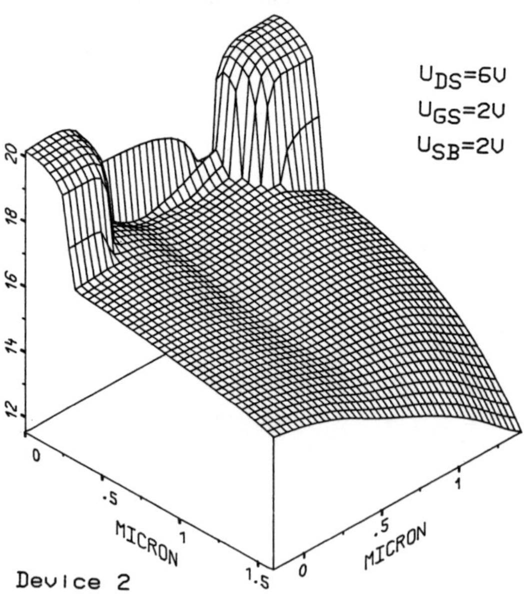

Fig. 9.1-12. Concentration of electrons [log., cm^{-3}] ($U_{DS}=6$ V, $U_{GS}=2$ V, $U_{SB}=2$ V)

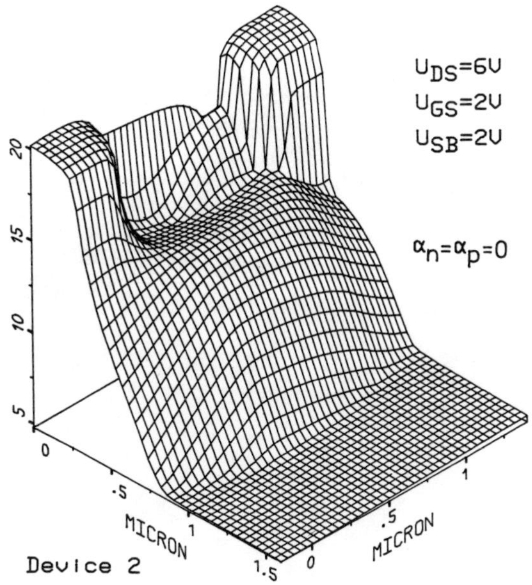

Fig. 9.1-13. Concentration of electrons [log., cm^{-3}] ($U_{DS}=6\,V$, $U_{GS}=2\,V$, $U_{SB}=2\,V$), avalanche generation neglected

The electron concentration with and without avalanche generation is shown in Fig. 9.1-12 and Fig. 9.1-13, respectively. One can see an enormous increase of the electron concentration in Fig. 9.1-12, which is partly caused by avalanche generation and partly by the rigid lowering of the source-bulk barrier (cf. Fig. 9.1-10). Note that the electron concentration is far above the intrinsic concentration in the whole figure. The distribution obtained by neglecting avalanche generation pretends a properly operating device. The behavior of the electron concentration in the inversion channel and in the pinch-off region is not influenced by neglecting impact ionization.

Fig. 9.1-14 and Fig. 9.1-15 show the hole concentration obtained with and without avalanche generation, respectively. First one has to note that the scale is significantly different for these figures. With avalanche generation the hole concentration is extraordinarily high with a maximum of about $4 \cdot 10^{16}\,cm^{-3}$ close to the pinch-off region. The hole concentration in Fig. 9.1-15 reaches only a very small maximal value. The part of the device shown in this figure is fully depleted. The plateau in the bulk and the various little peaks and dents are typical for the assumed operating condition if impact ionization is neglected. A discussion is irrelevant since the underlying model without avalanche generation is unrealistic.

Fig. 9.1-16 shows the net generation/recombination rate; impact ionization is included. A graphical representation of this quantity is not straightforward indeed since it may have large values with either sign and steep gradients. I usually apply the following transformation:

$$z = -\operatorname{sign}(R) \cdot \log\left(1 + \frac{|R|}{10^{18}\,cm^{-3}\,s^{-1}}\right) \tag{9.1-2}$$

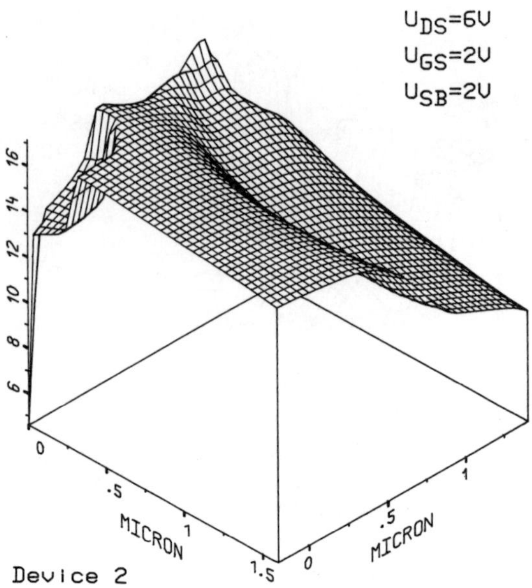

Fig. 9.1-14. Concentration of holes [log., cm^{-3}] ($U_{DS}=6$ V, $U_{GS}=2$ V, $U_{SB}=2$ V)

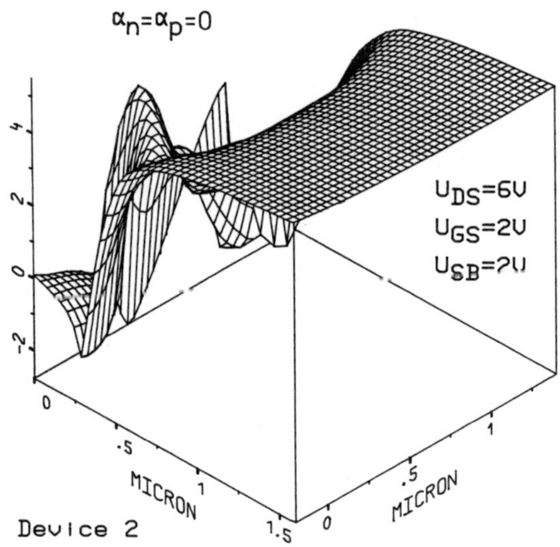

Fig. 9.1-15. Concentration of holes [log., cm^{-3}] ($U_{DS}=6$ V, $U_{GS}=2$ V, $U_{SB}=2$ V), avalanche generation neglected

The quantity obtained by the transformation (9.1-2) is positive if generation is dominant and negative if recombination is dominant. For values sufficiently large compared to 10^{18} cm^{-3} s^{-1} the representation becomes asymptotically logarithmic with proper sign adjustment. Thus, the value 10 in Fig. 9.1-16, which is about the maximum, corresponds to a generation rate of 10^{28} cm^{-3} s^{-1}. Note that generation is restricted to a small part of the shown region. In most parts recombination is

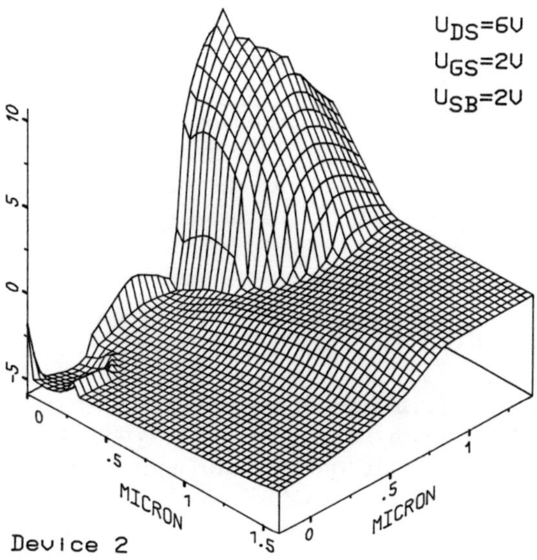

Fig. 9.1-16. Net generation/recombination rate [cf. description of figure] ($U_{DS}=6$ V, $U_{GS}=2$ V, $U_{SB}=2$ V)

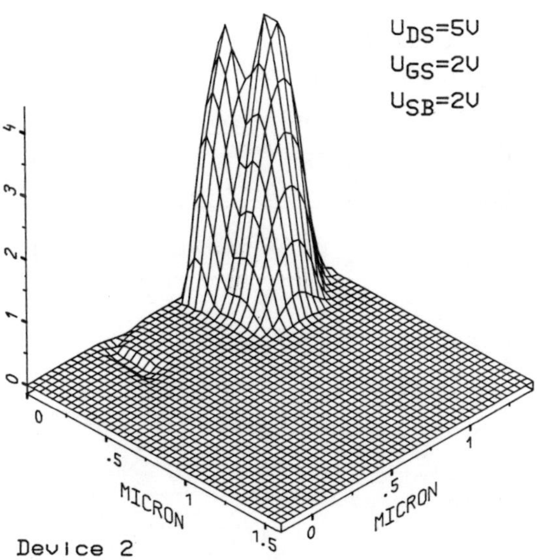

Fig. 9.1-17. Net generation/recombination rate [cf. description of figure] ($U_{DS}=5$ V, $U_{GS}=2$ V, $U_{SB}=2$ V)

dominant since the electron and the hole concentration are well above the intrinsic concentration (cf. Fig. 9.1-12, Fig. 9.1-14). This interesting phenomenon which essentially determines all device parameters relevant for the characterization of avalanche breakdown (e.g. snap-back voltage, sustain voltage) has been thoroughly discussed in [9.14]. Some further comments can be found in [9.12]. The maximal value of the recombination rate is about $10^{23}\,cm^{-3}\,s^{-1}$ which is quite large. By reducing the drain bias to 5 V the influence of impact ionization is significantly decreased. From Fig. 9.1-17 we see that the maximum generation rate has dropped by about six orders of magnitude. The reason for the two peaks in the generation rate can be found in the local distribution of the electric field and the current densities whose inner product has two maxima for this biasing condition. The moderately pronounced dent at the source-channel diode indicates recombination in this forward biased diode.

9.2 The Rate Effect in Thyristors

In this section I present a quasi-three-dimensional transient simulation of a thyristor which has been performed with the program SCDSS [9.7]. The simulation is termed quasi-three-dimensional because the thyristor under consideration exhibits cylindrical symmetry which allows a two-dimensional treatment after a reformulation of the basic semiconductor equations in cylindric coordinates. The discretization of the basic semiconductor equations in cylindric coordinates is analogous to the procedures given in Chapter 6 for cartesian coordinates. Some remarks on this subject and implementation details can be found in, e.g., [9.1], [9.8].
Fig. 9.2-1 and Fig. 9.2-2 show the geometry and the doping profile of the thyristor structure which will be considered here. The cathode contact is assumed to be a circular area with 96 μm radius. The *pn*-junction between the cathode and the *p*-base is very shallow, i.e. 3 μm, considering the total dimensions of the device.

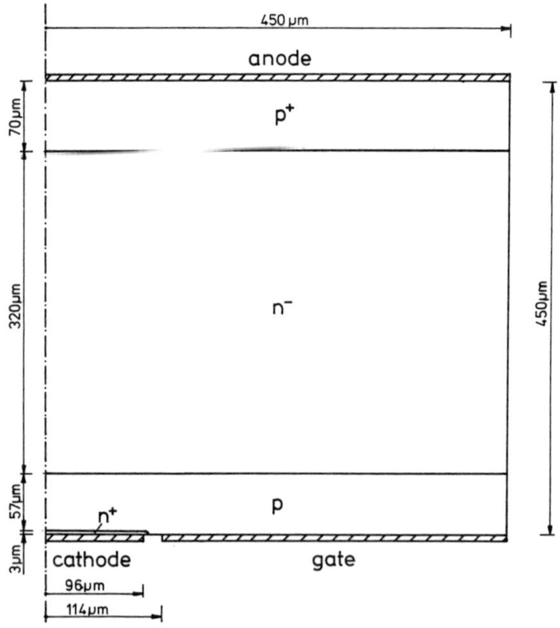

Fig. 9.2-1. Simulation domain for thyristor

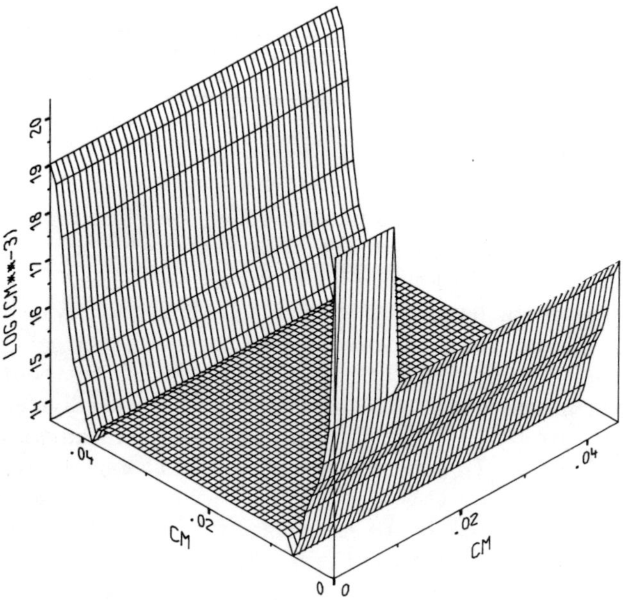

Fig. 9.2-2. Doping concentration [log., cm^{-3}] for thyristor

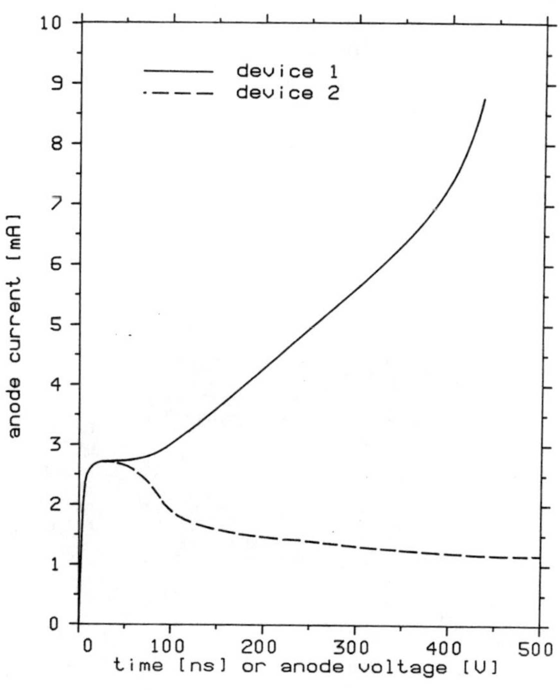

Fig. 9.2-3. Transient anode current

In the following we discuss the influence of a rapid rate, i.e. dUA/dt, upon the performance of the thyristor. For that purpose we apply a steep ramp voltage to the anode:

$$UA(t) = 1000 \frac{\text{V}}{\mu\text{s}} \cdot t \tag{9.2-1}$$

Two devices are considered; device 1 is assumed to have a floating gate and device 2 has a short between cathode and gate which is an established method to reduce the sensitivity with respect to unwanted dUA/dt triggering. Fig. 9.2-3 shows the anode current versus time characteristic for device 1 (solid line) and device 2 (dashed line). The initial phase of exponentially increasing current, which represents the loading of the depletion capacitances, is identical for both devices. At about 50 ns (which corresponds to an anode bias of 50 V) the characteristics visually depart from each other. For device 1 we can first observe a linear increase of the anode current with a slope of about 15 mA/μs. This increase tends to become exponential and leads to triggering at about $t = 450$ ns ($UA = 450$ V!). In the following we discuss snapshots of the electrostatic potential, the electron concentration and the hole concentration for both devices. Fig. 9.2-4, Fig. 9.2-5 and Fig. 9.2-6 show the electrostatic potential, the electron concentration and the hole concentration at $t = 0$, i.e. at equilibrium. The electrostatic potential is basically determined by the built-in potential caused by the doping concentration. The electron concentration and, complementary, the hole concentration follow, obviously, the electrostatic potential. The equilibrium solution is certainly identical for device 1 and device 2.
The first snapshot is taken at $t = 25$ ns ($UA = 25$ V). Fig. 9.2-7 shows the electrostatic potential which looks identical in the bird's-eye-view presentation for both devices. The total voltage drop occurs at the *pn*-junction between the *p*-base and the *n*-bulk. Fig. 9.2-8, Fig. 9.2-9 and Fig. 9.2-10, Fig. 9.2-11 show the electron, hole concentration for device 1 and device 2, respectively. We observe that the electron

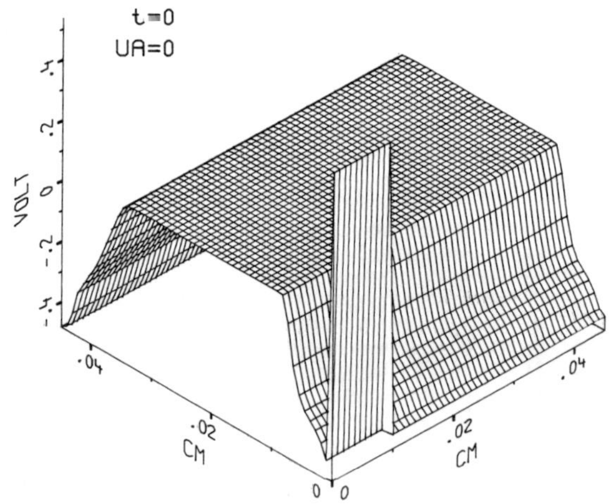

Fig. 9.2-4. Electrostatic potential [V] in thermal equilibrium

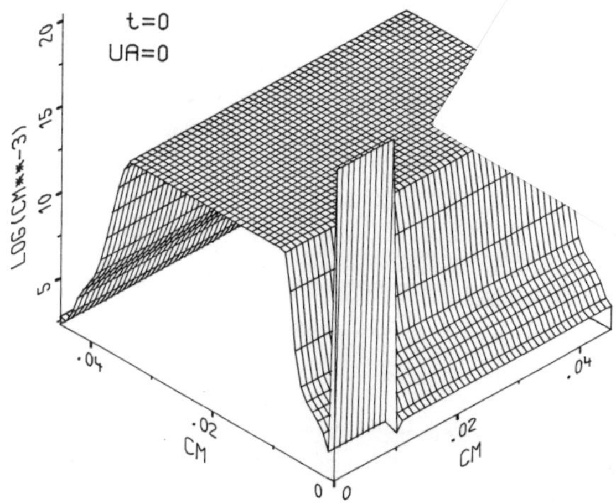

Fig. 9.2-5. Concentration of electrons [log., cm^{-3}] in thermal equilibrium

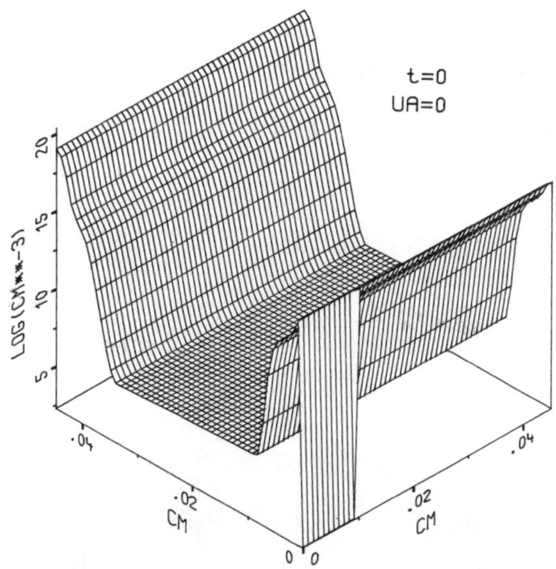

Fig. 9.2-6. Concentration of holes [log., cm^{-3}] in thermal equilibrium

concentration for device 1 is significantly higher in the *pn*-junction which is intended to block compared to the same quantity for device 2. From the hole concentrations we can deduce that the injection from the anode is markedly higher for device 1 than for device 2. Note that for device 2 the hole concentration close to the blocking *pn*-junction is not affected at this time by the injected holes. However, the total current is identical at time $t = 25$ ns (cf. Fig. 9.2-3).

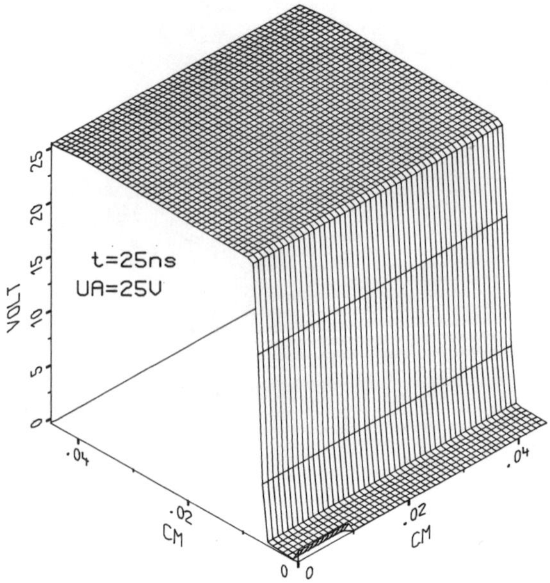

Fig. 9.2-7. Electrostatic potential [V] at $t = 25$ ns

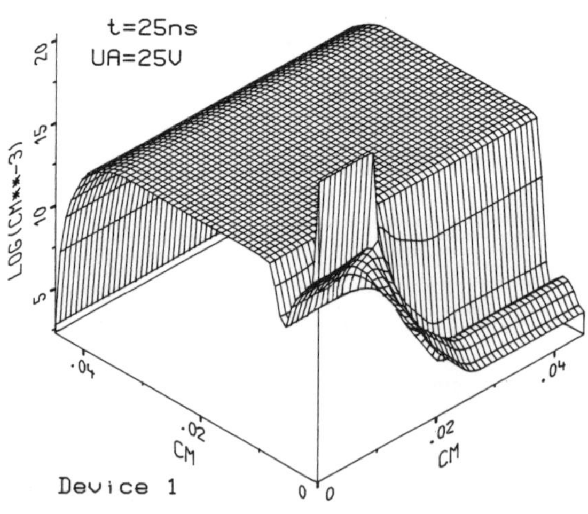

Fig. 9.2-8. Concentration of electrons [log., cm^{-3}] for device 1 at $t = 25$ ns

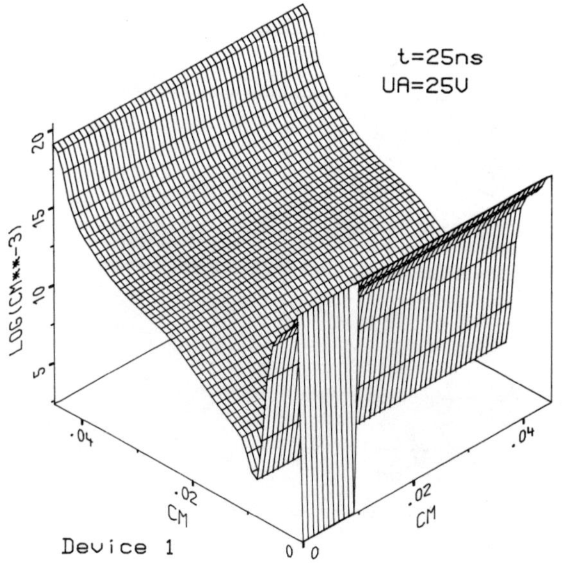

Fig. 9.2-9. Concentration of holes [log., cm^{-3}] for device 1 at $t=25$ ns

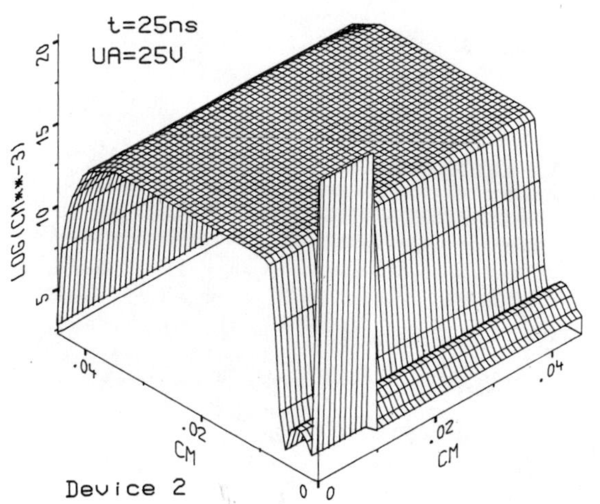

Fig. 9.2-10. Concentration of electrons [log., cm^{-3}] for device 2 at $t=25$ ns

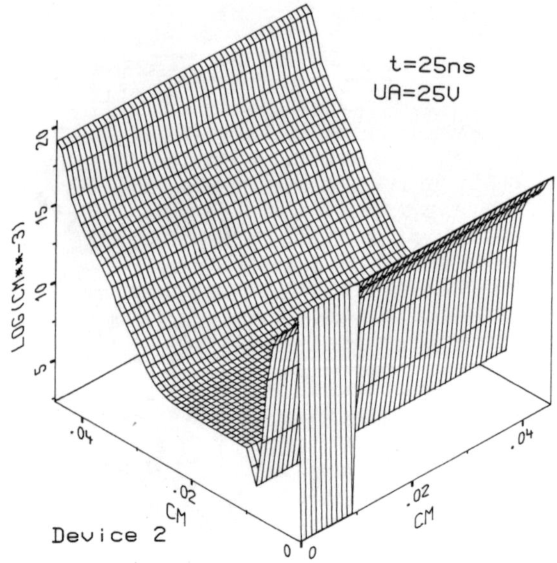

Fig. 9.2-11. Concentration of holes [log., cm^{-3}] for device 2 at $t=25$ ns

The next snapshot is taken at $t=170$ ns ($UA=170$ V). The electrostatic potential (Fig. 9.2-12) looks qualitatively similar to the one at the previous snapshot. The scale, however, has changed by more than a factor of six. The electron concentration for device 1 (Fig. 9.2-13) shows a slightly more extended depletion region.

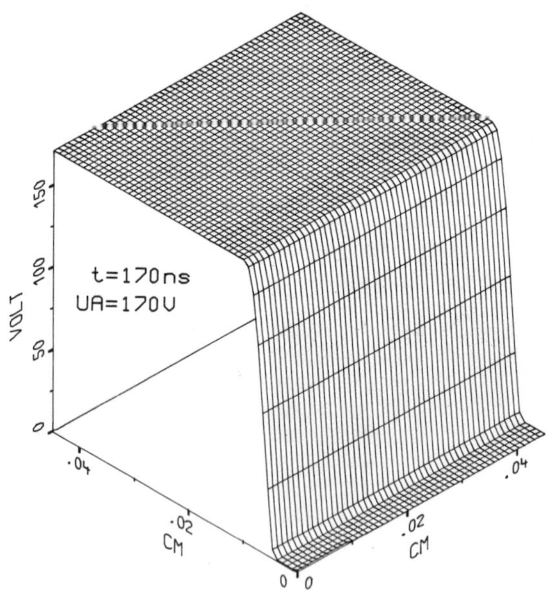

Fig. 9.2-12. Electrostatic potential [V] at $t=170$ ns

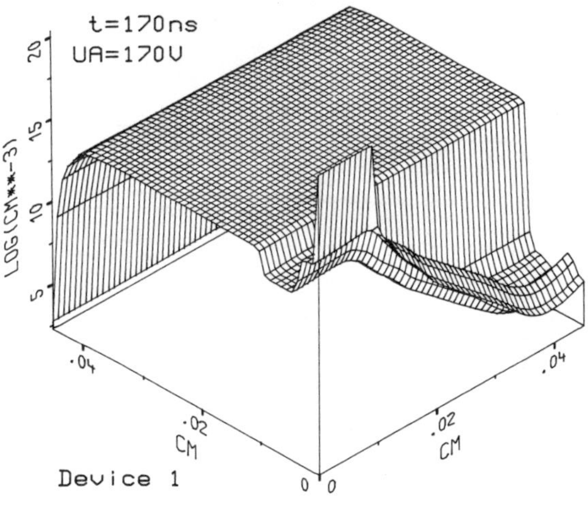

Fig. 9.2-13. Concentration of electrons [log., cm^{-3}] for device 1 at $t=170$ ns

Nevertheless, close to the cathode the electron concentration is already above the intrinsic concentration such that one should not speak of a depletion region. The injection of holes (Fig. 9.2-14) from the anode is now so strong that the entire *n*-bulk is flooded. The electron concentration for device 2 (Fig. 9.2-15) has only insignificantly changed compared to the previous snapshot. The holes injected from the anode have now reached the depletion region (Fig. 9.2-16).

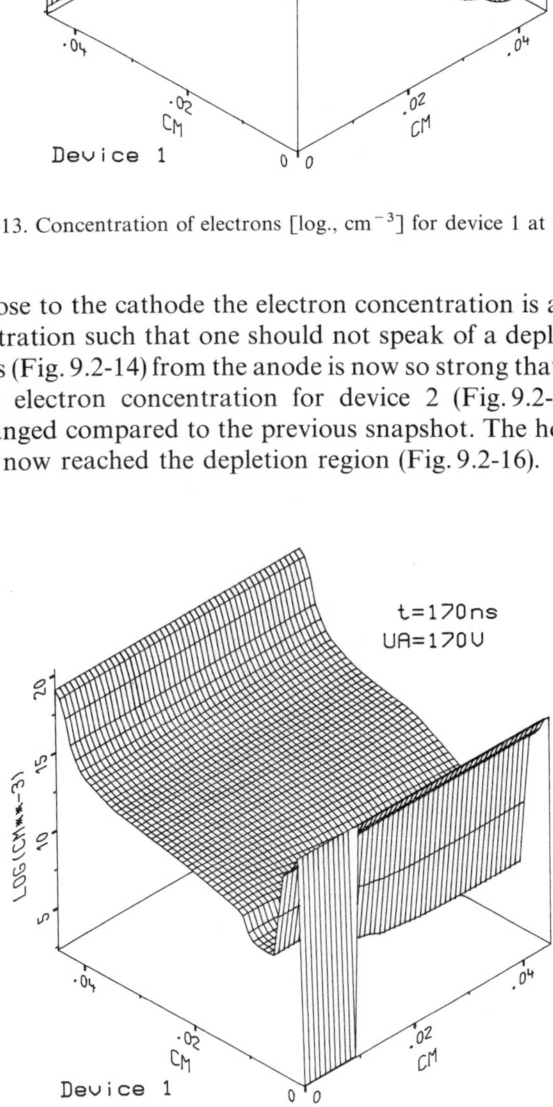

Fig. 9.2-14. Concentration of holes [log., cm^{-3}] for device 1 at $t=170$ ns

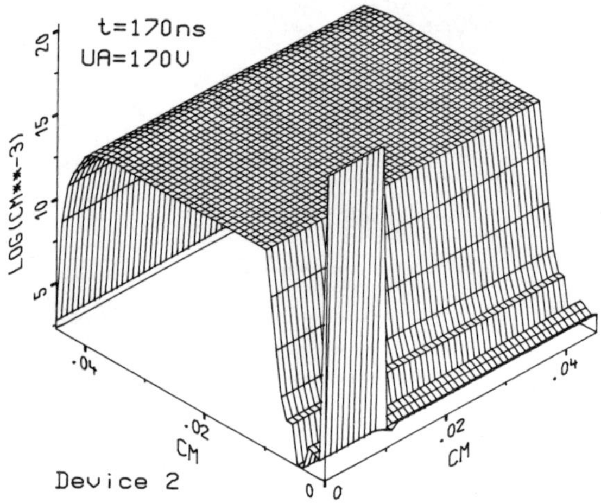

Fig. 9.2-15. Concentration of electrons [log., cm^{-3}] for device 2 at $t=170$ ns

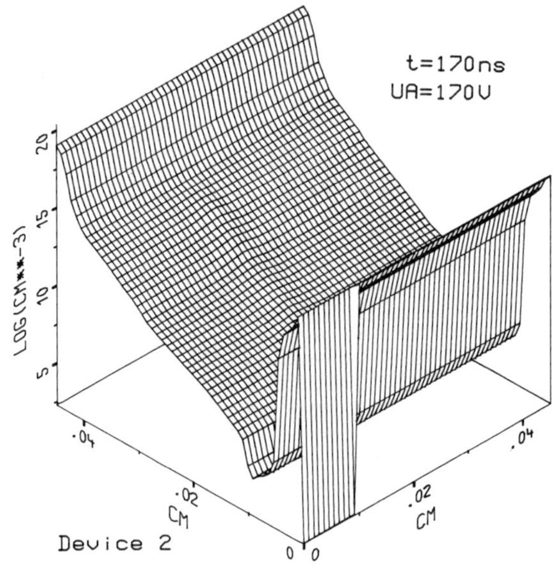

Fig. 9.2-16. Concentration of holes [log., cm^{-3}] for device 2 at $t=170$ ns

The last snapshot is taken close to the triggering of device 1 at $t=430$ ns ($UA=430$ V). The electrostatic potential is given in Fig. 9.2-17. The total voltage drop certainly takes place at the *pn*-junction between the *p*-base and the *n*-bulk. The carrier concentrations in device 1 (Fig. 9.2-18, Fig. 9.2-19) have reached enormously high values at this operating condition. There is almost no barrier left considering in particular the hole concentration. In device 2 we can see the formation of a wide depletion region instead (Fig. 9.2-20, Fig. 9.2-21). The level of the hole concentration

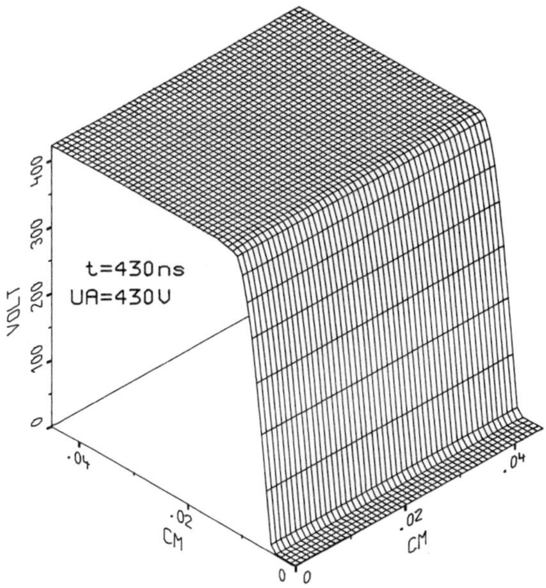

Fig. 9.2-17. Electrostatic potential [V] at $t = 430$ ns

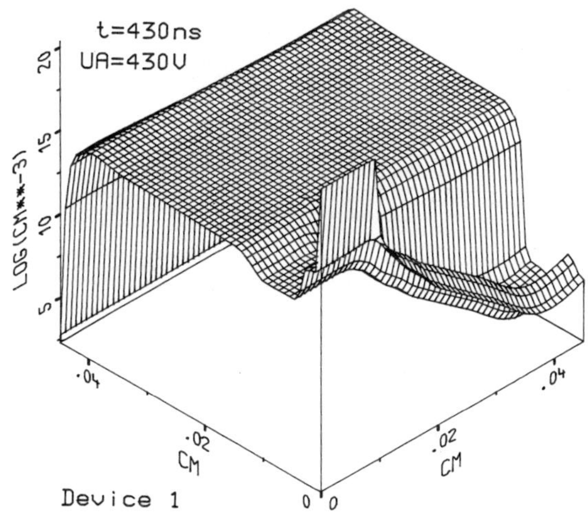

Fig. 9.2-18. Concentration of electrons [log., cm^{-3}] for device 1 at $t = 430$ ns

due to injection from the anode did not markedly increase compared to the distribution at the previous snapshot (cf. Fig. 9.2-16). Further computations for device 1 have been stopped here because an enormous increase of the total current for $t > 430$ μs necessitated impractically short time steps for the actual computation.

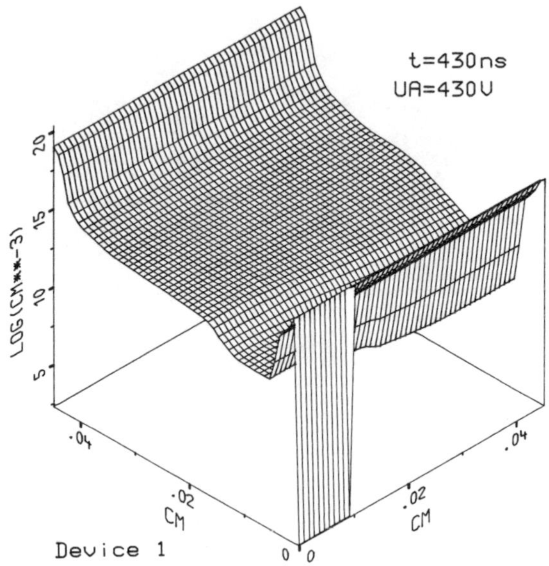

Fig. 9.2-19. Concentration of holes [log., cm^{-3}] for device 1 at $t=430$ ns

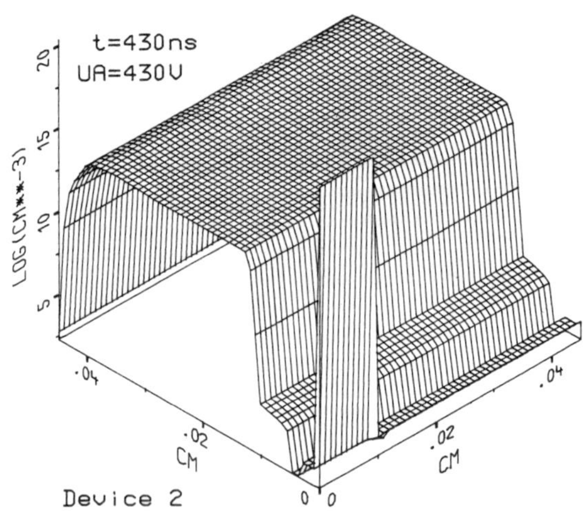

Fig. 9.2-20. Concentration of electrons [log., cm^{-3}] for device 2 at $t=430$ ns

The transient characteristic for device 2 has been calculated without difficulties up to $t=1\ \mu s$ which corresponds to an anode voltage of 1 kV. The electrostatic potential, the electron concentration and the hole concentration for this final snapshot are given in Fig. 9.2-22, Fig. 9.2-23 and Fig. 9.2-24, respectively. One can observe nicely the wide depletion region which is capable to block 1000 V reverse bias. Note that the niveaus of the carrier concentrations in the depletion region have increased due to relatively strong thermal generation compared to the niveaus at $UA = 430$ V.

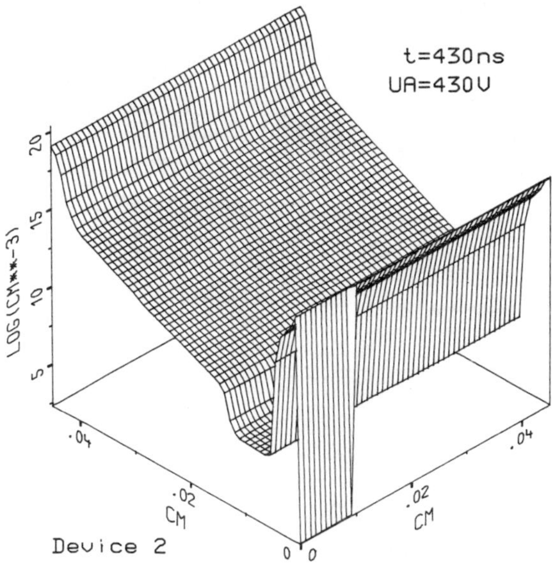

Fig. 9.2-21. Concentration of holes [log.,cm^{-3}] for device 2 at $t=430$ ns

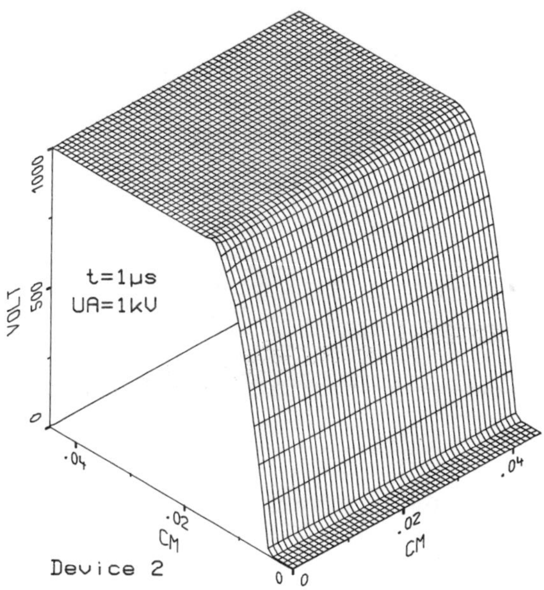

Fig. 9.2-22. Electrostatic potential [V] for device 2 at $t=1$ μs

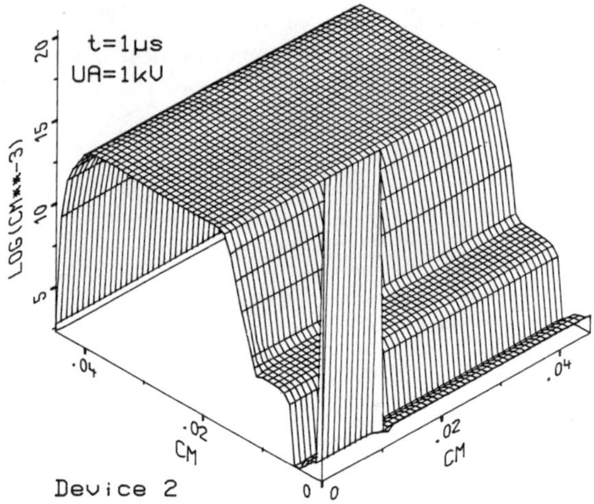

Fig. 9.2-23. Concentration of electrons [log., cm^{-3}] for device 2 at $t = 1\,\mu s$

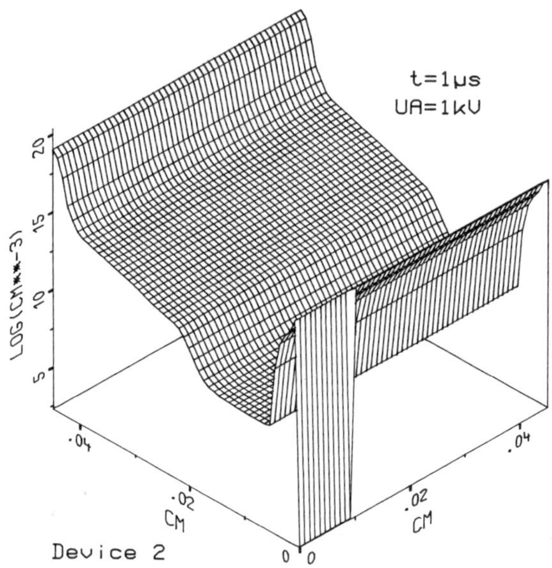

Fig. 9.2-24. Concentration of holes [log., cm^{-3}] for device 2 at $t = 1\,\mu s$

The SCDSS program uses the finite boxes method for the spatial discretization as outlined in Section 6.2. For the time discretization the backward Euler method (cf. Section 6.4) is implemented. The mesh is updated during the time integration in order to equidistribute the local truncation error of the difference approximations (cf. Section 6.5). A typical mesh used in the computations given above is shown in

Fig. 9.2-25. One can see that the cathode region appears only as a black stripe since many mesh points have been required there. Fig. 9.2-26 and Fig. 9.2-27 show enlarged details of the mesh. The resolution of the steeply graded *pn*-junction between the cathode region and the *p*-base becomes nicely apparent.

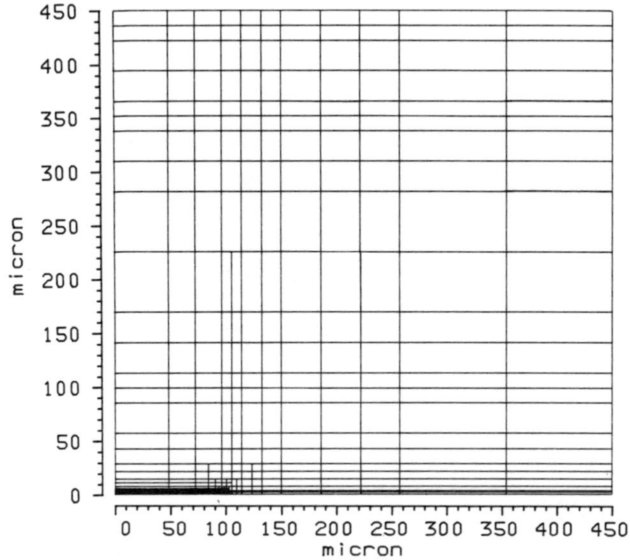

Fig. 9.2-25. Typical finite boxes mesh for thyristor simulation

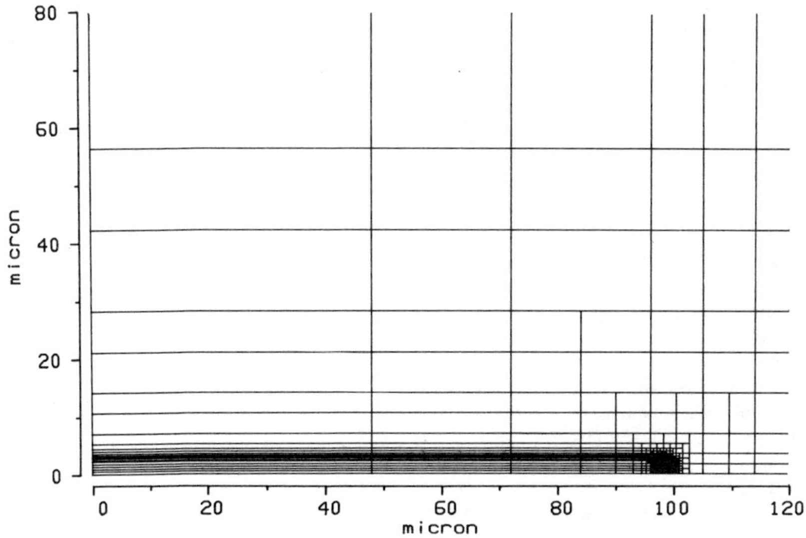

Fig. 9.2-26. Enlarged detail of Fig. 9.2-25

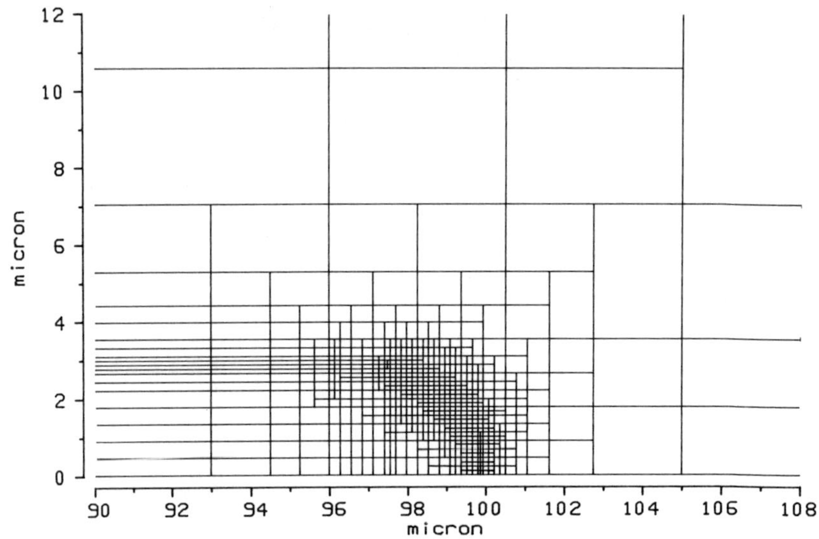

Fig. 9.2-27. Enlarged detail of Fig. 9.2-26

9.3 References

[9.1] Adler, M. S., Temple, V. A. K., Rustay, R. C.: Theoretical Basis for Field Calculations on Multi-Dimensional Reverse Biased Semiconductor Devices. Solid-State Electron. *25,* No. 12, 1179 – 1186 (1982).

[9.2] Antoniadis, D. A., Hansen, S. E., Dutton, R. W.: SUPREM II – a Program for IC Process Modeling and Simulation. Report 5019-2, Stanford University, 1978.

[9.3] Barnes, J. J., Shimohigashi, K., Dutton, R. W.: Short-Channel MOSFET's in the Punchthrough Current Mode. IEEE Trans. Electron Devices *ED-26,* 446 – 453 (1979).

[9.4] Chamberlain, S. G., Husain, A.: Three-Dimensional Simulation of VLSI MOSFET's. Proc. Int. Electron Devices Meeting, pp. 592 – 595 (1981).

[9.5] Chynoweth, A. G.: Ionization Rates for Electrons and Holes in Silicon. Physical Review *109,* 1537 – 1540 (1958).

[9.6] Demoulin, E., Greenfield, J. A., Dutton, R. W., Chatterjee, P. K., Tasch, A. F.: Process Statistics of Submicron MOSFET's. Proc. Int. Electron Devices Meeting, pp. 34 – 37 (1979).

[9.7] Franz, A. F., Franz, G. A., Selberherr, S., Ringhofer, C., Markowich, P.: Finite Boxes – A Generalization of the Finite Difference Method Suitable for Semiconductor Device Simulation. IEEE Trans. Electron Devices *ED-30,* No. 9, 1070 – 1082 (1983).

[9.8] Franz, G. A., Franz, A. F., Selberherr, S., Markowich, P.: A Quasi Three Dimensional Semiconductor Device Simulation Using Cylindrical Coordinates. Proc. NASECODE III Conf., pp. 122 – 127. Dublin: Boole Press 1983.

[9.9] Greenfield, J. A., Dutton, R. W.: Nonplanar VLSI Device Analysis Using the Solution of Poisson's Equation. IEEE Trans. Electron Devices *ED-27,* 1520 – 1532 (1980).

[9.10] Kotani, N., Kawazu, S.: Computer Analysis of Punch-Through in MOSFET's. Solid-State Electron. *22,* 63 – 70 (1979).

[9.11] Kotani, N., Kawazu, S.: A Numerical Analysis of Avalanche Breakdown in Short-Channel MOSFET's. Solid-State Electron. *24,* 681 – 687 (1981).

[9.12] Müller, W., Risch, L., Schütz, A.: Analysis of Short Channel MOS Transistors in the Avalanche Multiplication Regime. IEEE Trans. Electron Devices *ED-29,* No. 11, 1778 – 1784 (1982).

[9.13] Schütz, A., Selberherr, S., Pötzl, H. W.: A Two-Dimensional Model of the Avalanche Effect in MOS Transistors. Solid-State Electron. *25,* 177 – 183 (1982).

[9.14] Schütz, A., Selberherr, S., Pötzl, H. W.: Analysis of Breakdown Phenomena in MOSFET's. IEEE Trans. Computer-Aided-Design of Integrated Circuits *CAD-1,* 77–85 (1982).

[9.15] Selberherr, S., Schütz, A., Pötzl, H. W.: MINIMOS – a Two-Dimensional MOS Transistor Analyzer. IEEE Trans. Electron Devices *ED-27,* 1540–1550 (1980).

[9.16] Selberherr, S., Schütz, A., Pötzl, H.: Investigation of Parameter Sensitivity of Short Channel MOSFET's. Solid-State Electron. *25,* 85–90 (1982).

[9.17] Selberherr, S., Schütz, A., Pötzl, H.: Two Dimensional MOS-Transistor Modeling. In: Process and Device Simulation for Integrated Circuit Design, pp. 490–581. The Hague: Martinus Nijhoff 1983.

[9.18] Toyabe, T., Yamaguchi, K., Asai, S., Mock, M. S.: A Numerical Model of Avalanche Breakdown in MOSFET's. IEEE Trans. Electron Devices *ED-25,* 825–832 (1978).

[9.19] Troutman, R. R.: VLSI Limitations from Drain-Induced Barrier Lowering. IEEE Trans. Electron Devices *ED-26,* 461–469 (1979).

[9.20] VanOverstraeten, R., DeMan, H.: Measurement of the Ionization Rates in Diffused Silicon p-n Junctions. Solid-State Electron. *13,* 583–608 (1970).

Author Index

Subject Index

Table Index

Satz: Austro-Filmsatz Richard Gerin, A-1021 Wien
Druck: Paul Gerin, A-1021 Wien

CISM

International Centre
for Mechanical
Sciences, Udine

Courses and Lectures

No. 279

Secure Digital Communications

Edited by **G. Longo,** Università di Trieste

1983. 118 figures. V, 332 pages.
ISBN 3-211-81784-0

The development of communications and the increasing exchange of data between man and computers makes it necessary to protect the information from unauthorized users . . . This book covers the vast area of cryptology, with an accent to applications, as well as other aspects of secure communications: communication in the presence of jamming, multiple-access communication systems and protocols for conflict resolution . . . This volume is of interest both to students of mathematics, computer science and communication theory and to the practitioners in communication engineering. Its wide coverage makes it useful for getting a good appreciation of the secure digital communication area.

No. 265

Multi-User Communication Systems

Edited by **G. Longo,** Università di Trieste

1981. 30 figures. IV, 259 pages.
ISBN 3-211-81612-7

The traditional object of information theory is the one-source one-channel system, but since Shannon's pioneering work also more complicated two-source two-channel systems have been considered. These systems model realistic communication networks which are often met in the practice of data transmission and computer communication. The course is devoted to the information-theoretical analysis of the multi-user systems as well as to the practical implementation of such systems. Moreover, multiple-access communication systems will be dealt with in detail. The emphasis is on the possibility of eliminating the noise and interference effects from communication and data processing.

Springer-Verlag
Wien New York